PRINCIPLES OF MECHANICAL METALLURGY

PRINCIPLES OF MECHANICAL METALLURGY

Iain Le May
Department of Mechanical Engineering
University of Saskatchewan
President, Metallurgical Consulting Services, Ltd.

ELSEVIER
New York · Oxford

Elsevier North Holland, Inc.
52 Vanderbilt Avenue, New York, New York 10017

Distributors outside the United States and Canada:

Edward Arnold (Publishers), Ltd.
41 Bedford Square, London WC1B 3DQ

Library of Congress Cataloging in Publication Data

Le May, Iain.
Principles of mechanical metallurgy.

Includes bibliographical references and index.
1. Physical metallurgy. 2. Strength of materials.
I. Title.
TN690.L28 620.1′6 81-1356
ISBN 0-444-00612-5 AACR2

Manufactured in the United States of America

PREFACE

This text has been developed from senior undergraduate and graduate courses in mechanical metallurgy given primarily at the University of Saskatchewan, Canada, as well as a graduate course given at the Escóla Politécnica, Universidade de São Paulo, Brazil. An attempt has been made to build on the base offered by these courses to develop a text which provides a broad and reasonably complete survey of the principles involved in mechanical metallurgy: it should be useful to senior undergraduates in mechanical engineering and metallurgy, graduate students embarking on detailed study of specific topics in this subject area, and practicing engineers. Wherever possible, topics in the areas of fracture, fatigue and creep, in particular, have been related to the needs of the designer, not by providing materials data as such — these being available in handbooks and reference material — but rather by emphasizing principles as they relate to design and service applications.

Many people contribute to the writing of a book of this kind. First, there are the authors of other major works in the field, together with the authors of original papers: these are identified by the referencing in the text. A significant contribution has been made by my former and current graduate students and colleagues who have worked on specific topics in mechanical metallurgy: again, their names are identified by references to their published work in the appropriate parts of the text. I have also benefitted greatly from close contacts with many fellow engineers and research workers while serving as the first Technical Editor and, subsequently, as an Associate Editor of ASME's *Journal of Engineering Materials and Technology*. These contacts have, I believe, helped to broaden my outlook on the subject matter and, hopefully, may have aided me in presenting a balanced view of a complex subject.

The comments of the following, who have reviewed particular sections of the material in draft form, are gratefully acknowledged: Dr. Ney Quadros, Instituto de Energia Atômica, São Paulo, for comments on basic deformation mechanisms; Professor K. K. Chawla, Instituto Militar de Engenharia, Rio de Janeiro, for discussions concerning composite materials; and Professor W. E. White, University of Calgary, for discussions concerning high temperature creep. However, any errors or omissions are entirely my responsibility, as is also the choice of topics covered, which is very much a matter of personal judgement and, indeed, prejudice.

The assistance of Mr. Alex Kozlow in preparing the figures has been indispensible to the completion of this work. Sincere thanks is also offered to Miss Dawn Korchinski who did the phototypesetting, and to the staff of the Printing Services Department, University of Saskatchewan, for their work in laying out the pages for the printers. The support of the Natural Sciences and Engineering Research Council of Canada and of the International Copper Research Association Inc. for studies in subjects of mechanical metallurgy is gratefully acknowledged: this has helped to provide the background for this book.

Finally, sincere thanks is extended to my wife, Shona, and to my family for their forebearance and support throughout the writing of this text.

Iain Le May
Saskatoon, Canada
1980

CONTENTS

PRINCIPLES OF MECHANICAL METALLURGY

1

STRESS AND STRAIN

1.1 INTRODUCTION

When a solid body is subjected to a load, deformation results. Experience has shown that up to some limiting load value the component will return to its original dimensions after the load is removed: in such a case the deformation has been *elastic* and *recoverable.* If a load greater than this *elastic limit* is applied, permanent deformation will be observed on unloading: this permanent deformation is termed *plastic deformation.*

For the majority of engineering materials, and specifically for metals, the elastic deformation is directly proportional to applied load up to the elastic or *proportional limit,* and such materials are said to obey Hooke's law, which requires that load (or *stress*) and elastic extension (or *strain*) be proportional.

In this chapter we shall examine the definitions of stress and strain, and the relations between them, as a necessary preliminary to examining the loading criteria which lead to a material's or component's inability to continue to carry an applied load.

1.1.1 Concepts of Stress and Strain

Consider a cylindrical bar of initial length l_0 and initial cross-sectional area A_0 under the action of a uniaxial tensile load P, as shown in Fig. 1.1. The applied force is spread across the section of the bar and is balanced by the stress within the bar, σ, acting over the cross-section. The internal force is $\int \sigma dA$, and assuming that the stress is uniformly distributed, the *average*

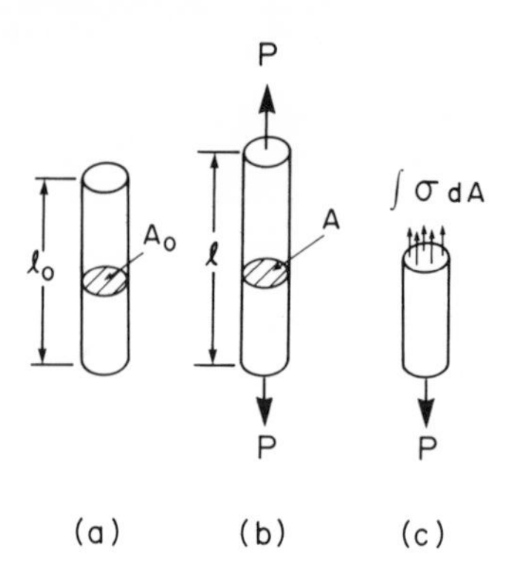

Fig. 1.1 Circular bar of initial length l_o and initial cross-sectional area A_o loaded in tension by load P

normal stress is given by P/A where A is the current cross-sectional area of the bar. In many engineering situations we can ignore any changes in cross-sectional area upon loading (except where the plastic deformations are large — see below), and base the value of the normal stress on the original cross-sectional area, A_o. Thus, the *engineering stress* is given by

$$\sigma = P/A_o \tag{1.1}$$

When subjected to the load P,the bar will extend to length l, and the *engineering (normal) strain*, e, is defined as

$$e = (l - l_o)/l_o \tag{1.2}$$

True normal stress, which takes account of the changes in geometry with load, and which may differ substantially from engineering stress for large plastic deformations, is defined as

$$\sigma = P/A, \tag{1.3}$$

while the *true* or *natural (normal) strain*, ϵ, is defined on the basis of considering infinitesimal incremental strains and integrating these between the initial and final lengths. Thus

$$\epsilon = \int_{l_o}^{l} (1/l)\, dl$$

or

$$\epsilon = \ln (l/l_o) \tag{1.4}$$

In addition to normal stress and strain, we must define *shear stress* and *shear strain*. Figure 1.2 shows a body subjected to pure shearing forces of magnitude S, and it is seen that the application of such shear forces does not lead to extension, but to shape change only. It should also be noted that

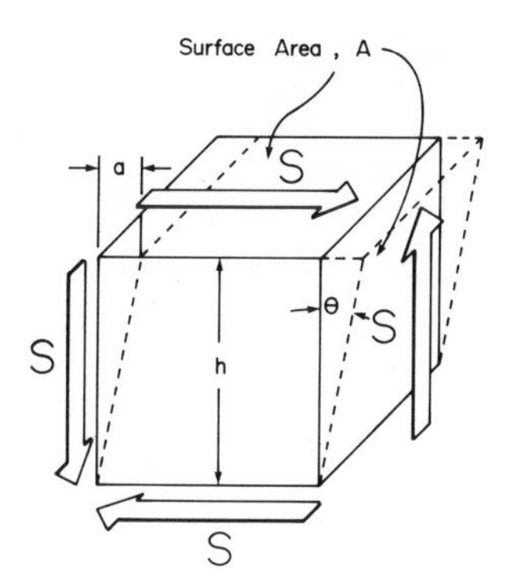

Fig. 1.2 Body subjected to pure shearing forces of magnitude *S*

complementary shearing forces are required in order to maintain the body in equilibrium. The magnitude of the shear stress, τ, on the surface of the body is given by

$$\tau = S/A \tag{1.5}$$

where A is the surface area of the face over which S is applied. The *shear strain*, γ, is defined in terms of the displacement, a, divided by the distance between the planes, h. This ratio is also the tangent of the angle through which the body (or element) has been rotated, θ, and for small angles this is equal to the angle expressed in radians. Hence, shear strain can be expressed as

$$\gamma = a/h = \tan\theta \simeq \theta \tag{1.6}$$

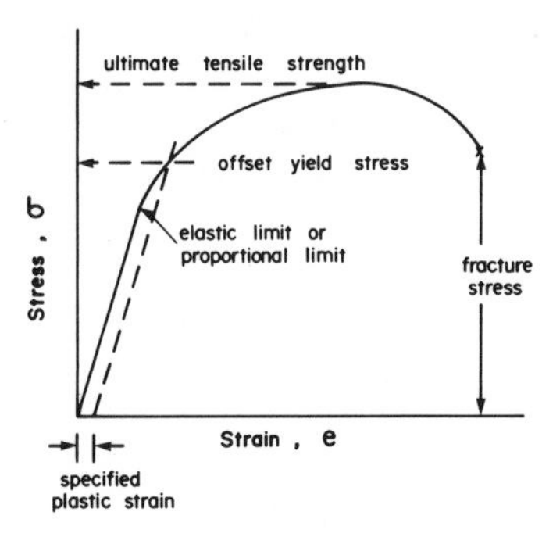

Fig. 1.3 Stress-strain curve for a ductile metal under uniaxial tensile loading

1.1.2 Engineering Stress-Strain Behavior

The uniaxial tensile stress-strain curve for a ductile metal is shown in Fig. 1.3, and the salient points are marked on this: these will be discussed more fully in Chapter 3.

During elastic deformation, normal stress and normal strain are related by

$$\sigma = E\,e \tag{1.7}$$

where E is a constant, termed *Young's modulus* or the *modulus of elasticity*. For shear stress and shear strain a similar relation holds:

$$\tau = G\,\gamma \tag{1.8}$$

where G is the *modulus of rigidity* or *shear modulus of elasticity*.

1.2 STRESS

Having set out the concept of stress, we shall now examine this more rigorously, beginning with the stress acting at a point on a body and then examining the stresses in a body subject to multiaxial loading.

1.2.1 Stress at a Point

Consider a region of surface of a solid body of area ΔA surrounding a point *0*, and let the force acting on this be ΔP. The stress, σ, acting at point *0* is defined as the limiting value of $\Delta P/\Delta A$ as ΔA is reduced to zero, i.e.,

$$\sigma = \lim_{\Delta A \to 0} (\Delta P/\Delta A) \tag{1.9}$$

The force, ΔP, and the stress defined as above will act in some specific direction and we can resolve them into components in various ways, but most conveniently into components normal and parallel to the surface being considered. Let x, y and z be a set of orthogonal axes as shown in Fig. 1.4. The stress is resolved into components σ_{zz} normal to the z-plane in the z-direction, τ_{zx} on the z-plane in the x-direction, and τ_{zy} on the z-plane in the y-direction. The first component is the *normal stress* on the z-plane, and the latter two are components of *shear stress*.

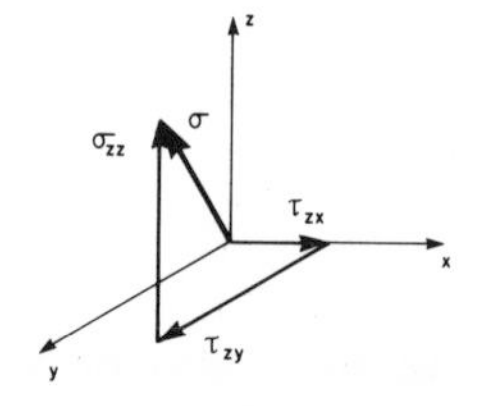

Fig. 1.4 Stress σ resolved into shear and normal stress components

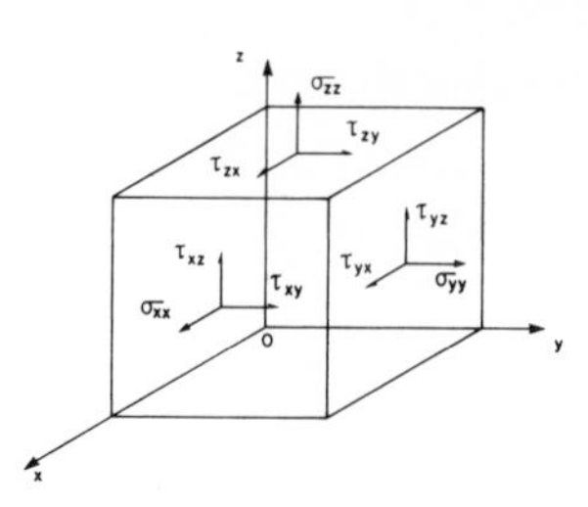

Fig. 1.5 Cartesian components of stress at a point

In the foregoing the stress at a point was defined on a particular plane, but in the general case we do not refer to any specific plane. Consider three mutually perpendicular planes passing through the point in question. Figure 1.5 illustrates this by means of a small rectangular element, the element being sufficiently small that the stresses are homogeneous within it and have the same magnitude and direction at every point in the cross-section. It may be seen that nine quantities must be defined to establish the state of stress at a point. However, we may simplify matters by summing the moments about the axes, to give

$$\tau_{zx} = \tau_{xz}\ ,\ \tau_{xy} = \tau_{yx}\ ,\ \tau_{yz} = \tau_{zy}\ .$$

Thus, only six independent quantities, σ_{xx}, σ_{yy}, σ_{zz}, τ_{xy}, τ_{xz} and τ_{yz}, are required to define the state of stress at a point.

The full state of stress (or the *stress tensor*) at the point may be written for convenience in the form σ_{ij}, which represents all nine components with i and j taking appropriate values of x, y, and z, and may be displayed as

$$\begin{matrix} \sigma_{xx} & \tau_{xy} & \tau_{xz} \\ \tau_{yx} & \sigma_{yy} & \tau_{yz} \\ \tau_{zx} & \tau_{zy} & \sigma_{zz} \end{matrix}$$

Stresses are frequently referred to coordinates other than Cartesian, and in cylindrical polar coordinates (z, r and θ) the components are as shown in Fig. 1.6. The stress tensor, σ_{ij}, may be used as before, with i and j taking appropriate values of z, r and θ. Similarly, Fig. 1.7 shows stresses in curvilinear coordinates, which are useful in the stress analysis of cracks and notches. For the plane $z = 0$,

$$x = c \cosh\alpha \cos\beta$$

$$y = c \sinh\alpha \sin\beta$$

where c is a constant. The lines of constant α are confocal ellipses with β varying from 0 to 2π, while lines of constant β are confocal hyperbolae. By

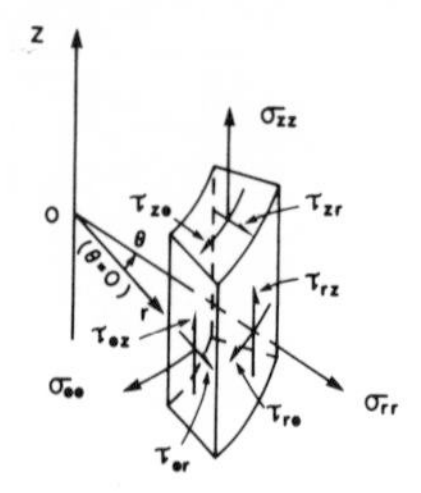

Fig. 1.6 Components of stress in cylindrical polar coordinates

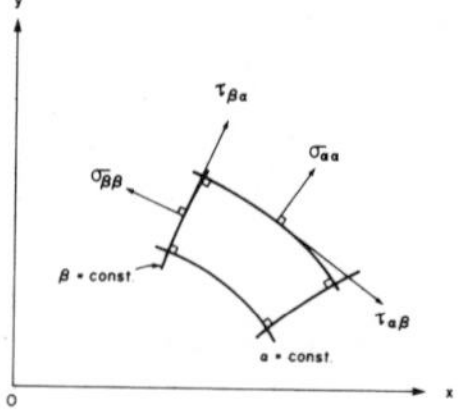

Fig. 1.7 Components of stress in curvilinear coordinates

adjustment of the value of the constant, an ellipse can be elongated to conform to the shape of an internal crack, or else the hyperbolae can be adjusted to conform to the shape of external notches.

1.2.2 Principal Stresses

Consider a stress acting on an oblique plane *ABC* of area *A* (Fig. 1.8), and let the plane be such that the stress acting normal to *ABC*, σ_i, is a *principal normal stress*, i.e., there is no shear stress acting on the plane. Let *l, m* and *n* be the direction cosines of σ_i, i.e., the cosines of the angles between σ_i and the *x, y* and *z*-axes, respectively. The components of σ_i which act in the *x, y* and *z*-directions are $S_x (= \sigma_i l)$, $S_y (= \sigma_i m)$ and $S_z (= \sigma_i n)$, and these act on areas *Al, Am* and *An*, respectively. For equilibrium, the summation of the forces in the *x*-direction gives

$$\sigma_i Al - \sigma_{xx} Al - \tau_{yx} Am - \tau_{zx} An = 0$$

or

$$(\sigma_i - \sigma_{xx})l - \tau_{yx} m - \tau_{zx} n = 0 \qquad (1.10a)$$

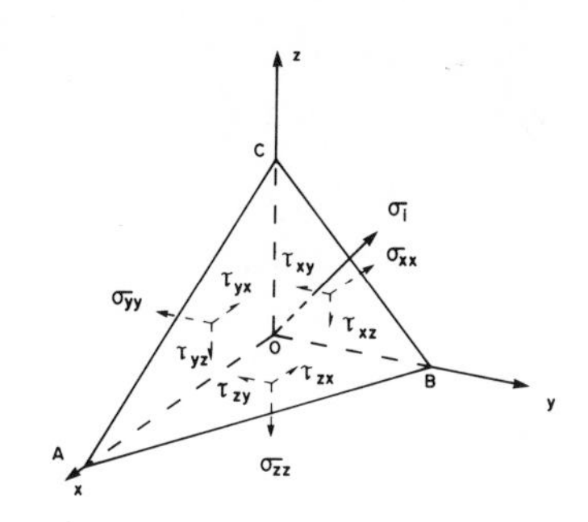

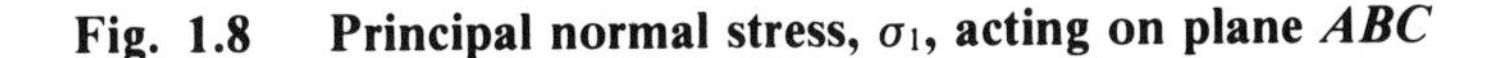

Fig. 1.8 Principal normal stress, σ_1, acting on plane *ABC*

Similarly,

$$-\tau_{xy}\, l + (\sigma_i - \sigma_{yy})m - \tau_{zy}\, n = 0 \qquad (1.10b)$$

$$-\tau_{xz}\, l - \tau_{yz}\, m + (\sigma_i - \sigma_{zz}) = 0 \qquad (1.10c)$$

A solution for Eqs. (1.10) for σ_i can be obtained by setting the determinant of *l*, *m* and *n* equal to zero, i.e.,

$$\begin{vmatrix} \sigma_i - \sigma_{xx} & -\tau_{yx} & -\tau_{zx} \\ -\tau_{xy} & \sigma_i - \sigma_{yy} & -\tau_{zy} \\ -\tau_{xz} & -\tau_{yz} & \sigma_i - \sigma_{zz} \end{vmatrix} = 0$$

From which we obtain

$$\sigma_i^3 - I_1\, \sigma_i^2 + I_2 \sigma_i - I_3 = 0 \qquad (1.11)$$

where

$$\left.\begin{aligned} I_1 &= \sigma_{xx} + \sigma_{yy} + \sigma_{zz} \\ I_2 &= \sigma_{xx}\,\sigma_{yy} + \sigma_{yy}\,\sigma_{zz} + \sigma_{zz}\,\sigma_{xx} + \tau_{xy}^2 - \tau_{yz}^2 - \tau_{zx}^2 \\ I_3 &= \begin{vmatrix} \sigma_{xx} & \tau_{xy} & \tau_{xz} \\ \tau_{xy} & \sigma_{yy} & \tau_{yz} \\ \tau_{xz} & \tau_{yz} & \sigma_{zz} \end{vmatrix} \end{aligned}\right\} \qquad (1.12)$$

The three roots of Eq. (1.11) are the three principal normal stresses, σ_1, σ_2 and σ_3, and their directions with respect to the *x*, *y* and *z*-axes can be found by substitution of σ_1, σ_2 and σ_3 into Eqs. (1.10), the resulting expressions being solved with the aid of the additional relationship, $l^2 + m^2 + n^2 = 1$. The principal normal stresses act on the *principal planes*, and these are mutually orthogonal. By convention, we choose σ_1, σ_2 and σ_3 such that algebraically $\sigma_1 > \sigma_2 > \sigma_3$.

The quantities I_1, I_2 and I_3 are independent of the choice of axes in the original coordinate system, since the principal stresses are independent of this. Thus, they are termed *invariants* of the stress tensor.

Consider an oblique plane (not a principal plane), given the principal normal stresses acting on the body, and let l, m and n be the direction cosines of the plane with respect to the principal axes (see Fig. 1.9). The total stress, S, on the plane is given by

$$S^2 = \sigma_1^2 l^2 + \sigma_2^2 m^2 + \sigma_3^2 n^2 \tag{1.13}$$

The normal stress on the plane is

$$\sigma = \sigma_1 l^2 + \sigma_2 m^2 + \sigma_3 n^2 \tag{1.14}$$

Thus, the shear stress on the plane is given by

$$\tau^2 = S^2 - \sigma^2 = \sigma_1^2 l^2 + \sigma_2^2 m^2 + \sigma_3^2 n^2 - (\sigma_1 l^2 + \sigma_2 m^2 + \sigma_3 n^2)$$

which reduces to

$$\tau = [(\sigma_1 - \sigma_2)^2 l^2 m^2 + (\sigma_2 - \sigma_3)^2 m^2 n^2 + (\sigma_3 - \sigma_1)^2 n^2 l^2]^{1/2} \tag{1.15}$$

on substitution of $l^2 + m^2 + n^2 = 1$.

It may be seen from Eq. (1.15) that if $\sigma_1 = \sigma_2 = \sigma_3$ (the condition of *hydrostatic stress*) the shear stress disappears, regardless of orientation. The maximum values of shear stress occur on planes for which the direction cosines of their normals bisect the angle between two of the principal axes. These are planes on which the *principal shear stresses* act, and these stresses are

$$\left.\begin{aligned} \tau_{12} &= (\sigma_2 - \sigma_3)/2 \\ \tau_{23} &= (\sigma_1 - \sigma_3)/2 \\ \tau_{31} &= (\sigma_1 - \sigma_2)/2 \end{aligned}\right\} \tag{1.16}$$

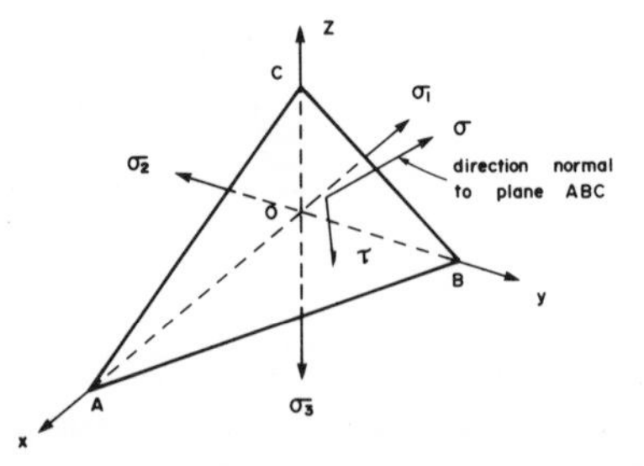

Fig. 1.9 Normal and shear components of stress acting on an oblique plane *ABC*, the body being subjected to principal stresses σ_1, σ_2, σ_3

Since σ_1 is algebraically the largest and σ_3 is the smallest principal normal stress, τ_2 is the *maximum shear stress*, τ_{max}, and this quantity is of importance in the theories of yielding.

Another plane of importance in failure theory is that cutting the three principal stress axes at equal distances from the origin, and illustrated in Fig. 1.10. This is one of the eight faces of a regular octahedron, and is termed the *octahedral plane*. The direction cosines of its normal are equal and of value $1/\sqrt{3}$. Thus, the *octahedral shear stress*, τ_{oct}, is given by

$$\tau_{oct} = [(\sigma_1 - \sigma_2)^2 + (\sigma_2 - \sigma_3)^2 + (\sigma_3 - \sigma_1)^2]^{1/2}/3 \tag{1.17}$$

and the normal stress on this plane is

$$\sigma_{oct} = (\sigma_1 + \sigma_2 + \sigma_3)/3 \tag{1.18}$$

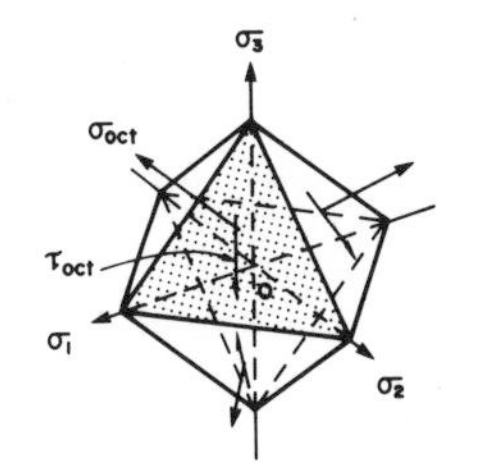

Fig. 1.10 Octahedral planes cutting principal stress axes at equal distances from the origin

1.2.3 Mohr's Circle for Stress

In addition to the use of Eqs. (1.14) and (1.15) for the determination of stresses on an oblique plane, a graphical method due to Mohr may be used. This method of representing the stress state provides a useful tool for the visualization of stresses in a body, and in it a triaxial state of stress, defined in terms of the three principal stresses, is represented by three Mohr's circles.

Figure 1.11(a) shows a cubic element cut by an oblique plane, and subject to principal stresses, σ_1, σ_2 and σ_3. The normal and shear stresses on the plane *abcd* are respectively σ and τ as shown, and they act on this at point Q. The plane is drawn such that at Q it is a tangent to a spherical surface lying within the cube. PQ is the line of intersection of the planes A_2PB_3Q and A_3B_1Q which are shown shaded, and these planes are oriented with respect to σ_1 by the angles θ and ϕ, respectively, Thus, the line PQ is defined with respect to the principal axes by θ and ϕ.

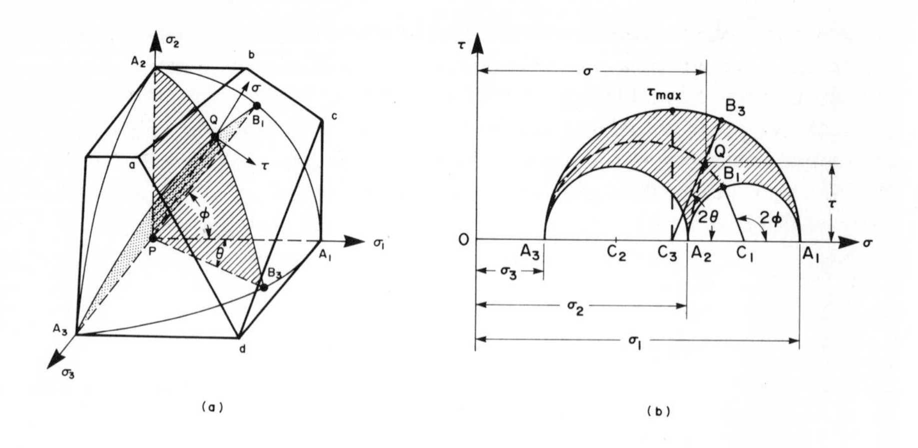

Fig. 1.11 Mohr's circle for three-dimensional stress

Referring to Fig. 1.11(b), we can apply the following procedure to determine the values of σ and τ [*1*]:

(1) On a σ - τ coordinate system, mark off σ_1, σ_2 and σ_3 as shown.
(2) Draw three Mohr semicircles centered at C_1, C_2 and C_3, and with diameters A_1A_2, A_2A_3 and A_1A_3.
(3) At C_1 draw the line C_1B_1 at an angle 2ϕ to the σ-axis; at C_3 draw C_3B_3 at 2θ. B_1 and B_3 are the points of intersection with the circles centered on C_1 and C_3, respectively.
(4) Construct arcs through A_3 and B_1, and A_2 and B_3, both being centered on the σ-axis. Their point of intersection defines Q.

All values of Q, covering all possible stress conditions within the body, fall within the shaded area or along the circumferences of the three circles [*2*]. The extreme value of τ is achieved when $\theta = 45°$ and $\phi = 0$, and is

$$\tau_{max} = (\sigma_1 - \sigma_3)/2$$

The octahedral plane is defined by $\theta = \phi = 45°$, and thus σ_{oct} and τ_{oct} may be determined graphically.

Figure 1.12 shows Mohr's circles for several stress states, and several points may be noted. First, we see that biaxial tensile loading does not reduce the value of the maximum shear stress which is produced under uniaxial tensile loading: this would *not* be clear if the more commonly used two-dimensional Mohr's circle had been used[1]. We see that the application of triaxial tensile stress, however, does reduce the maximum shear stress, and in the extreme

---

[1] For treatment of the two-dimensional case, see, e.g., Refs. [*1*] or [*3*].

case of equal triaxial tensile stresses, the Mohr's circle reduces to a point, and the material would behave in a brittle manner, there being no shear stresses to produce plastic flow. The application of compressive stresses in the σ_2 and σ_3 directions to a body under uniaxial tension, σ_1, is seen to increase greatly the maximum shear stress, and provides an explanation for the good ductility obtained in metals when they are drawn through a die.

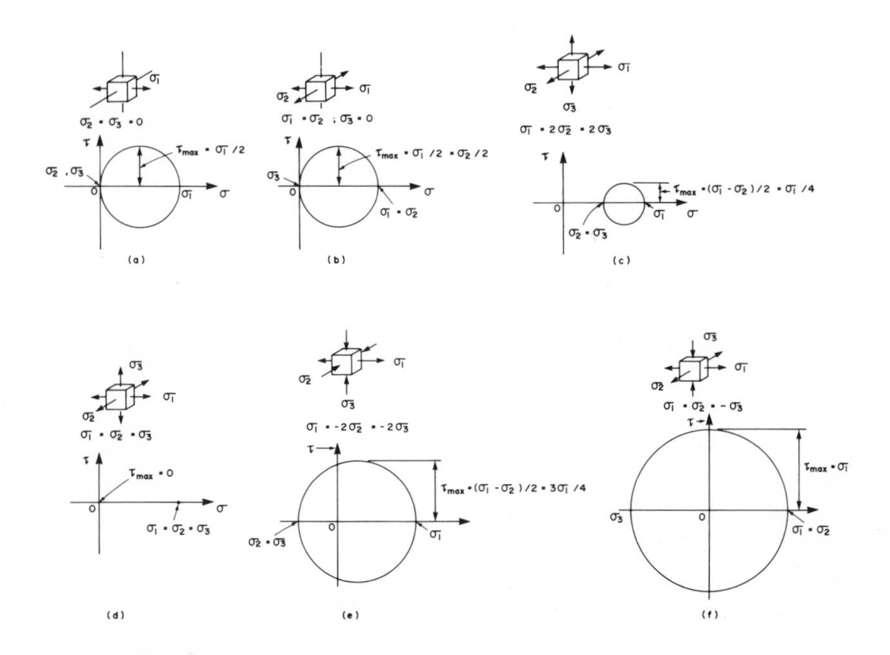

Fig. 1.12 Mohr's circles for various stress states: (a) uniaxial tension; (b) equibiaxial tension; (c) triaxial tension; (d) hydrostatic tension; (e) uniaxial tension plus equibiaxial compression; (f) equibiaxial tension plus uniaxial compression

1.3 STRAIN

The concepts of strain were described in Section 1.1.1. Here we are concerned with the generalized description of strain at a point, the determination of principal strains in a body, and the equations of compatibility.

1.3.1 Strain at a Point

We may define the strain at a point in a manner similar to the definition of stress. Consider a body which has been deformed such that a length originally of magnitude Δx has changed to $\Delta x + \Delta u$. The normal strain at a point may be defined as the limit of $\Delta u / \Delta x$ as Δx tends to zero, or

$$e_{xx} = \lim_{\Delta x \to 0} (\Delta u / \Delta x) = du/dx \tag{1.19}$$

It is conventional to denote elongations as positive strains and contractions as negative ones.

Now consider a two-dimensional body in which all deformation is confined to the xy-plane, as shown in Fig. 1.13. The normal strains in the x and y-directions are given by

$$e_{xx} = \partial u / \partial x \tag{1.20a}$$

$$e_{yy} = \partial v / \partial y \tag{1.20b}$$

For small deformations, $\tan \alpha_x \simeq \alpha_x \simeq \partial v / \partial x$, and $\tan \alpha_y \simeq \alpha_y \simeq \partial u / \partial y$, and the total shear strain, γ_{xy}, which is also equal to γ_{yx}, is given by the change in the angle of the two elements originally parallel to the x and y-axes. Thus,

$$\gamma_{xy} = \partial v / \partial x + \partial u / \partial y = \gamma_{yx} \tag{1.20c}$$

If we extend the analysis to three dimensions, it is simply seen that

$$e_{zz} = \partial w / \partial z \tag{1.20d}$$

$$\gamma_{xz} = \partial w / \partial x + \partial u / \partial z = \gamma_{zx} \tag{1.20e}$$

$$\gamma_{yz} = \partial w / \partial y + \partial v / \partial z = \gamma_{zy} \tag{1.20f}$$

Thus, we see that six terms are required to define strain at a point, as given in Eqs. (1.20 a to f).

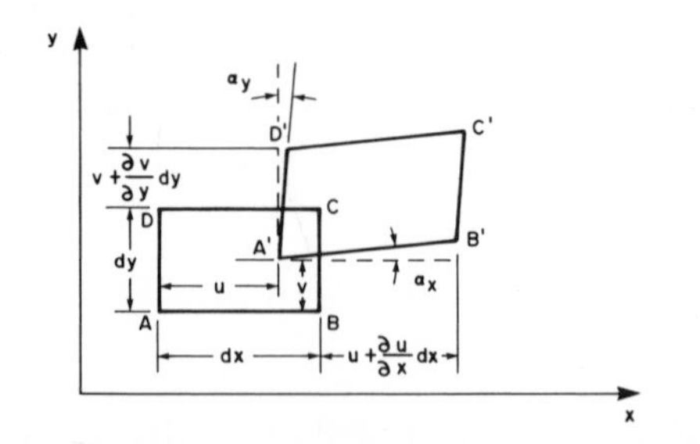

Fig. 1.13 Two-dimensional body *ABCD* subjected to strain

The strain at a point may be more conveniently defined in terms of the *strain tensor*, e_{ij}, which may be set in the form

$$\begin{matrix} e_{xx} & e_{xy} & e_{xz} \\ e_{yx} & e_{yy} & e_{yz} \\ e_{zx} & e_{zy} & e_{zz} \end{matrix}$$

or by

$$e_{ij} = (\partial u_i/\partial x_j + \partial u_j/\partial x_i)/2 \tag{1.21}$$

where $i, j = x, y, z$. The normal strains are obtained from this when $i = j$.

It is important to note that the shear strain components obtained from Eq. (1.21), which have the physical meaning of pure rotationless shears, are half the total shear strain components γ_{ij}, which are physically equivalent to simple shear as in torsion testing.

1.3.2 Principal Strains

By analogy with stress, a set of orthogonal coordinate axes may be defined along which there are no shear strains. For an isotropic body these *principal axes of strain* are found to be identical to the principal stress axes [*4*]. Thus, an element aligned along one of these axes will be subject to simple extension or compression only, without any rotation or shear.

The determination of planes of principal strains, $e_1 > e_2 > e_3$, and of the principal shearing strains, follows that for stress, and we may use the same equations, substituting appropriate strain components for the stress components. In the substitution, σ is replaced by e, and τ by $\gamma/2$. The three principal strains are the roots of the following equation obtained from Eq. (1.11):

$$e_i^3 - I_1 e_i^2 + I_2 e_i - I_3 = 0 \tag{1.22}$$

where

$$\left.\begin{aligned} I_1 &= e_{xx} + e_{yy} + e_{zz} \\ I_2 &= e_{xx}e_{yy} + e_{yy}e_{zz} + e_{zz}e_{xx} - (\gamma_{xy}/2)^2 - (\gamma_{yz}/2)^2 - (\gamma_{zx}/2)^2 \\ I_3 &= \begin{vmatrix} e_{xx} & \gamma_{xy}/2 & \gamma_{xz}/2 \\ \gamma_{xy}/2 & e_{yy} & \gamma_{yz}/2 \\ \gamma_{xz}/2 & \gamma_{yz}/2 & e_{zz} \end{vmatrix} \end{aligned}\right\} \tag{1.23}$$

The *principal shearing strains* are obtained from Eq. (1.16) as

$$\left.\begin{aligned} \gamma_{12} &= e_2 - e_3 \\ \gamma_{23} &= e_1 - e_3 \\ \gamma_{31} &= e_1 - e_2 \end{aligned}\right\} \tag{1.24}$$

γ_2 being the maximum shearing strain.

1.3.3 Mohr's Circle for Strain

If one set of strain components is given, another set may be required for different axes. If these new axes are derived from rotation about one of the old, the problem can be solved once more using the Mohr circle diagram. In this, normal strain, e_{ii} is plotted on the abscissa, and shear strain represented by $e_{ij} = \gamma_{ij}/2$ is plotted on the ordinate. Figure 1.14 shows the general case for three-dimensional strain, and solution of problems is as for the stress diagram.

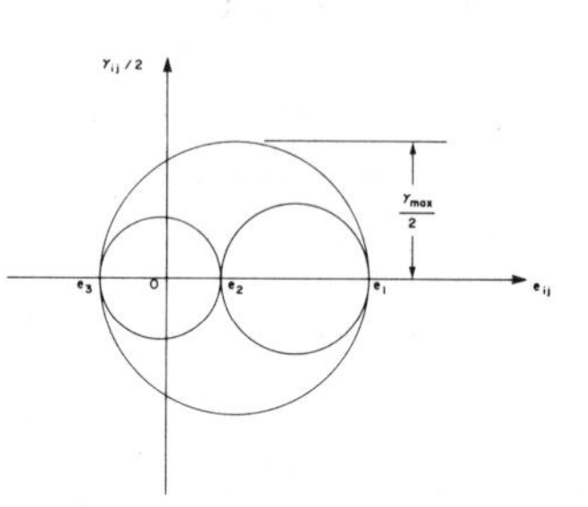

Fig. 1.14 Mohr's circle plot for three-dimensional strain

1.3.4 Compatibility Equations

As seen in Eq. (1.21), six strain components are connected to three components of displacement only. Hence, there must be restrictions on how the strain components are related. If we consider a two-dimensional strain with e_{xx}, e_{yy} and γ_{xy}, and differentiate the expressions for these, as given in Eqs. (1.20 a to c), appropriately with respect to x and y, we obtain

$$\partial^2 e_{xx}/\partial y^2 = \partial^3 u/\partial x \partial y^2, \quad \partial^2 e_{yy} = \partial^3 u/\partial x^2 \partial y,$$
$$\partial^2 \gamma_{xy}/\partial x \partial y = \partial^3 u/\partial x \partial y^2 + \partial^3 u/\partial x^2 \partial y$$

or

$$\partial^2 e_{xx}/\partial y^2 + \partial^2 e_{yy}/\partial x^2 = \partial^2 \gamma_{xy}/\partial x \partial y \tag{1.25}$$

Similarly, we obtain the three-dimensional *compatibility equations* as

$$\left.\begin{aligned}
\frac{\partial^2 e_{xx}}{\partial y^2} + \frac{\partial^2 e_{yy}}{\partial x^2} = \frac{\partial^2 \gamma_{xy}}{\partial x\, \partial y}, \quad 2\frac{\partial^2 e_{xx}}{\partial y\, \partial z} = \frac{\partial}{\partial x}\left(-\frac{\partial \gamma_{yz}}{\partial x} + \frac{\partial \gamma_{xz}}{\partial y} + \frac{\partial \gamma_{xy}}{\partial z}\right) \\
\frac{\partial^2 e_{yy}}{\partial z^2} + \frac{\partial^2 e_{zz}}{\partial y^2} = \frac{\partial^2 \gamma_{yz}}{\partial y\, \partial z}, \quad 2\frac{\partial^2 e_{yy}}{\partial z\, \partial x} = \frac{\partial}{\partial y}\left(\frac{\partial \gamma_{yz}}{\partial x} - \frac{\partial \gamma_{xz}}{\partial y} + \frac{\partial \gamma_{xy}}{\partial z}\right) \\
\frac{\partial^2 e_{zz}}{\partial x^2} \times \frac{\partial^2 e_{xx}}{\partial z^2} = \frac{\partial^2 \gamma_{zx}}{\partial z\, \partial x}, \quad 2\frac{\partial^2 e_{zz}}{\partial x\, \partial y} = \frac{\partial}{\partial x}\left(\frac{\partial \gamma_{yz}}{\partial x} + \frac{\partial \gamma_{xz}}{\partial y} - \frac{\partial \gamma_{xy}}{\partial z}\right)
\end{aligned}\right\} \tag{1.26}$$

For a body to be deformed without fracturing, there cannot be discontinuities in its strain functions, and the components of strain at every point must satisfy the compatibility relations in Eq. (1.26).

1.4 STRESS-STRAIN RELATIONSHIPS

In uniaxial elastic deformation of a metal, stress and strain are directly related as shown in Eq. (1.7), the modulus of elasticity, E, being termed the *elastic coefficient* relating stress to strain. The coefficient relating strain to stress $(1/E)$ is termed the *elastic compliance*, and *Poisson's ratio* (ν), is the ratio of the lateral contraction to longitudinal expansion. For application of the tensile stress in the x-direction, we have

$$e_{yy} = e_{zz} = -\nu\, e_{xx} = -\nu\, \sigma_{xx}/E \tag{1.27}$$

For a perfectly isotropic elastic material, $\nu = 0.25$, but for most metals it lies in the range 0.28 - 0.33.

In the general case for three-dimensional stressing, and where the elastic constants may vary with orientation, the six independent stress components must be related to the six independent strain components by means of six linear equations, which involve 36 coefficients. Thus, we may write

$$\left.\begin{aligned}
\sigma_{xx} &= C_{11}\, e_{xx} + C_{12}\, e_{yy} + C_{13}\, e_{zz} + C_{14}\, e_{xy} + C_{15}\, e_{xz} + C_{16}\, e_{yz}\\
\sigma_{yy} &= C_{21}\, e_{xx} + C_{22}\, e_{yy} + C_{23}\, e_{zz} + C_{24}\, e_{xy} + C_{25}\, e_{xz} + C_{26}\, e_{yz}\\
\sigma_{zz} &= C_{31}\, e_{xx} + C_{32}\, e_{yy} + C_{33}\, e_{zz} + C_{34}\, e_{xy} + C_{35}\, e_{xz} + C_{36}\, e_{yz}\\
\tau_{xy} &= C_{41}\, e_{xx} + C_{42}\, e_{yy} + C_{43}\, e_{zz} + C_{44}\, e_{xy} + C_{45}\, e_{xz} + C_{46}\, e_{yz}\\
\tau_{yz} &= C_{51}\, e_{xx} + C_{52}\, e_{yy} + C_{53}\, e_{zz} + C_{54}\, e_{xy} + C_{55}\, e_{xz} + C_{56}\, e_{yz}\\
\tau_{zx} &= C_{61}\, e_{xx} + C_{62}\, e_{yy} + C_{63}\, e_{zz} + C_{64}\, e_{xy} + C_{65}\, e_{xz} + C_{66}\, e_{yz}
\end{aligned}\right\} \tag{1.28}$$

or

$$\left.\begin{aligned}
e_{xx} &= S_{11}\, \sigma_{xx} + S_{12}\, \sigma_{yy} + S_{13}\, \sigma_{zz} + S_{14}\, \tau_{xy} + S_{15}\, \tau_{xz} + S_{16}\, \tau_{yz}\\
e_{yy} &= S_{21}\, \sigma_{xx} + S_{22}\, \sigma_{yy} + S_{23}\, \sigma_{zz} + S_{24}\, \tau_{xy} + S_{25}\, \tau_{xz} + S_{26}\, \tau_{yz}\\
e_{zz} &= S_{31}\, \sigma_{xx} + S_{32}\, \sigma_{yy} + S_{33}\, \sigma_{zz} + S_{34}\, \tau_{xy} + S_{35}\, \tau_{xz} + S_{36}\, \tau_{yz}\\
e_{xy} &= S_{41}\, \sigma_{xx} + S_{42}\, \sigma_{yy} + S_{43}\, \sigma_{zz} + S_{44}\, \tau_{xy} + S_{45}\, \tau_{xz} + S_{46}\, \tau_{yz}\\
e_{yz} &= S_{51}\, \sigma_{xx} + S_{52}\, \sigma_{yy} + S_{53}\, \sigma_{zz} + S_{54}\, \tau_{xy} + S_{55}\, \tau_{xz} + S_{56}\, \tau_{yz}\\
e_{zx} &= S_{61}\, \sigma_{xx} + S_{62}\, \sigma_{yy} + S_{63}\, \sigma_{zz} + S_{64}\, \tau_{xy} + S_{65}\, \tau_{xz} + S_{66}\, \tau_{yz}
\end{aligned}\right\} \tag{1.29}$$

where C_{11}, C_{12}, ..., C_{ij} are the *elastic coefficients*, and S_{11}, S_{12}, ..., S_{ij} are the *elastic compliances*.

We can reduce the number of constants required to calculate stress from strain (or *vice versa*) as, from considerations of symmetry, those elastic constants whose subscripts are equal are equivalent when the order of the subscripts in reversed, or $C_{ij} = C_{ji}$, and $S_{ij} = S_{ji}$. For isotropic solids the number of constants can be reduced still further on the basis of the coincidence of the principal axes of stress and strain, and the symmetry of displacements about these axes. For the principal axes, we may write

$$\left.\begin{aligned} \sigma_1 &= (\lambda + 2G)e_1 + \lambda e_2 + \lambda e_3 \\ \sigma_2 &= \lambda e_1 + (\lambda + 2G)e_2 + \lambda e_3 \\ \sigma_3 &= \lambda e_1 + \lambda e_2 + (\lambda + 2G)e_3 \end{aligned}\right\} \quad (1.30)$$

where λ and G are termed the Lamé constants. Substituting the dilatation, $\Delta = e_1 + e_2 + e_3$, we obtain

$$\left.\begin{aligned} \sigma_1 &= \lambda\Delta + 2Ge_1 \\ \sigma_2 &= \lambda\Delta + 2Ge_2 \\ \sigma_3 &= \lambda\Delta + 2Ge_3 \end{aligned}\right\} \quad (1.31)$$

Referring to the x, y and z axes, we have

$$\left.\begin{aligned} \sigma_{xx} &= \lambda\Delta + 2Ge_{xx}\,, \quad \tau_{xy} = G\gamma_{xy} \\ \sigma_{yy} &= \lambda\Delta + 2Ge_{yy}\,, \quad \tau_{yz} = G\gamma_{yz} \\ \sigma_{zz} &= \lambda\Delta + 2Ge_{zz}\,, \quad \tau_{zx} = G\gamma_{zx} \end{aligned}\right\} \quad (1.32)$$

Thus, we see that G is the shear modulus of elasticity, as defined in Eq. (1.8).

Consider a thin member in uniaxial tension, i.e., $\sigma_2 = \sigma_3 = 0$. We may write the stresses as

$$\sigma_1 = \lambda\Delta + 2Ge_1 = Ee_1 \quad (1.33)$$

$$\sigma_2 = \sigma_3 = 0 = \lambda\Delta + 2Ge_2 = \lambda\Delta + 2G\nu e_1 \quad (1.34)$$

From Eqs. (1.33) and (1.34), we obtain

$$G = E/2(\nu + 1) \quad (1.35)$$

It is seen that the three commony used material constants, G, E, and ν are related, and that for an isotropic material there are only two independent constants. The second of Lamé's constants, λ, is given by

$$\lambda = \nu E/[(1 + \nu)(1 - 2\nu)] \quad (1.36)$$

For a body acted upon by a generalized stress system, the strain along a principal axis is given by the strain produced by the stress acting along that axis plus the superimposed strains arising from the Poisson effect of the other two principal stresses. Hence, we may write

$$\left.\begin{aligned} e_1 &= [\sigma_1 - \nu(\sigma_2 + \sigma_3)]/E \\ e_2 &= [\sigma_2 - \nu(\sigma_3 + \sigma_1)]/E \\ e_3 &= [\sigma_3 - \nu(\sigma_1 + \sigma_2)]/E \end{aligned}\right\} \quad (1.37)$$

For biaxial or *plane stress*, $\sigma_3 = 0$, and Eqs. (1.37) reduce to

$$\left.\begin{aligned} e_1 &= (\sigma_1 - \nu\sigma_2)/E \\ e_2 &= (\sigma_2 - \nu\sigma_1)/E \\ e_3 &= -\nu(\sigma_1 + \sigma_2)/E \end{aligned}\right\} \quad (1.38)$$

For *plane strain*, where $e_2 = 0$, we obtain

$$\left.\begin{aligned} e_1 &= (1 + \nu)[(1 - \nu)\sigma_1 - \nu\sigma_3]/E \\ e_3 &= (1 + \nu)[(1 - \nu)\sigma_3 - \nu\sigma_1]/E \end{aligned}\right\} \quad (1.39)$$

Setting $e_2 = 0$ rather than $e_3 = 0$, enables us to preserve the convention of $\sigma_1 > \sigma_2 > \sigma_3$.

It is useful to be able to determine the values of principal stresses in a plane stress situation from measured values of strain, for example from strain gages on the surface of a component. From Eqs. (1.38) for e_1 and e_2, we obtain

$$\left.\begin{aligned} \sigma_1 &= E(e_1 + \nu e_2)/(1 - \nu^2) \\ \sigma_2 &= E(e_2 + \nu e_1)/(1 - \nu^2) \end{aligned}\right\} \quad (1.40)$$

Hence, from measured or computed values of the principal strains in the plane of a surface, the principal stresses can be found.

1.5 ELEMENTS OF ELASTICITY THEORY

The *theory of elasticity* involves a more rigorous consideration of stresses and strains in a body than is done from an elementary analysis using strength of materials theory. The two principles on which it is based are that the conditions of equilibrium are satisfied and that the compatibility equations are obeyed. Figure 1.15 shows an element in a body with the nine pairs of force components acting on it. Considering force equilibrium, it may be seen that the following relationships must be satisfied:

$$\left.\begin{aligned} \Sigma P_x &= \partial\sigma_{xx}/\partial x + \partial\tau_{yx}/\partial y + \partial\tau_{zx}/\partial z = 0 \\ \Sigma P_y &= \partial\tau_{xy}/\partial x + \partial\sigma_{yy}/\partial y + \partial\tau_{zy}/\partial z = 0 \\ \Sigma P_z &= \partial\tau_{xz}/\partial x + \partial\tau_{yz}/\partial y + \partial\sigma_{zz}/\partial z = 0 \end{aligned}\right\} \quad (1.41)$$

ignoring any body forces which may act on the element (e.g., its weight).

Eqs. (1.41) must be satisfied at every point throughout the body. In addition, the boundary conditions at the surface of the body must be satisfied, i.e., the stresses at the surface must balance the externally applied forces.

The compatibility equations [Eqs. (1.26)] must also be satisfied so that there is elastic continuity throughout the body. They may be rewritten in terms of stress by differentiating Eqs. (1.32) and substituting appropriately. To simplify matters, we consider a two-dimensional plane stress system with

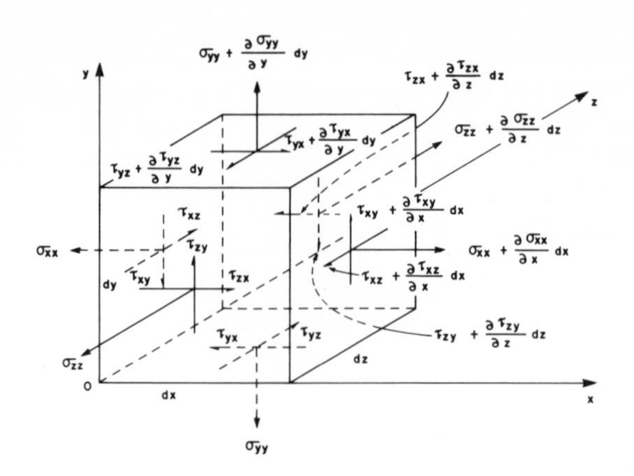

Fig. 1.15 Element of a body under equilibrium with nine pairs of force components on it

stresses of the type σ_{zi} all equal to zero. Thus, we obtain

$$\partial^2\sigma_{xx}/\partial y^2 - \nu\,\partial^2\sigma_{yy}/\partial y^2 + \partial^2\sigma_{yy}/\partial x^2 - \nu\,\partial^2\sigma_{xx}/\partial x^2 = 2\,(1+\nu)\partial^2\tau_{yx}/\partial x\partial y \quad (1.42)$$

The solution of a problem in the theory of elasticity requires that expressions for the stress components be found which satisy the equilibrium equations [Eqs. (1.41)], the compatibility equations [Eq. (1.42)] and the boundary conditions. The solution of such problems frequently involves considerable mathematical complication, most of which arises from the requirement of continuity. For tractable solutions it is preferable that the body can be described by simple mathematical functions, but even so such solutions are not simple unless rotational symmetry or similar simplifications can be applied.

1.5.1 The Airy Stress Function

In order to solve elasticity problems, we require a function ϕ in x and y which satisfies Eqs. (1.41) and (1.42) and enables the stresses to be related to the external loads. Such a *stress function* was introduced by Airy [5] for two-dimensional stressing, and he showed that the stresses could be derived from it as follows:

$$\sigma_{xx} = \partial^2\phi/\partial y^2,\ \sigma_{yy} = \partial^2\phi/\partial x^2,\ \tau_{xy} = -\,\partial^2\phi/\partial x\partial y \quad (1.43)$$

These equations satisfy the conditions of equilibrium [Eqs. (1.41)] with stresses of the type $\sigma_{zi} = 0$. Substitution of them into the compatibility equation (Eq. 1.42) gives us

$$\partial^4\phi/\partial x^4 + 2\partial^4\phi/\partial x^2\partial y^2 + \partial^4\phi/\partial y^4 = \nabla^4\phi = 0 \quad (1.44)$$

If a stress function can be found to satisfy this equation, the stresses can be found from Eqs. (1.43) provided the boundary conditions are also satisfied. Determination of the stress function for a particular situation is normally difficult, and finite element methods or finite difference equations are frequently used. For more detailed consideration of this topic, the reader should consult the relevant standard reference texts, e.g. [*6, 7*].

1.6 STRESS CONCENTRATION

A important matter is the determination of the stress distribution at a discontinuity in a member. Adjacent to the discontinuity the stress will be higher than the average stress value. This is of particular relevance in components containing holes, notches or sharp changes in section, and such points are likely starting points for failure to occur, if care is not taken to minimize the local stresses by providing a gradual change is section. Such local stresses are normally specified in terms of the *theoretical stress concentration factor, k_t*. This is specified on the basis of the ratio of the maximum stress, determined on the basis of elasticity theory, to the nominal stress based on the net section.

Stress concentration factors have been calculated from elasticity theory for many simple cases, while for more complex situations, which are not tractable analytically, the techniques of photoelasticity are frequently used [*8*]. In addition, finite element analysis is being used increasingly to solve more difficult geometries, although it is largely used in two-dimensional and axisymmetric cases.

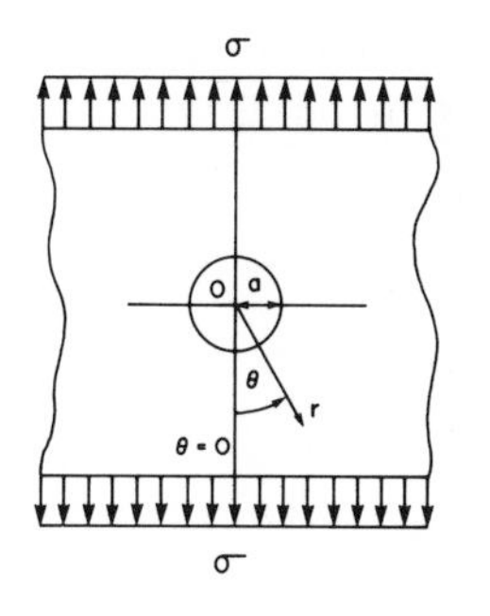

Fig. 1.16 Thin rectangular sheet containing a circular hole and which is subjected to axial loading

As an example, consider a circular hole in a thin rectangular sheet whose boundaries are an infinite distance from the center of the hole, and which is subjected to axial loading as shown in Fig. 1.16. We assume that the situation is one of plane stress, and we consider the case of polar coordinates r and θ.[2]

Neglecting body forces, the equilibrium equations may be written as

$$\left.\begin{aligned} \partial\sigma_{rr}/\partial r + (1/r)\partial\tau_{r\theta}/\partial\theta + (\sigma_{rr} - \sigma_{\theta\theta})/r = 0 \\ (1/r)\,\partial\sigma_{\theta\theta}/\partial\theta + \partial\tau_{r\theta}/\partial r + 2\tau_{r\theta}/r = 0 \end{aligned}\right\} \quad (1.45)$$

and the compatibility equation is

$$\nabla^4\phi = \left(\frac{\partial^2}{\partial r^2} + \frac{1}{r}\frac{\partial}{\partial r} + \frac{1}{r^2}\frac{\partial^2}{\partial\theta^2}\right)\left(\frac{\partial^2\phi}{\partial r^2} + \frac{1}{r}\frac{\partial\phi}{\partial r} + \frac{1}{r^2}\frac{\partial^2\phi}{\partial\theta^2}\right) = 0 \quad (1.46)$$

The stresses are given by the following relations, in terms of an appropriate stress function:

$$\left.\begin{aligned} \sigma_{rr} &= (1/r)\partial\phi/\partial r + (1/r^2)\partial^2\phi/\partial\theta^2 \\ \sigma_{\theta\theta} &= \partial^2\phi/\partial r^2 \\ \tau_{r\theta} &= (1/r)\partial\phi/\partial\theta - (1/r)\partial^2\phi/\partial r\partial\theta = -\frac{\partial}{\partial r}\left(\frac{1}{r}\frac{\partial\phi}{\partial\theta}\right) \end{aligned}\right\} \quad (1.47)$$

The boundary conditions at the circumference of the hole are

$$\sigma_{rr} = \tau_{r\theta} = 0 \quad \text{at } r = a$$

At large distances from the hole, $\sigma_{\theta\theta}$, σ_{rr} and $\tau_{r\theta}$ are given by their values for an infinite plate in the absence of a hole. These are

$$\sigma_{rr} = \tfrac{1}{2}\sigma(1 + \cos 2\theta),\ \sigma_{\theta\theta} = \tfrac{1}{2}\sigma\,(1 - \cos 2\theta),\ \tau_{r\theta} = -\tfrac{1}{2}\sigma \sin 2\theta,$$

and on solution we obtain the stress function satisfying the compatibility equation and these boundary conditions as,

$$\phi = [-(\sigma/4)r^2 - (a^4/4)(\sigma/r^2) + (a^2/2)\sigma]\cos 2\theta \quad (1.48)$$

Substituting this in Eq. (1.47) we obtain

$$\left.\begin{aligned} \sigma_{rr} &= (\sigma/2)\,[(1 - a^2/r^2) + (1 + 3a^4/r^4 - 4a^2/r^2)\cos 2\theta] \\ \sigma_{\theta\theta} &= (\sigma/2)\,[(1 + a^2/r^2) - (1 + 3a^4/r^4)\cos 2\theta] \\ \tau_{r\theta} &= -(\sigma/2)\,(1 - 3a^4/r^4 + 2a^2/r^2)\sin 2\theta \end{aligned}\right\} \quad (1.49)$$

The maximum value of $\sigma_{\theta\theta}$ occurs at $\theta = \pi/2$ or $3\pi/2$, and for $r = a$. For this case

$$\sigma_{\theta\theta} = 3\sigma \quad (1.50)$$

this being the maximum stress. Hence the theoretical stress concentration factor for a circular hole in a plate is 3. We may also see that a radial stress has

[2] The solution of this problem is considered in some detail by Ugural and Fenster [1], for example. The main steps only are presented here.

been produced close to the hole. For $\theta = \pi/2$ or $3\pi/2$,

$$\sigma_{rr} = (3/2)\,(a^2/r^2 - a^4/r^4) \tag{1.51}$$

which has a maximum value at $r = \sqrt{2}a$ of $3\sigma/8$. The variation in both σ_{rr} and $\sigma_{\theta\theta}$ with r, along the line for $\theta = \pi/2$ or $3\pi/2$, is shown in Fig. 1.17, and it may be seen that the peak stress value ($\sigma_{\theta\theta}$ at $r = a$) falls off rapidly, being only some 7% above σ at $r = 3a$.

For the plane strain situation, where $e_{zz} = 0$, we have σ_{zz} from Eq. (1.37) as

$$\sigma_{zz} = \nu\,(\sigma_{rr} + \sigma_{\theta\theta}) \tag{1.52}$$

Thus, a state of triaxial tension will be produced in this situation at a small distance from the surface of the hole. This will inhibit yielding by reducing the maximum shear stress as discussed briefly in Section 1.2.3.

Another particularly useful analysis was made by Inglis [*9*], who solved the problem of an elliptical hole in an infinite plate stressed uniaxially as before in Fig. 1.16, and with the major axis of the ellipse lying at right angles to the applied stress. The maximum stress (at the end of the major axis on the surface of the hole) is given by

$$\sigma_{max} = \sigma(1 + 2a/b) = \sigma[1 + 2(a/\rho)^{1/2}] \tag{1.53}$$

where $2a$ and $2b$ are respectively the major and minor axes of the ellipse, and ρ is the tip radius at the end of the major axis. For a circular hole ($a = b$), Eq. (1.53) reduces to Eq. (1.50).

Data on stress concentration factors for a wide range of geometries are available in the literature in the form of graphs, tables and formulae, and the reader is referred to the appropriate sources, e.g. [*10-12*].

So far, we have discussed the *theoretical* rather than the *actual* stress concentration. In a ductile metal, yielding will take place at the point of maximum stress when the condition for yielding is exceeded (see Chapter 2). The peak stresses will be reduced and there will be stress redistribution across

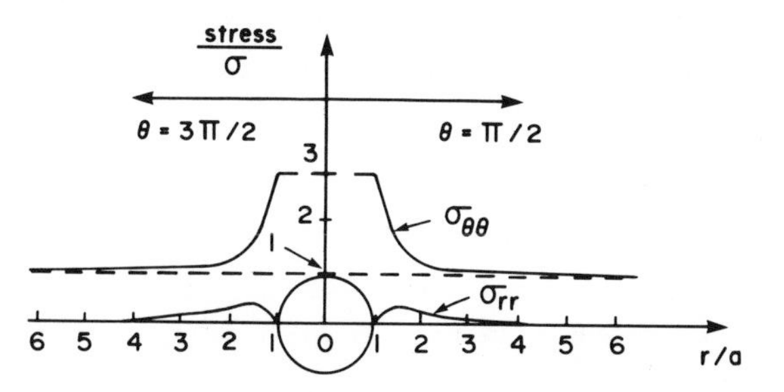

Fig. 1.17 Variation in radial stress, σ_{rr}, and hoop stress, $\sigma_{\theta\theta}$, along the line for $\theta = \pi/2$ or $3\pi/2$ for the loading situation shown in Fig. 1.16

the section because of plastic flow, to produce an actual stress concentration factor less than the theoretical value. In brittle solids, however, no such redistribution can take place, and they are much more sensitive to stress concentrations in promoting failure than are ductile solids.

1.7 PLASTIC FLOW

The criteria for yield or the onset of plastic deformation in multiaxial stressing will be set out in some detail in Chapter 2. Here, our purpose is to describe the basic relations governing plastic flow in metals. These are of importance in the area of metal forming, which is not considered specifically in this volume, in creep deformation, which is discussed in Chapter 11, and in the relief of stresses at points of concentration as discussed in the preceding section.

For small strains ($< \sim 5\%$) true strain and engineering strain differ little: however, at larger strains considerable differences arise, and it is general in plastic deformation to use true strains. True and engineering strain are related as follows. From Eq. (1.2), engineering strain is

$$e = l/l_o - 1 \quad \text{or} \quad e + 1 = l/l_o$$

True strain is [from Eq. (1.4)]

$$\epsilon = \ln (l/l_o) = \ln (e + 1) \tag{1.54}$$

During plastic deformation the volume remains approximately constant (neglecting elastic strains and any internal cracking or void formation), thus we can write Eq. (1.54) as[3]

$$\epsilon = \ln (l/l_o) = \ln (A_o/A) \tag{1.55}$$

and

$$\epsilon_1 + \epsilon_2 + \epsilon_3 = 0 \tag{1.56}$$

Consider now a rod deformed in uniaxial tensile loading. The principal true strains are ϵ_1, $-\epsilon_2$ and $-\epsilon_3$, and from symmetry $\epsilon_2 = \epsilon_3$. From Eq. (1.56), $\epsilon_1 = 2\epsilon_2 = 2\epsilon_3$. Hence, from the definition of Poisson's ratio [Eq. (1.27)], we see that $\nu = 0.5$ for fully plastic flow[4].

---

[3] In this and the remainder of this section, we are taking ϵ to be identically the total and plastic strains. Likewise, $\dot{\epsilon}$ represents the total or plastic strain rate.

---

[4] It should be noted, however, that ν increases from its elastic value ($\sim$0.25 to 0.33) to 0.5 for the *fully* plastic condition in a progressive manner, so that for a work-hardening material it is *not* as much as 0.5 during plastic deformation.

In obtaining Eq. (1.56), we have assumed that the principal strain axes do not rotate during deformation, and this is a reasonable assumption in most cases. In consequence, we may differentiate Eq. (1.56) directly with respect to time to give

$$\dot{\epsilon}_1 + \dot{\epsilon}_2 + \dot{\epsilon}_3 = 0 \tag{1.57}$$

St. Venant [*13*] originally proposed that the principal axes of strain *increment* coincided with the principal stress axes. Lévy [*14*] and von Mises [*15*] independently introduced general relations between strain increment and *reduced* stress in conformity with this, as

$$\frac{d\epsilon_{xx}}{\sigma'_{xx}} = \frac{d\epsilon_{yy}}{\sigma'_{yy}} = \frac{d\epsilon_{zz}}{\sigma'_{zz}} = \frac{d\gamma_{yz}}{\tau_{yz}} = \frac{d\gamma_{zx}}{\tau_{zx}} = \frac{d\gamma_{xy}}{\tau_{xy}} = \text{constant} \tag{1.58}$$

where σ'_{xx} has the value $(\sigma_{xx} - \sigma'')$, where σ'' is the hydrostatic component of stress [see Eq. (2.2)] = $(\sigma_1 + \sigma_2 + \sigma_3)/3$. Eq. (1.58) also implies that the Mohr circles for stress and strain increment are geometrically similar.

We can rewrite Eq. (1.58) for the principal strain increments only, as

$$\left.\begin{aligned} d\epsilon_1 &= d\lambda\,[\sigma_1 - (\sigma_2 + \sigma_3)/2] \\ d\epsilon_2 &= d\lambda\,[\sigma_2 - (\sigma_3 + \sigma_1)/2] \\ d\epsilon_3 &= d\lambda\,[\sigma_3 - (\sigma_1 + \sigma_2)/2] \end{aligned}\right\} \tag{1.59}$$

which are the Lévy-von Mises relationships. They are plastic counterparts of Hooke's law with $d\lambda$ replacing $1/E$ [cf. Eqs. (1.37)] and Poisson's ratio = 1/2.

The problem is now one of determining the constant $d\lambda$ which depends on the criteria for flow and the stress-strain characteristics of the material. For a material to follow a given flow theory, the stress-strain relationship can be defined in terms of *effective stress* and *effective strain* as $\epsilon_{\text{eff}} = f(\sigma_{\text{eff}})$. Various invariant stress and strain parameters may be used which will define the plastic stress-strain curve independent of the state of stress, including the octahedral shear stress and octahedral shear strain [*2*]. However, a useful quantity is the effective stress defined in the form

$$\sigma_{\text{eff}} = \frac{1}{\sqrt{2}}[(\sigma_1 - \sigma_2)^2 + (\sigma_2 - \sigma_3)^2 + (\sigma_3 - \sigma_1)^2]^{1/2} \tag{1.60}$$

and the corresponding effective strain increment is

$$d\epsilon_{\text{eff}} = \frac{\sqrt{2}}{3}[(d\epsilon_1 - d\epsilon_2)^2 + (d\epsilon_2 - d\epsilon_3)^2 + (d\epsilon_3 - d\epsilon_1)^2]^{1/2} \tag{1.61}$$

It is well demonstrated by experiment that σ_{eff} determines the ability of multiaxial stressing to induce yielding and plastic flow [*2*], and it may be identified as the tensile yield stress, Y, when $\sigma_2 = \sigma_3 = 0$. Similarly, $d\epsilon_{\text{eff}} = d\epsilon_1$ for pure tension. In the particular case of *proportional loading* or *proportional straining*, in which there are no changes in the directions of the principal strain increments and all increments are related in constant ratio, we can write $d\epsilon_2/d\epsilon_1 = A$, and from Eq. (1.56), we obtain $d\epsilon_3/d\epsilon_1 = -(A+1)$. Thus

$$d\epsilon_{\text{eff}} = \frac{2}{\sqrt{3}}(1 + A + A^2)^{1/2}\, d\epsilon \tag{1.62}$$

For A constant, $d\epsilon_1$, $d\epsilon_2$, $d\epsilon_3$ and $d\epsilon_{\text{eff}}$ can be replaced by ϵ_1, ϵ_2, ϵ_3 and ϵ_{eff}, respectively. Thus, we may write the effective strain as

$$\epsilon_{\text{eff}} = \frac{\sqrt{2}}{3}[(\epsilon_1 - \epsilon_2)^2 + (\epsilon_2 - \epsilon_3)^2 + (\epsilon_3 - \epsilon_1)^2]^{1/2} \tag{1.63}$$

When plastic deformation takes place, work is done. It is easily shown [*16*] that the incremental work term, generalized for all components of stress and strain, is given by[5]

$$dw = \sigma_{ij}\, d\epsilon_{ij} \tag{1.64}$$

We may also write the plastic work increment as

$$dw = \sigma_{\text{eff}}\, d\epsilon_{\text{eff}} \tag{1.65}$$

Now, if we define $d\lambda$ as

$$d\lambda = d\epsilon_{\text{eff}}/\sigma_{\text{eff}} \tag{1.66}$$

and substitute this in Eqs. (1.59), we obtain the quantity dw in Eq. (1.64) from summing its components in identical form to Eq. (1.65). For a strain hardening material we see that $\lambda = \epsilon_{\text{eff}}/\sigma_{\text{eff}}$, and this is shown in Fig. 1.18 to illustrate its physical meaning.

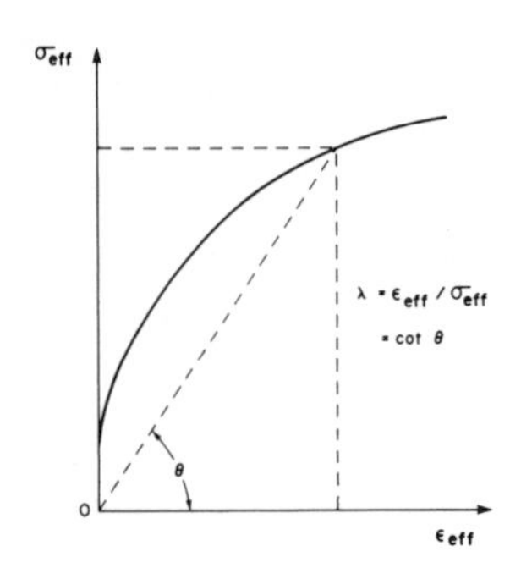

Fig. 1.18 Variation of effective strain with effective stress in a strain hardening material, showing the definition of λ

[5] Eq. (1.64) summarizes the nine component terms. However, since $d\epsilon_{ij} = d\gamma_{ij}/2$ from Section 1.3.1, we must include both $\sigma_{ij}d\epsilon_{ij}$ and $\sigma_{ji}d\epsilon_{ji}$ for $i \neq j$, or else only include a single $\tau_{ij}d\gamma_{ij}$ term in their place.

For proportional loading, we may replace the incremental strains in Eqs. (1.59) with the integrated strains. Hence,

$$\left.\begin{aligned} \epsilon_1 &= \frac{\epsilon_{eff}}{\sigma_{eff}}[\sigma_1 - (\sigma_2 + \sigma_3)/2] \\ \epsilon_2 &= \frac{\epsilon_{eff}}{\sigma_{eff}}[\sigma_2 - (\sigma_3 + \sigma_1)/2] \\ \epsilon_3 &= \frac{\epsilon_{eff}}{\sigma_{eff}}[\sigma_3 - (\sigma_1 + \sigma_2)/2] \end{aligned}\right\} \quad (1.67)$$

If the relationship between σ_{eff} and ϵ_{eff} is known, we can determine the plastic strains from a knowledge of the principal stresses.

1.7.1 Isotropic Rigid Ideal Plastic Material

The discussion so far has concerned plastic strains achieved rather than rates of deformation. For an isotropic rigid plastic material we can determine the flow rates under multiaxial loading.

In such a material, whose stress strain curve is shown in Fig. 1.19, the principal shear strain rates are proportional to the principal shear stresses [*2*]. Thus,

$$\dot{\gamma}_{12}/\tau_{12} = \dot{\gamma}_{23}/\tau_{23} = \dot{\gamma}_{31}/\tau_{31} = \text{constant} \quad (1.68)$$

These may also be written in terms of the principal strain rates and principal stresses as[6]

$$\dot{\epsilon}_1/\sigma_1' = \dot{\epsilon}_2/\sigma_2' = \dot{\epsilon}_3/\sigma_3' = \text{constant} \quad (1.69)$$

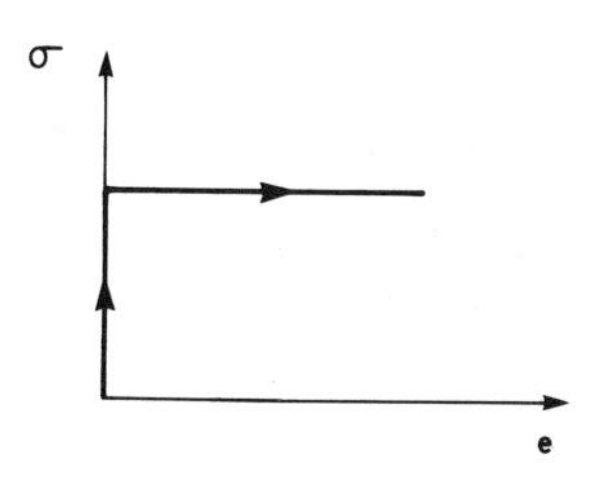

Fig. 1.19 Stress-strain behavior of a rigid ideal plastic material

---

[6] This may be simply demonstrated by evaluating $\dot{\epsilon}_2$ and $\dot{\epsilon}_3$ from Eq. (1.69) and subtracting them to give γ_{12}, and identity with Eq. (1.68) is found.

As before, in deriving Eqs. (1.59), we can rewrite these in the form

$$\dot{\epsilon}_1 = \lambda[\sigma_1 - (\sigma_2 + \sigma_3)/2], \text{ etc}$$

and we may substitute for λ as $\dot{\epsilon}_{eff}/\sigma_{eff}$ [Eq. (1.66)], where

$$\dot{\epsilon}_{eff} = \frac{\sqrt{2}}{3}[(\dot{\epsilon}_1 - \dot{\epsilon}_2)^2 + (\dot{\epsilon}_2 - \dot{\epsilon}_3)^2 + (\dot{\epsilon}_3 - \dot{\epsilon}_1)^2]^{1/2} \tag{1.70}$$

this being determined from the relation

$$\dot{\epsilon}_{eff} = f(\sigma_{eff})$$

which can be established for the simple uniaxial case. Hence, we have

$$\left.\begin{aligned} \dot{\epsilon}_1 &= \frac{\dot{\epsilon}_{eff}}{\sigma_{eff}}[\sigma_1 - (\sigma_2 + \sigma_3)/2] \\ \dot{\epsilon}_2 &= \frac{\dot{\epsilon}_{eff}}{\sigma_{eff}}[\sigma_2 - (\sigma_3 + \sigma_1)/2] \\ \dot{\epsilon}_3 &= \frac{\dot{\epsilon}_{eff}}{\sigma_{eff}}[\sigma_3 - (\sigma_1 + \sigma_2)/2] \end{aligned}\right\} \tag{1.71}$$

From Eqs. (1.71) and a knowledge of the relationship between flow rate and effective stress, we can predict the principal strain rates in an ideal "flowing" solid subject to multiaxial stressing. When elastic deformation must also be taken into account or when the material is non-ideal and strain hardens during deformation, the analysis is made more complex, but it is hoped that the simple analysis presented here will introduce the basic principles of plastic flow and its prediction. We shall apply these principles in Chapter 11 when examining multiaxial creep deformation, in which the material can be treated as an ideal rigid plastic solid when subject to steady state deformation, at least as a first approximation.

REFERENCES

1. UGURAL, A. C., and FENSTER, S. K., *Advanced Strength and Applied Elasticity,* Elsevier, New York (1975).
2. NADAI, A., *Theory of Flow and Fracture of Solids,* 2nd ed., McGraw-Hill, New York (1950).
3. DIETER, G. E., *Mechanical Metallurgy*, McGraw-Hill, New York (1961).
4. WANG, C. T.,*Applied Elasticity*, McGraw-Hill, New York (1953).
5. AIRY, G. B., *British Assoc. Advance. Sci. Report*, London (1862).
6. TIMOSHENKO, S., and GOODIER, J., *Theory of Elasticity,* McGraw-Hill, New York (1970).
7. ZIENKIEOWICZ, O.C., *The Finite Element Method in Engineering Science,* McGraw-Hill, New York (1971).
8. HEYWOOD, R. B., *Designing by Photoelasticity*, Chapman and Hall, London (1952).
9. INGLIS, C. E., *Trans. Inst. Naval Architects,* **55**, 219 (1913).

10. NEUBER, H. P., *Kerbspannungslere,* 2nd ed., Springer, Berlin (1958).
11. PETERSON, R. E., *Stress Concentration Design Factors*, 2nd ed., Wiley, New York (1974).
12. *Data Sheets*, The Royal Aeronautical Society, London.
13. ST. VENANT, B. DE, *Compt. rend., Acad. Sci., Paris,* **70**, 473 (1870).
14. LÉVY, M., *Compt. rend., Acad. Sci., Paris,* **70**, 1323 (1870).
15. VON MISES, R., *Nachr. Ges. Wiss. Göttingen, Math.-physik, Klasse,* 582 (1913).
16. BACKOFEN, W. A., *Deformation Processing,* Addison-Wesley, Reading, Mass. (1972).

2

CRITERIA FOR FAILURE

2.1 INTRODUCTION

"Failure" may be defined in several different ways when applied to the inability of a member or component to continue to carry the required load. First, it may refer to the onset of yielding in ductile materials so that permanent deformation occurs in the component; second, it may relate to the occurrence of elastic instability of a member such as a column or shell (Euler buckling), so that collapse will take place; or third, it may be considered as the condition of fracturing in a material, e.g., in a brittle solid under static tensile loading, where it may be taken as the ultimate tensile strength (UTS), or it may refer to a more ductile material failing either under cyclic loading (by fatigue), or under a steady load under complex stress conditions where sudden failure may take place at a stress well below the UTS.

In this chapter, we shall consider the criteria for failure under the action of a three-dimensional stress system where the material is (a) ductile, and (b) brittle, and then examine the criteria for failure under multiaxial cyclic loading.

Discussion of the mechanisms of fracture and of the fracture mechanics approach to failure will not be dealt with here, but are considered in Chapters 8 and 9; neither is creep failure considered, being dealt with separately in Chapter 11. Essentially, the approach here is one which considers solids as continuous, structureless media, having idealized elastic-plastic behavior, an appropriate one for many design problems.

As will become apparent in the various sections dealing with failure criteria, there is a lack of good and reliable data concerning the strength of materials under biaxial or triaxial loading, and it is to be hoped that, with increased interest in this area, such data will be more readily available in the not too distant future, so that the various theories which have been formulated can be examined more critically.

2.2 THE YIELD SURFACE

The stress-strain curve for a ductile metal was shown in Fig. 1.3, and the stress at which plastic deformation begins, or the yield stress (σ_Y), has been discussed briefly. However, we may broaden this definition to make σ_Y the *current* yield stress, i.e., the yield stress after whatever plastic straining has taken place. Hence we can plot yield stress versus plastic strain as shown in Fig. 2.1. Now for zero Bauschinger effect[1], the material is elastic when $-\sigma_Y < \sigma < \sigma_Y$.

Consider an element under biaxial stressing as shown in Fig. 2.2. Yielding will occur when σ_1 and σ_2 reach some critical combination, giving rise to a series of points as shown in Fig. 2.3. For all combinations of σ_1 and σ_2 giving rise to yielding, there will be a *yield locus*. Inside this line the material will be elastic; outwith it the applied stresses will cause yielding. It may be seen that we may have an *initial yield locus* or a *subsequent yield locus*, depending on whether or not prior yielding has taken place.

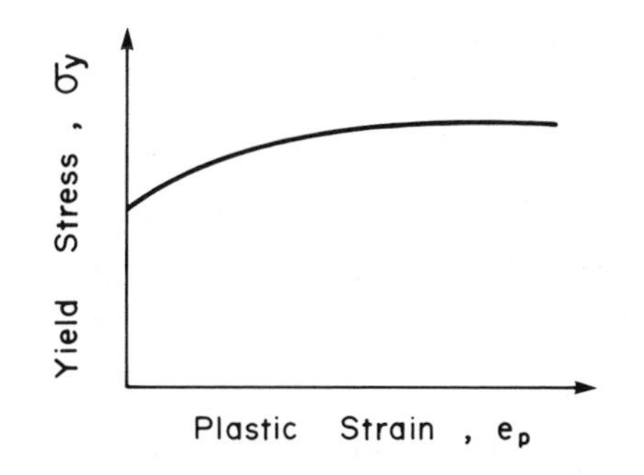

Fig. 2.1 Plot of yield stress versus plastic strain

[1] That is, it is assumed that it does not matter whether prior loading was tensile or compressive, the yield stress will be altered to the same degree as shown in Fig. 2.1, by the amount of plastic strain which has been given. The Bauschinger effect is discussed in more detail in Section 4.4.4.

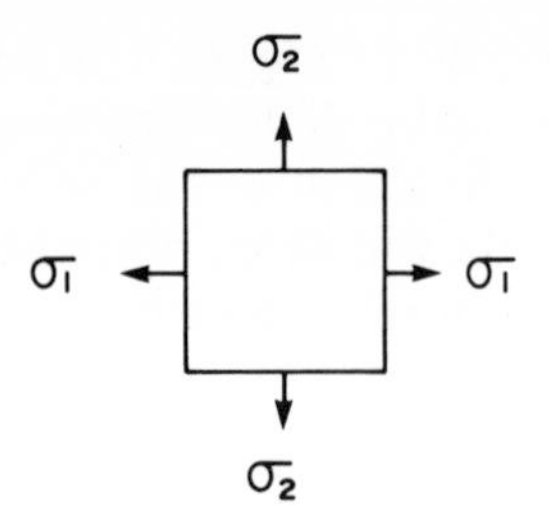

Fig. 2.2 Biaxially stressed element

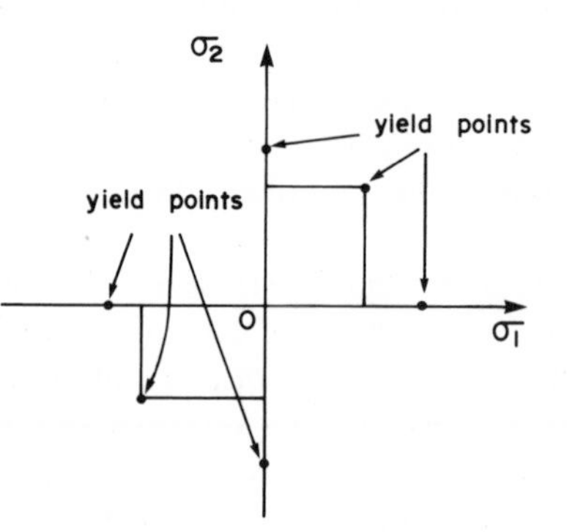

Fig. 2.3 Yield points for different combinations of σ_1 and σ_2

For triaxial stress, the first point to be noted is that hydrostatic pressure does *not* affect yield behavior. This is obvious when one considers that there is no change in shape, hence no plastic deformation (due to dislocation movement - see Chapter 4) has taken place.

Consider an element under triaxial principal stresses, σ_1, σ_2, σ_3, as shown in Fig. 2.4. Our statement above says that if the point $(\sigma_1, \sigma_2, \sigma_3) = (a, b, c)$ is on the yield surface, then the point $(a + h, b + h, c + h)$ also falls on this surface for all values of h. Thus any point (a, b, c) on the yield surface generates a line through it and parallel to the line $\sigma_1 = \sigma_2 = \sigma_3$ as shown in Fig. 2.5. Hence the *yield surface* in three-dimensional principal stress space is a prism formed by sliding a curve along the line $\sigma_1 = \sigma_2 = \sigma_3$, and the cross-section of this prism defines the yield stress.

Any plane perpendicular to $\sigma_1 = \sigma_2 = \sigma_3$ has the equation

$$\sigma_1 + \sigma_2 + \sigma_3 = \text{constant}$$

Hence, consider the plane $\sigma_1 + \sigma_2 + \sigma_3 = 0$, i.e., the plane passing through the origin, *0*. This is the most convenient plane to choose, and is termed the *π-plane.* It intersects the yield prism to form the *C-curve.*

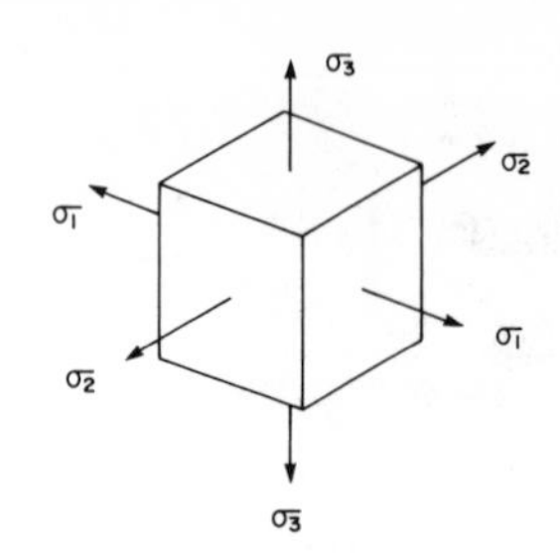

Fig. 2.4 Triaxially stressed element

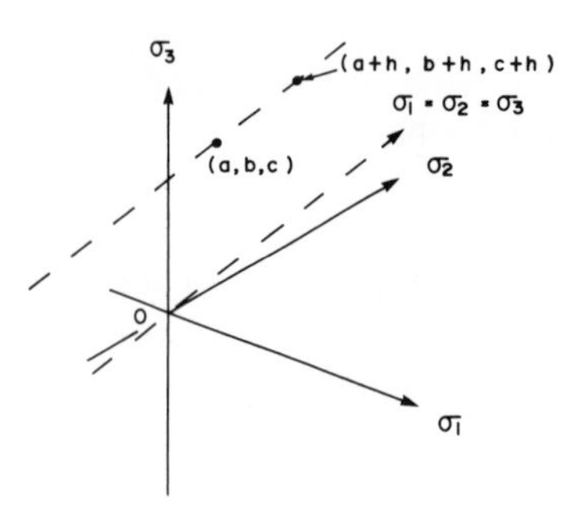

Fig. 2.5 Three-dimensional principal stress space showing a point (*a, b, c*) where yield occurs. Yield also occurs at (*a+h, b+h, c+h*) on the dotted line through (*a, b, c*) and parallel to $\sigma_1 = \sigma_2 = \sigma_3$

Consider the point (a, b, c) on the yield surface. It can be projected onto the C-curve to give a point

$$(a\text{-}h, b\text{-}h, c\text{-}h)$$

and, since it lies on the π-plane, the sum of the three principal stresses must be zero. Hence

$$h = (a + b + c)/3$$

Thus, for any point on the yield surface

$$(\sigma_1, \sigma_2, \sigma_3) \equiv (\sigma_1', \sigma_2', \sigma_3') + (\sigma'', \sigma'', \sigma'') \tag{2.1}$$

where

$$\left.\begin{aligned} \sigma'' &= (\sigma_1 + \sigma_2 + \sigma_3)/3 \\ \sigma_1' &= \sigma_1 - \sigma'' \\ \sigma_2' &= \sigma_2 - \sigma'' \\ \sigma_3' &= \sigma_3 - \sigma'' \end{aligned}\right\} \quad (2.2)$$

Note, from our definition of the π-plane, that

$$\sigma_1' + \sigma_2' + \sigma_3' = 0 \quad (2.3)$$

The stress, $(\sigma_1', \sigma_2', \sigma_3')$, is termed the *deviatoric stress*, and $(\sigma'', \sigma'', \sigma'')$ is the *hydrostatic stress*. The deviatoric stress lies in the π-plane, while the hydrostatic stress is perpendicular to this.

The form of the C-curve on the π-plane is shown in Fig. 2.6 for a material which is isotropic and has no Bauschinger effect, and it is seen that in such a case the complete C-curve can be specified by defining the C-curve in a typical 30° sector [1]. It may be seen that three of the six axes of symmetry represent projections of uniaxial tension/compression axes; the other three correspond to projections of states of pure shear plus hydrostatic pressure. The angular range of one sector may be covered by tests of thin tubes under combined tensional and torsional loading [1], which was the basis of the experimental work of Taylor and Quinney [2] discussed in Section 2.3.3.

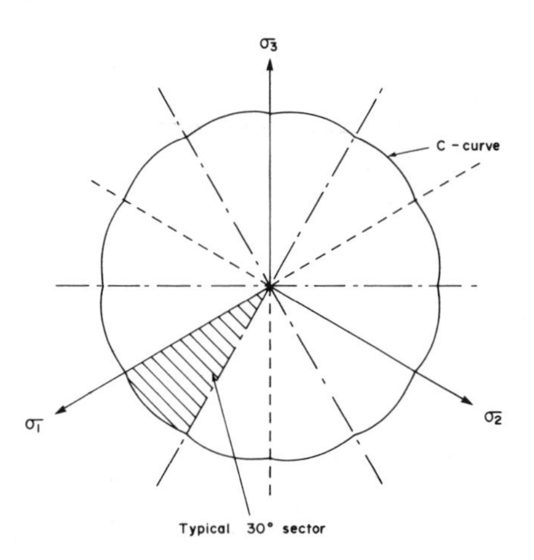

Fig. 2.6 The *C*-curve on the π-plane in principal stress space

2.3 YIELD CRITERIA

The two criteria for yielding which are most commonly used are the Tresca [3] or maximum shear stress criterion, and the von Mises [4] criterion. They will be described separately, and then compared in terms of their relative validity.

2.3.1 The Tresca Criterion

A single crystal deforms plastically when the resolved shear stress on a specific plane in a specific direction reaches a critical value, this being discussed more fully in Chapter 4. In a polycrystalline metal, composed of many individual crystals, general plastic flow involving many individual crystals occurs approximately when the shear stress exceeds some critical value [5].

The Tresca criterion postulates that yielding will occur when the shear stress τ on *any* plane reaches some critical value k, which is a characteristic of the material, and which must be evaluated for the material in the same condition as it will be used. The criterion for yielding can be written as

$$\tau_{crit} = (\sigma_1 - \sigma_3)/2 = k \quad (2.4)$$

where σ_1 and σ_3 are respectively the maximum and minimum principal stresses.

The value of k can be determined experimentally by determining the values of σ_1 and σ_3 at which yielding occurs, and substituting these in Eq. (2.4). However, this presupposes the validity of the criterion, and strictly speaking, k should be determined from a torsion test, which gives rise to conditions of pure shear.

Plotted on the π-plane for $\sigma_1 > \sigma_2 > \sigma_3$, the stress point for yielding lies between the σ_1(+ve) and σ_3(-ve) axes, and the complete C-curve is shown in Fig. 2.7. It may be seen that the complete yield surface has the form of a hexagonal prism in principal stress space.

The yield criterion

$$\sigma_1 - \sigma_3 \; (= \sigma_1' - \sigma_3') = 2\,k \quad (2.5)$$

is a widely used and *safe* criterion for the design of structural members in ductile materials.

2.3.2 The von Mises Criterion

Von Mises [4] replaced the hexagonal C-curve with a circle, for reasons of simplicity. The C-curve is then the intersection of a sphere of radius R with the π-plane. The equation of the sphere is

$$\sigma_1^2 + \sigma_2^2 + \sigma_3^2 = R^2 \quad (2.6)$$

and the equation of the circle in the π-plane is given by

$$(\sigma_1')^2 + (\sigma_2')^2 + (\sigma_3')^2 = R^2 \quad (2.7)$$

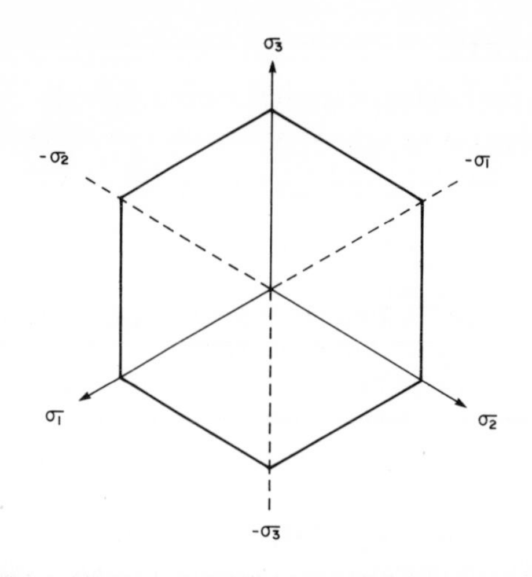

Fig. 2.7 The *C*-curve for the Tresca yield criterion

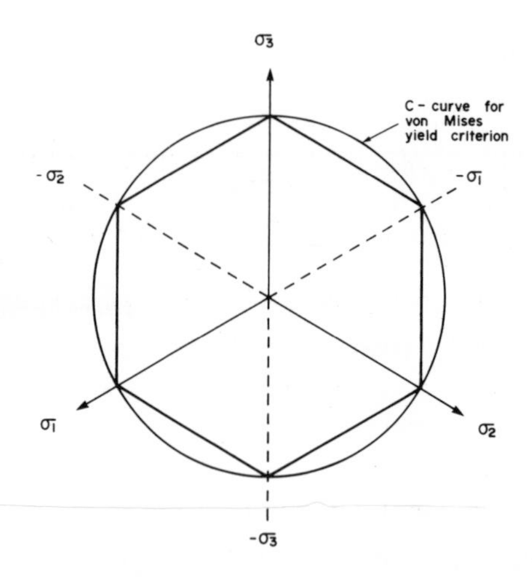

Fig. 2.8 The *C*-curve for the von Mises yield criterion with the Tresca *C*-curve drawn to touch at conditions of principal stress

This is also the equation of a cylinder representing the yield surface in three-dimensional principal stress space.

Let us assume that the uniaxial tensile yield stress has the value Y. Each of the principal stress axes lies at an angle of $\cos^{-1}(1/\sqrt{3})$ to the π-plane, and R must be assigned the value $Y/\sqrt{3}$ for the von Mises circle and the Tresca hexagon to touch at the conditions of principal stress shown in Fig. 2.8.

Hence, we may write the von Mises criterion for yield in the form

$$(\sigma_1')^2 + (\sigma_2')^2 + (\sigma_3')^2 = Y^2/3 \tag{2.8}$$

which can be rearranged to give

$$(\sigma_1 - \sigma_2)^2 + (\sigma_2 - \sigma_3)^2 + (\sigma_3 - \sigma_1)^2 = 2Y^2 \tag{2.9}$$

a useful form for many engineering problems.

A particular advantage of the von Mises over the Tresca criterion, is that it is not necessary, in applying the former, to evaluate the principal stresses, so that they can be ranked in order. Hence, we can express the von Mises criterion in terms of orthogonal stresses, the form being

$$\tfrac{1}{2}\,[(\sigma_{xx} - \sigma_{yy})^2 + (\sigma_{yy} - \sigma_{zz})^2 + (\sigma_{zz} - \sigma_{xx})^2 + 6(\tau_{xy}^2 + \tau_{yz}^2 + \tau_{zx}^2)] - Y^2 \geqslant 0 \tag{2.10}$$

for yielding to occur.

It is possible to evaluate the Tresca criterion in terms of orthogonal stress components, but the procedure is more complex, and there are more steps involved. For more detailed discussion see Sines [*6*] (pp. 63-64).

It is appropriate to consider the physical basis of the von Mises criterion, and Eq. (2.9) can be derived on several semi-physical bases. In particular, it can be derived on the basis that yielding occurs when the shear stress on the (macroscopic) octahedral plane reaches a critical value; when the elastic distortion energy reaches a critical level; or when the root mean square value of the shear stress averaged over all planes exceeds a certain critical value, this being discussed by Sines [*6*]. Certainly, this last-mentioned approach appears to have the greatest physical meaning, as plastic deformation is always related to shear stress. Intuitively, too, one may expect there to be an effect of the intermediate principal stress, this being omitted in the Tresca criterion.

2.3.3 Relative Validity of Tresca and von Mises Criteria

The question must obviously be raised as to which of the two yield criteria outlined in the foregoing discussion corresponds more closely with observed yield under complex stress. From the *C*-curves shown in Figs. 2.7 and 2.8, there are obvious differencies under certain loading conditions.

As shown in Fig. 2.8, we consider the two yield surfaces or *C*-curves to touch at the principal stress axes, and we have also specified that the uniaxial tensile yield stress has the value Y. Thus, the maximum separation between the two surfaces will occur under conditions of pure shear.

Substituting the conditions for pure shear of value k, we have principal stresses (k, $-k$, 0), and inserting these into the von Mises criterion (Eq. 2.9) we obtain

$$k = Y/\sqrt{3} = 0.577\,Y \tag{2.11}$$

From the Tresca relation (Eq. 2.5), we have

$$k = Y/2 \tag{2.12}$$

Thus the greatest discrepancy is about 15%, with the von Mises criterion predicting a greater stress being required for yield. This discrepancy can be reduced to ~7½ % by altering the hexagon by this amount, but there is normally little point in attempting "dodges" such as this, and it makes good sense to maintain the yield surfaces in contact at the uniaxial tensile conditions, at which most materials evaluations are made.

Experimental tests have shown the shear yield stress to lie between 0.5 Y and 0.6 Y, with an average value of around 0.57 Y [*7*], indicating that the von Mises criterion fits most experimental data better. The results of experiments by Taylor and Quinney [*2*], using isotropic polycrystalline copper, mild steel and aluminum loaded in combined bending and torsion, are shown in Fig. 2.9, and it may be seen that the von Mises line is closer to the experimental results.

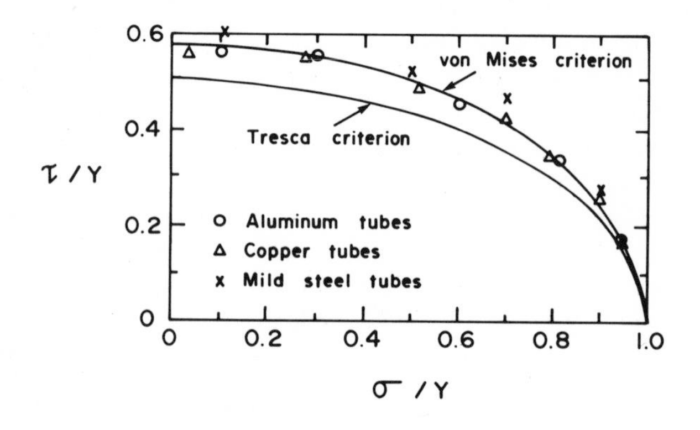

Fig. 2.9 The results of Taylor and Quinney [*2*] for combined tension and torsion of tubes, showing the better agreement between observed yielding and the von Mises criterion rather than the Tresca criterion

2.4 YIELD UNDER CONDITIONS OF PLANE STRESS OR PLANE STRAIN

In evaluating stresses in components and in determining whether or not yield will take place under given loading conditions, it is important to consider stresses and strains operating in all three principal directions.

2.4.1 Plane Stress

For conditions of plane stress, as may be found in thin sheet material under biaxial loading, $\sigma_3 = 0$, and the von Mises criterion (Eq. 2.9) becomes

$$\sigma_1^2 + \sigma_2^2 - \sigma_1\sigma_2 = Y^2 \qquad (2.13)$$

which is the equation of an ellipse. This is shown in Fig. 2.10, together with the Tresca condition for plane stress conditions.

Lode [8] adopted a sensitive method to evaluate the validity of the two criteria by determining the effect of the intermediate principal stress on yielding. As previously noted, the Tresca theory predicts that this will have no effect: hence

$$(\sigma_1 - \sigma_3)/Y = 1 \qquad (2.14)$$

Lode's stress parameter, μ, was defined as

$$\mu = (2\sigma_2 - \sigma_3 - \sigma_1)/(\sigma_1 - \sigma_3) \qquad (2.15)$$

Solving Eq. (2.15) for σ_2, and substituting in the von Mises criterion (Eq. 2.9), we obtain

$$(\sigma_1 - \sigma_3)/Y = 2/(3 + \mu^2)^{1/2} \qquad (2.16)$$

Experimental data plot better against this equation than the Tresca relation (Eq. 2.14), showing that σ_2 does indeed have an effect on yielding [5, 7].

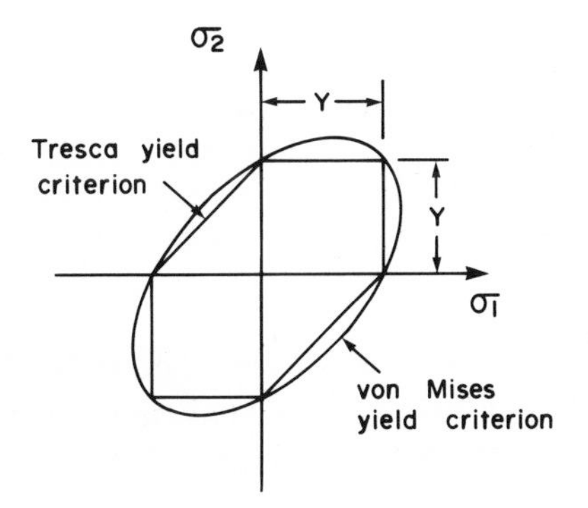

Fig. 2.10 Tresca and von Mises yield criteria for plane stress conditions

2.4.2 Plane Strain

Plane strain deformation is common in metalworking operations such as rolling or forging, as well as during the propagation of a crack, or in the "brittle" fracture of materials. In this situation all strain takes place in two principal directions only, hence we can consider that e_3 is zero. In consequence, the principal stress, σ_3, is not zero, but is given by

$$\sigma_3 = E\,[e_3 + \nu(\sigma_1 + \sigma_2)/E] \tag{2.17}$$

where E is the modulus of elasticity, and ν is Poisson's ratio. Also

$$e_3 = [\sigma_3 - \nu(\sigma_1 + \sigma_2)]/E = 0 \tag{2.18}$$

For plastic deformation, the equations are analogous with $\nu = 0.5$ (see Section 1.7). Hence

$$\sigma_3 = (\sigma_1 + \sigma_2)/2 \tag{2.19}$$

Substituting this in Eq. (2.9), the von Mises criterion, we obtain

$$\sigma_1 - \sigma_2 = 2Y/\sqrt{3} = 1.15Y \tag{2.20}$$

For the Tresca criterion

$$\sigma_1 - \sigma_2 = 2\,k \tag{2.21}$$

σ_2 being defined as the minimum principal stress. Now the uniaxial tensile yield stress, Y, is related to the pure shear yield stress, k, through the von Mises relation by Eq. (2.11). Hence the two criteria are equivalent for plane strain conditions.

Eq. (2.20) is an important and useful one in many metal deformation problems. For example, it is important to realize that when compressive yield stress is measured by means of the plane strain compression test [9], the value obtained will be greater than the tensile yield stress (Y). Similarly, the flow curve obtained is this way will be above the tensile flow curve in terms of magnitude. This arises because, in plane strain compression testing, $\sigma_2 \simeq 0$, and the critical value of σ_1 for yielding is then $1.15Y$.

2.5 EFFECTS OF TEXTURE ON THE YIELD LOCUS

Texture, or the preferred crystallographic orientation of grains in sheet or rods of metal, may arise both from mechanical working (deformation texture) and from annealing subsequent to cold working (annealing texture), the development of such textures being discussed briefly in Section 4.4.5. Textures give rise to anisotropy in the yield stress and in elastic moduli. Here, we shall consider briefly the effect of texture in a sheet metal on the shape of the yield locus. We assume that there is planar isotropy, that is, the crystallographic texture is rotationally similar about the sheet normal; that the yield stress in the *plane* of the sheet is constant; that there is no Bauschinger effect; and that yielding is unaffected by hydrostatic stress.

Consider the sheet compressed through its thickness until it yields at $\sigma_{Y(3)}$ [Fig. 2.11(a)]. If a hydrostatic tension of magnitude $h = \sigma_{Y(3)}$ is applied, the yielding condition will not be affected, but we now have a specimen yielding under balanced biaxial tension, as shown in Fig. 2.11(b).

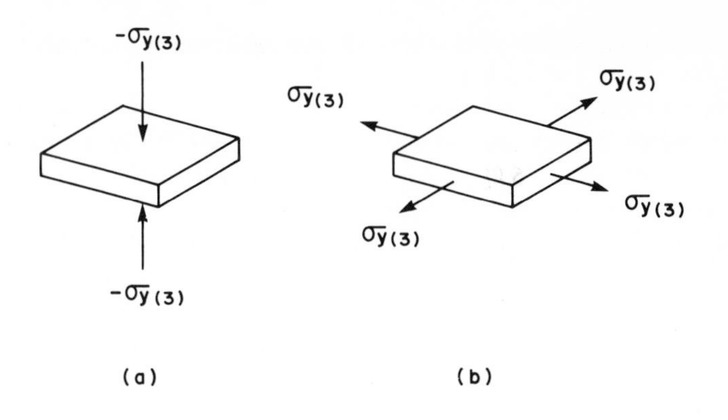

Fig. 2.11 Yielding under balanced biaxial tension. In (a) a compressive stress is applied until yielding begins in the through-thickness direction at $\sigma_{Y(3)}$; in (b) hydrostatic tension of magnitude $\sigma_{Y(3)}$ has been added, so that it is seen that yielding under balanced biaxial tension begins at a stress equal to $\sigma_{Y(3)}$

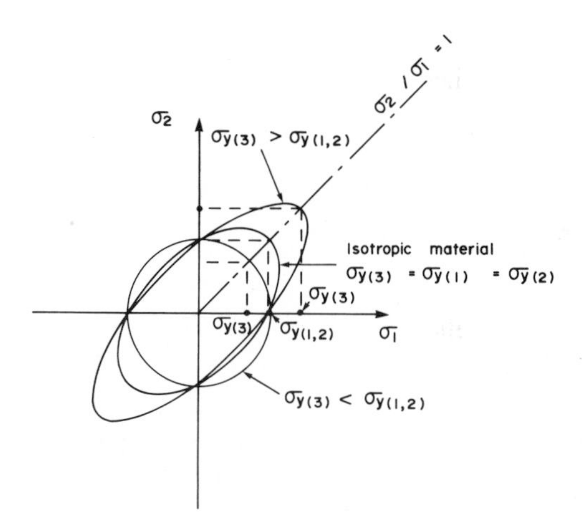

Fig. 2.12 Yield loci for textured sheet under biaxial stressing

Hence, we see that yielding on the load path, $\sigma_2/\sigma_1 = 1$, illustrated in Fig. 2.12, is controlled by the magnitude of $\sigma_{Y(3)}$. The condition for yielding is

$$\sigma_1 = \sigma_2 = \sigma_{Y(3)}$$

which is trivial for isotropic material where $\sigma_{Y(1)} = \sigma_{Y(2)} = \sigma_{Y(3)}$. However, if through-thickness yield stress is not equal to the planar yield stress, the isotropic yield locus will be distorted as shown in Fig. 2.12. If there is not planar isotropy, then the distortion will not be along the line $\sigma_2/\sigma_1 = 1$, but along some other line such as $\sigma_2/\sigma_1 = 1.2$.

Where an increase in resistance to yield takes place in quadrant I of Fig. 2.12, i.e., where $\sigma_{Y(3)} > \sigma_{Y(1,\,2)}$, *texture hardening* [10] is said to take place. Conversely, for $\sigma_{Y(3)} < \sigma_{Y(1,\,2)}$, we have *texture softening*.

Hexagonal close packed metals can exhibit substantial preferred orientation after rolling or other mechanical working, and titanium is a good example of such a material where considerable texture hardening can be induced by rolling, facilitating the construction of lightweight pressure vessels. An increase of 55 % in the uniaxial through-thickness yield stress over that in the plane of the sheet is possible [11], providing a corresponding increase in the biaxial load carrying capacity of the sheet.

Empirical equations for the anisotropic yield locus of Fig. 2.12 are given by Backofen [12], although he points out that they also follow from Hill's general analysis of yielding in plastically anisotropic material [13]. They are as follows:

$$\sigma_1^2 + \sigma_2^2 - \sigma_1\sigma_2\{2 - [\sigma_{Y(1,2)}/\sigma_{Y(3)}]^2\} = \sigma_{Y(1,2)} \qquad (2.22)$$

and

$$\sigma_1/\sigma_{Y(1,\,2)} = \{1 + \beta^2 - \beta[2 - \sigma_{Y(1,\,2)}/\sigma_{Y(3)}]^2\}^{-1/2} \qquad (2.23)$$

where β is the stress ratio σ_2/σ_1.

Eq. (2.22) reduces to Eq. (2.13) when $\sigma_{Y(3)} = \sigma_{Y(1,\,2)} = Y$, and, on the $\sigma_2/\sigma_1 = 1$ path, yielding is predicted when $\sigma_1 = \sigma_{Y(3)}$, in agreement with our discussion of Fig. 2.11.

More detailed discussion of yielding under different conditions of plane stress and with different degrees of planar anisotropy is given by Backofen [12], together with detailed consideration of the structural basis for plastic anisotropy in cubic metals.

2.6 FAILURE CRITERIA FOR BRITTLE SOLIDS

There has recently been increased interest by mechanical engineers in the use of brittle materials having high compressive strength. Such materials, in the form of ceramics, carbon and graphite, are of interest in the aerospace industry where their excellent strength at high temperature can be utilized. In addition, there is continued and widespread use of brittle and relatively brittle cast irons, carbides and ceramics in many other engineering applications. In these materials, normally yielding does not occur before they fracture in a brittle manner, although it can be induced by subjecting them to a sufficiently large hydrostatic compressive stress, together with a sufficiently high deviatoric stress. This loading situation was exploited by von Kármán in deforming marble [14], and by Bridgman [15] in his classic studies, in which he plastically deformed many normally brittle materials.

The many theories which have been proposed to predict failure in brittle solids are basically of two types. The first group comprises essentially empirical equations, and includes these due to Coulomb [16], Marin [17] and Paul [18], while the other theories base their analyses on the stresses required to cause frature to initiate from pre-existing defects. This latter group includes the theories of Griffith [19], Fisher [20], Babel and Sines [21], and Frishmuth and McLaughlin [22], the last mentioned being specificially for cast irons which have a ductile matrix surrounding graphite flakes. The discussion here will relate primarily to those theories having a more physical basis.

In 1924, Griffith [19] proposed that fracture would take place from closed cracks in glass when a critical tensile stress σ^*, was reached at the surface of the flaw, this value being a characteristic of the material.[2] While closed cracks were a reasonable assumption for glass, this is not appropriate for ceramics and other materials where there will be a distribution of open flaws, and Babel and Sines [21] developed an extension of the Griffith analysis to cover this situation. The model for the flaw is an elliptical hole, variable between the two extremes of a circular hole and an infinitely sharp closed crack.

Under biaxial loading, the tensile stress at the surface of an elliptical hole is given by [23]

$$\sigma = \{(\sigma_1 + \sigma_2) \sinh 2u_o + (\sigma_1 - \sigma_2) [\exp(2u_o) \cos 2v - 1] \cos 2\theta + (\sigma_1 - \sigma_2) \exp(2u_o) \sin 2v \sin 2\theta\}/(\cosh 2u_o - \cos 2v) \tag{2.24}$$

where σ_1 and σ_2 are the principal stresses with $\sigma_1 \geqslant \sigma_2$.

Figure 2.13 shows the coordinate system and stresses on the elliptical hole, and v is seen to be a position coordinate on the hole, θ the angle between σ_2 and the major axis of the ellipse, while u_o relates to the sharpness of the crack, being defined as $u_o \equiv \tanh^{-1}(b/a)$, with a and b defining the ellipse. The maximum local tensile stress, σ, and its position on the hole, depend on the orientation of the ellipse with respect to the principal stress axes. By determining the conditions under which a maximum stress occurs, and then setting this equal to σ^*, the critical stress for fracture under the Griffith hypothesis, we obtain[3]

$$\sigma^* = (1 + 2r)\sigma_1 - \sigma_2 \tag{2.25}$$

and

$$\sigma^* = -(r + 1)^2(\sigma_1 - \sigma_2)^2/4r(\sigma_1 + \sigma_2) \tag{2.26}$$

where $r = a/b$.

---

2 See also Section 8.3.1.

---

3 The solution is detailed by Babel and Sines [21].

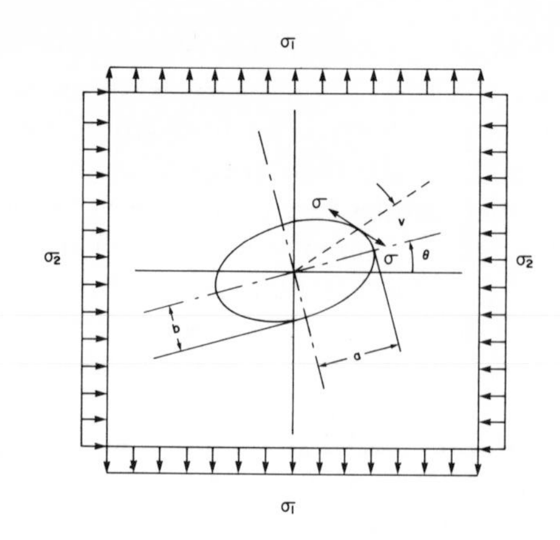

Fig. 2.13 Coordinate system and stresses on an elliptical hole

Thus, the behavior of the material is characterized by σ^* and $r \equiv a/b$, which are both microscopic parameters, and can be replaced by the macroscopic parameters σ_C (the uniaxial compressive strength) and σ_T (the uniaxial tensile strength), as follows. Inserting $\sigma_1 = \sigma_T$ and $\sigma_2 = 0$ into Eq. (2.25) we obtain

$$\sigma^* = (1 + 2\,r)\sigma_T \tag{2.27}$$

Inserting $\sigma_2 = -\sigma_C$ and $\sigma_1 = 0$ into Eq. (2.26), we obtain

$$\sigma^* = (r + 1)^2\sigma_C/4r \tag{2.28}$$

and solving Eqs. (2.27) and (2.28) simultaneously, we have

$$r = a/b = [-R + 2 - 2(R + 1)^{1/2}]/(R - 8) \tag{2.29}$$

where

$$R \equiv \sigma_C/\sigma_T$$

The linear part of the criterion expressed by Eq. (2.25) applies in the tension-tension quadrant and partially into the tension-compression one, while the curved part given by Eq. (2.26) applies for high ratios of compressive to tensile stress. The crossover point occurs at

$$\sigma_2/\sigma_1 = -(3r + 1)/(r - 1) \tag{2.30}$$

obtained by equating σ^* in the two equations. The criterion is represented in Fig. 2.14.

It may also be seen that the criterion shows that the ratio of tensile to compressive strength is dependent on crack sharpness, as, equating the right hand sides of Eqs. (2.27) and (2.28), we obtain:

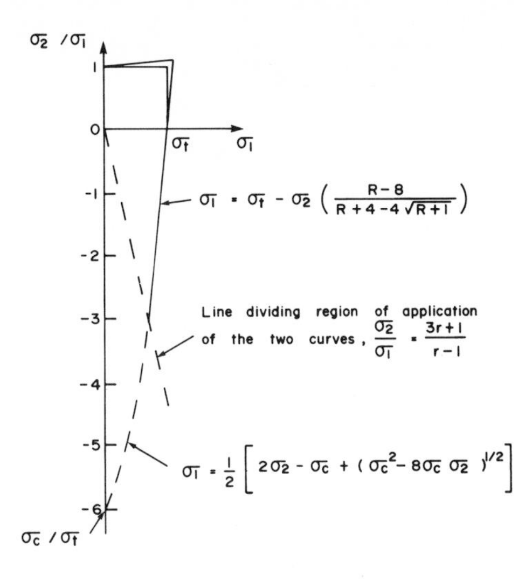

Fig. 2.14 Babel and Sines criterion for failure of a brittle solid containing an elliptical flaw under biaxial stressing [*21*]

$$R = \sigma_C/\sigma_T = 4r(1 + 2r)/(1 + r)^2 \tag{2.31}$$

As $a/b \equiv r \to \infty$, $R \to 8$ in Eq. (2.31), and the Babel and Sines criterion predicts the identical ratio of compressive to tensile strength as does the Griffith for this case. At the other extreme, when the flaws become circular, $R \to 3$, and the failure envelope becomes a straight line in agreement with the Coulomb-Mohr criterion for the tension-compression quadrant.

In the tension-tension quadrant, it is more conservative to utilize the Coulomb-Mohr criterion in the form

$$\sigma_1/\sigma_T - \sigma_2/\sigma_C - 1 \geqslant 0 \tag{2.32}$$

for fracture to occur. Even this criterion may not be sufficiently conservative, as some data show this region to be rounded off at the equibiaxial corner (see, e.g., Broutman *et al* [*24*]), but there is a lack of good experimental data for biaxial conditions, and the Coulomb-Mohr criterion may be used as a first approximation.

In the tension-compression quadrant, the Babel and Sines criterion agrees well with most experimental data, but underestimates the strength of cast irons [*22*]. It does seem that it is a reasonable criterion to use, although a more conservative approach would be to use the Coulomb-Mohr criterion for ceramics, this being illustrated in the failure surface of Fig. 2.15.

In the compression-compression quadrant there are only very limited reliable data available, but they appear to lie outside of the solid line of Fig.

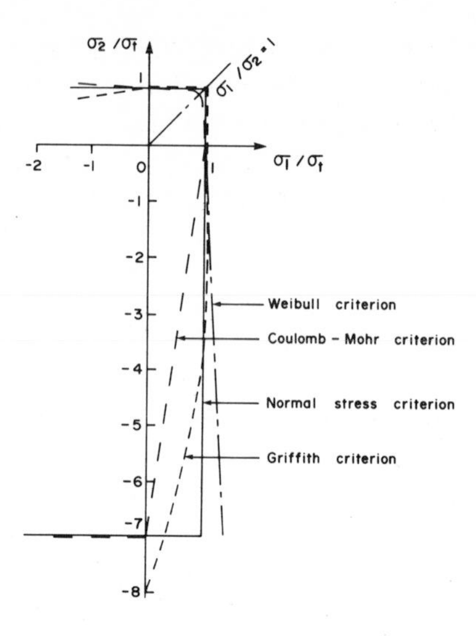

Fig. 2.15 Various failure criteria for a brittle solid under biaxial stressing

2.15, which represents the Griffith, extended Griffith, and Coulomb-Mohr theories, all of which predict fracture when the greatest compressive stress exceeds the uniaxial compressive strength. Using the Griffith-type model, which considers cylindrical elliptical flaws, whose tip stresses on an intersecting free surface are nearly the same as for penny-shaped cracks of the same cross-section, but are mathematically much simpler to handle, an argument may be given as to why the square line prediction of Fig. 2.15 is conservative [*25*].

Considering a distribution of cylindrical elliptical flaws, the critical flaw for uniaxial compression has its cylindrical axis perpendicular to the stress, and its major axis inclined at about 30° to it. The addition of another compressive stress at right angles to the first, and perpendicular to the cylindrical axis of the flaw, reduces the stress at the crack tip, and the Griffith parabola is seen to continue from the tension-compression quadrant into the compression-compression one, so predicting strengths higher than the uniaxial compressive strength [*26*]. For cracks whose cylindrical axes are parallel to the lesser compressive stress, no such benefit should occur, and we may expect that the biaxial strength would be the same as the uniaxial.

However, from careful analysis [*25*] it has been suggested that the biaxial strength should be intermediate to these two extremes. An interior penny-shaped crack of the latter orientation has a smaller stress concentration than the corresponding cylindrical elliptical one, and thus failure would not start from it until higher stresses were applied. When cut by a free surface, such a flaw will have the projection of the tip radius made larger, so reducing the stress concentration. These two effects will reduce the local stress by some 30% [*25, 27*]. Some data obtained by Sines and Adams [*25*], in the course of carefully conducted tests of ceramics, indicate that the failure surface does lie outside the square line prediction, and that it may be as much as 30% in error.

Some more recent data obtained by Sines and Adams [*28*] for a ceramic tested in the compression-compression quadrant suggest some contradictions to the previous results and discussion, and are shown in Fig. 2.16. The points to note are the large ratio between tensile and compressive strength (19.5), and the horizontal nature of the failure surface in the compression-compression quadrant[4]. The authors provide an explanation of these observations on the basis that, under compression, cracks which initiate from flaws become stable, increasing numbers of cracks being formed with increased load, until at failure 10^5-10^6 flaws will have extended. The Griffith criterion deals with the initiation of crack extension from flaws, thus these data suggest that much higher loads may be carried by brittle solids in the cracked condition than are

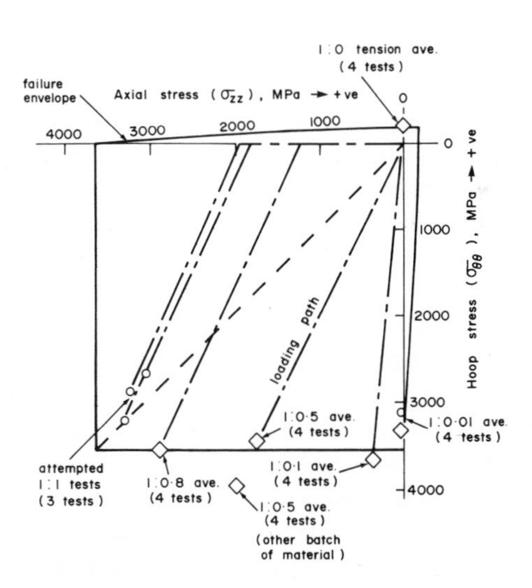

Fig. 2.16 Data of Sines and Adams [*28*] for failure of alumina

---

4 The anomalous results for 1:1 tests are explained in terms of cracking which took place during prior compression of the cylindrical specimens, before pressurization. The loading paths are shown in Fig. 2.16.

required to initiate cracking, and the previously discussed failure criterion and surface is probably still appropriate for design purposes where the intent is for crack formation to be avoided.

Sines [*6*] has discussed the extension of brittle fracture criteria to cover the triaxial stress state, and this is simply done, although there are not data to back up the analysis. Nevertheless, it is easily seen that extension of the conservative criteria of Fig. 2.15 to triaxial stress conditions will produce a cube in the tension-tension-tension octant; a set of three pyramids with square bases in the three tension-tension-compression octants; a set of three pairs of intersecting wedges in the tension-compression-compression octants; and in the compression-compression-compression octant the surface would intersect the biaxial stress planes to form three squares. Because hydrostatic compression should not cause fracture, but rather inhibit it, we can consider that the surface will have the form of a hexagonal prism with its axis having direction cosines $-1/\sqrt{3}$, $-1/\sqrt{3}$, $-1/\sqrt{3}$. In fact, we can expect the prism to have an expanding cross-section as we move away from the origin in a negative direction along this axis, as it has been found that uniaxial compressive strength in brittle materials increases with increasing hydrostatic stress (see, e.g., Bridgman [*15*]), and Fig. 2.17 shows the intersection of such an expanding hexagonal prism with the σ_1-σ_2 plane. Figure 2.18 is an attempt to summarize the shape of the various parts of the failure surface in the various octants.

In their analysis of fracture criteria for cast irons mentioned previously, Frishmuth and McLaughlin [*22*] consider a representative volume element (RVE), chosen as being typical of the material's microstructure, and analyze its behavior under three-dimensional macroscopic principal stress, to obtain upper and lower bounds for the RVE's failure surface. While the result is a lower bound close to that determined from other theories, including that of Babel and Sines [*21*], there is one potential advantage to Frishmuth and McLaughlin's approach, namely that the triaxial behavior of the cast iron can be predicted from microstructural data on the representative graphite flake or nodule size, the size of the eutectic cell (or the nodule spacing for ductile iron), and the unixial tensile yield strength of the *matrix*. Alternatively, if the uniaxial tensile and compressive failure strengths of a cast iron are determined, the theory allows estimation to be made of the microstructural parameters and matrix strength. Hence, the failure stresses under any other loading state can be predicted.

While failure under complex loading conditions can be predicted from a knowledge of tensile and compressive uniaxial strengths for essentially brittle materials using other theories of failure than that just discussed, Frishmuth and McLaughlin's work represents an interesting first attempt to predict the failure conditions for a common, but complex, group of engineering alloys on the basis of their microstructural characteristics.

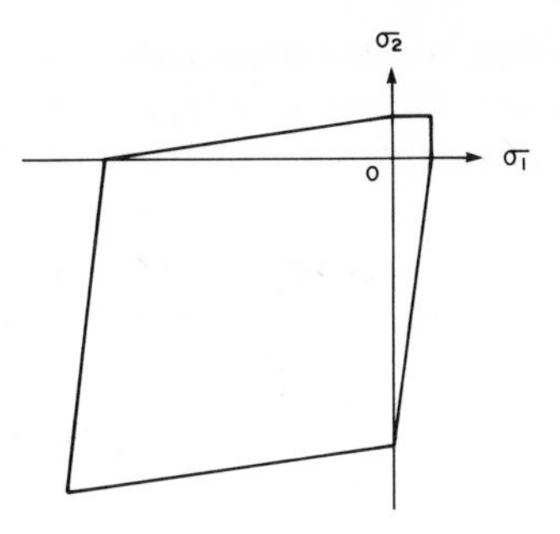

Fig. 2.17 Intersection of an expanding hexagonal prism with the σ_1-σ_2 plane. The prism has direction cosines - $1/\sqrt{3}$, $-1/\sqrt{3}$, $-1/\sqrt{3}$, and expands as one moves away from the origin in a negative direction

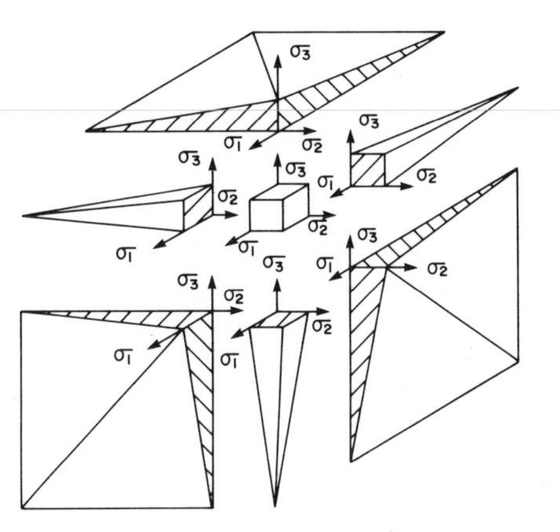

Fig. 2.18 The various parts of the failure surface for a brittle solid in principal stress space

2.7 FAILURE CRITERIA FOR FATIGUE UNDER MULTIAXIAL LOADING

Fatigue of metals is considered in some detail in Chapter 10, but it is appropriate here to consider the nature of the criteria for fatigue failure under multiaxial stressing, as the discussion follows naturally from the criteria for static failure already considered. First, it must be emphasized that fatigue failure is essentially a local phenomenon, involving damage and subsequent propagation of a fatigue crack under conditions of cyclic (or fluctuating) stress. When the member concerned is subjected to a regular sinusoidal fluctuating principal stress, variation with time can be represented by Fig. 2.19, the fluctuating principal stress being separated into alternating and static components.

When subjected to such a fluctuating stress, the member may fail after a certain number of stress cycles, N_f, and data are normally plotted in the form of a stress (S) versus log N_f (S-N) plot as shown in Fig. 2.20, which is drawn for a material (such as carbon steel) where those is a definite *fatigue limit* (see Section 10.2.1). The effect of a superimposed tensile mean stress is to decrease

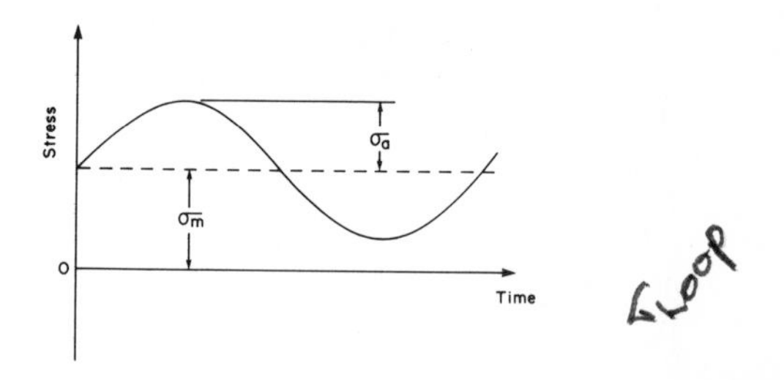

Fig. 2.19 Cyclic loading, showing the static component of stress, σ_m, and the alternating component, σ_a

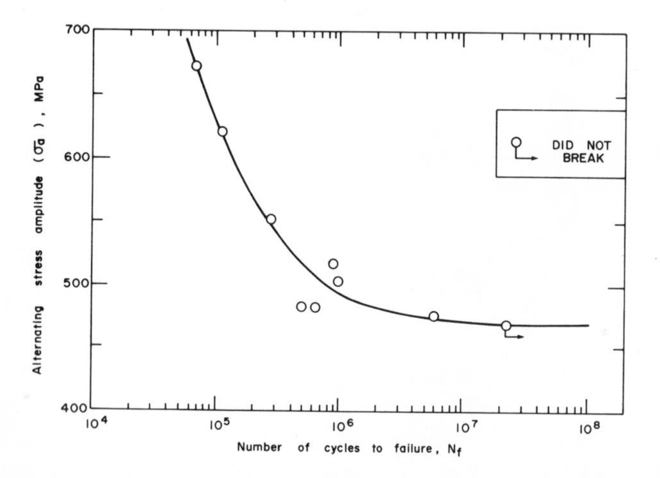

Fig. 2.20 *S-N* curve for high-strength, low-alloy, copper-containing steel

the life, or to lower the *S-N* curve in Fig. 2.20. Most fatigue data are available for simple loading situations only, e.g., as data for reversed bending or push-pull loading. However, many components undergoing cyclic loading and in which the possibility of fatigue failure must be considered are subject to multiaxial stressing in critical regions of high stress where fatigue fracture might initiate.

A number of analyses have been made and failure criteria proposed for fatigue under combined stresses, and we shall look first at the relatively simple analysis due to Sines [*6, 29*], which is based on the assumptions that the principal stresses are sinusoidal and can be described as in Fig. 2.19 by static and alternating components, that they are exactly in phase and are of the same frequency. Alternating stresses (or the alternating components of stress) can be combined by plotting their amplitudes to cause failure after a specified number of cycles, the data of Fig. 2.21 being for 10^7 cycles. In this figure, from the work of Sines [*29*], the data of Gough [*30*] and Sawert [*31*] are plotted, and it may seen that the failure locus has the form of an ellipse. The data for cast iron fall outside the ellipse, as do data for annealed hypereutectoid steel [*30*], and Sines [*29*] has argued convincingly that the relative independence of the effect of σ_2 on σ_1 in cast iron is due to the presence of graphite flakes which give rise to stress concentrations at their ends. Considering the idealized model shown in Fig. 2.22, the stress concentrations at A_1 and A_2 will be large under the action of σ_1 and may be expected to cause crack propagation from that flake, while there will be little or no effect when σ_2 is applied. Similarly, crack propagation at C_1 and C_2 will occur under the action of σ_2, with little effect of σ_1. Intermediately oriented flakes will have a lesser stress concentration (at B_1

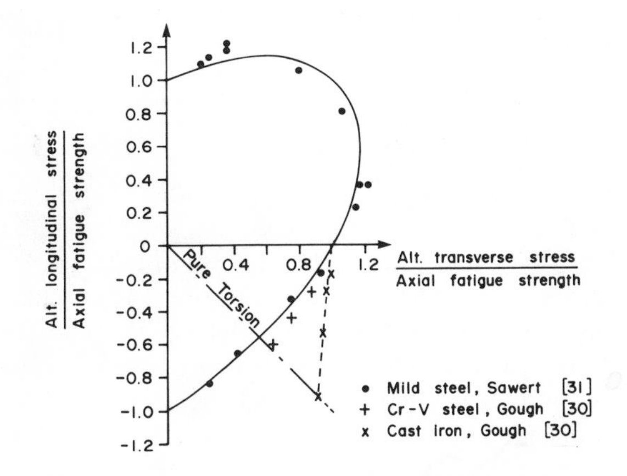

Fig. 2.21 Fatigue data for combined stresses. (After Sines [*29*])

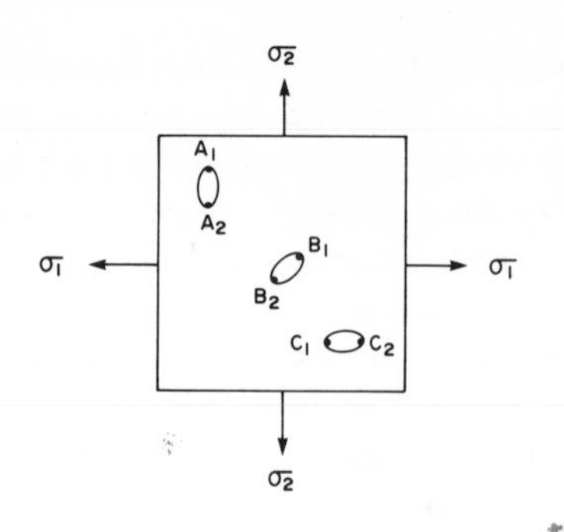

Fig. 2.22 Idealized model of graphite flakes in cast iron. (After Sines [*29*])

and B_2) under the action of either σ_1 or σ_2 and should contribute little to the failure.

Hence, a criterion may be proposed for fatigue failure under in-phase multiaxial stressing, having the same quadratic form as used for yielding by von Mises; namely that failure will occur when

$$(1/3)\,[(\sigma_1 - \sigma_2)^2 + (\sigma_2 - \sigma_3)^2 + (\sigma_3 - \sigma_1)^2]^{1/2} \geqslant \tau_{\text{oct-crit}} \tag{2.33}$$

where σ_1, σ_2 and σ_3 are the amplitudes of the fluctuating principal stresses, and $\tau_{\text{oct-crit}}$ is an experimental parameter for a given cyclic life. It has been observed that most fatigue data for metals do fall close to the ellipse defined in this way, except for the graphite-containing alloys already discussed [*6, 32*].

The effect of different combinations of static stress with alternating stress was examined by Sines [*29, 33*], and it was found that the permissible alteration of the octahedral shear stress is a linear function of the orthogonal normal static stresses (S_1, S_2 and S_3). Thus, the criterion for failure to occur can be expressed as

$$\underbrace{(1/3)\,[(\sigma_1 - \sigma_2)^2 + (\sigma_2 - \sigma_3)^2 + (\sigma_3 - \sigma_1)^2]^{1/2}}_{\text{(alternating)}} \geqslant A - \underbrace{\alpha(S_1 + S_2 + S_3)}_{\text{(static)}} \tag{2.34}$$

where α defines the variation of the permissible range of cyclic stress with the static stress, and A is a constant for the material, depending on the reversed stress fatigue strength. A and α are both functions of the lifetime required for design purposes, and their values can be evaluated for a particular material by determining any two fatigue curves for which the static stresses are appreciably different, commonly used tests being reversed axial loading or reversed bending, and zero-to-tension fluctuating stress. Using these tests, we obtain the relations [*29*]

$$\left.\begin{aligned} A &= (\sqrt{2}/3)f_1 \\ \alpha &= (\sqrt{2}/3)\,[(f_1/f_1') - 1] \end{aligned}\right\} \tag{2.35}$$

where f_1 is the amplitude of the stress in the reversed load test, and f_1' is the amplitude of the fluctuating stress to cause failure in the same lifetime as for the reversed stress f_1. Since fatigue failure normally starts at a free surface [*32*], one of the principal stresses can usually be taken as zero, hence we can plot the criterion in the form of a series of concentric ellipses on the plane containing the other two principal stresses. This is illustrated in Fig. 2.23, and the size of the ellipse will depend on the sum of the static stresses.

In the case of an interior source for fatigue cracking, we obtain a failure surface in the form of a cylinder similar to the von Mises yield surface in principal stress space. Such a criterion has not been proved for three-dimensional stressing, however, but it has been suggested that it is conservative because of the inherent weaknesses which arise at a free surface [*29*].

The Sines criterion has been criticized by Booth [*34*] for the case where stressing is biaxial with the mean stresses having opposite sign, but the alternating stresses are applied algebraically in phase. Eq. (2.34) predicts that the fatigue strength would be equal to the uniaxial strength with zero mean stress, but this is not in agreement with test results [*34*].

An effect which has been neglected in the foregoing discussion is anisotropy, which is of importance in forged or rolled material, and of particular importance when dealing with directionally solidified or composite materials. In order to take into account the effects of anisotropy on the fatigue failure surface in principal stress (or strain) space, Krempl [*35*] has proposed a generalized approach, which reduces to the more normal criteria (including

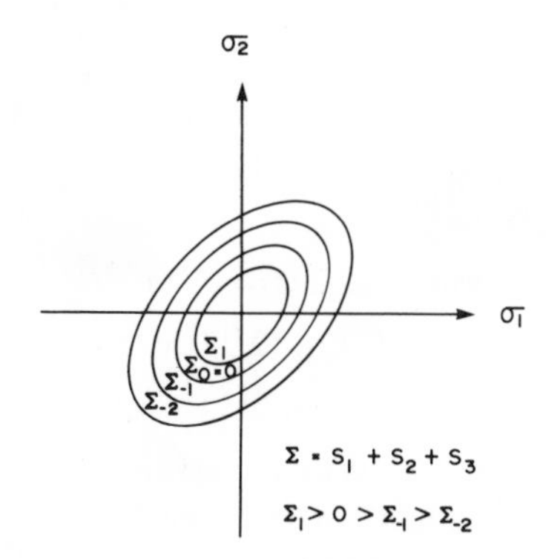

Fig. 2.23 Sines criterion [*29, 33*] for fatigue failure under multiaxial stressing. The size of the ellipse depends on the sum of the static components of stress

von Mises) upon making the appropriate substitutions. Caution must be exercised in applying von Mises-type relations to reduction of the fatigue strength of tubes subjected to an in-phase internal pressure and a fluctuating axial load. For a thin-walled tube this leads to each element in the wall being subjected to in-phase biaxial tension, and the results of such tests [*36, 37, 38*] show deviations from the von Mises-type relation, but this can be explained in most cases because the apparent circumferential fatigue limit was considerably less than the longitudinal one, indicating strong anisotropy of fatigue properties. It may be suggested that this reduction in circumferential fatigue strength is more apparent than real, as circumferential fatigue strengths were deduced from tube tests made in the presence of oil under fluctuating pressure, whereas the longitudinal data were determined in the normal way, and it is known that fatigue strength is reduced in the presence of a fluctuating oil pressure [*32*].This reduction can be explained by the pressurized oil entering into the cracks at an early stage, to give an increased driving force for their propagation, and by the increased ease of crack nucleation from surface discontinuities. Modification of the crack growth equation in the presence of oil has been discussed by Tomkins [*39*], and is considered further in Section 10.9 when discussing the prediction of fatigue life under service conditions.

An alternative approach to that due to Sawert [*31*] and Sines [*6, 29*] is the empirical one of McDiarmid [*40*] for long life in-phase multiaxial fatigue. He formulated a criterion based on a critical range of shear stress, modified for the effects of normal stress acting on the maximum shear stress plane, and of material anisotropy. For general reversed stress systems, the failure criterion is [*40*]

$$\tau_a = \tau_A - [(\tau_A - \sigma_A/2)/(\sigma_A/2)^{1.5}]\sigma_s^{1.5} \tag{2.36}$$

where τ_a is the shear, and σ_s the normal stress amplitude acting on the plane of maximum shear stress, and σ_A and τ_A are the reversed uniaxial and reversed pure shear fatigue strengths, respectively. Assuming $\tau_A/\sigma_A = 0.58$ for ductile materials, Eq. (2.36) gives

$$\tau_a = 0.58\sigma_A - (0.225/\sigma_A^{0.5})\sigma_s^{1.5} \tag{2.37}$$

When mean stresses are present, Eq. (2.37) is modified to give [*40*]

$$\tau_a = [0.58\sigma_A - (0.225/\sigma_A^{0.5})\sigma_s^{1.5}]\,[1 - \sigma_{Ms}/(\sigma_u/2)]^{0.5} \tag{2.38}$$

where σ_{Ms} is the mean normal stress on the maximum shear stress plane, and σ_u is the ultimate tensile strength. Thus, only one fatigue strength value, σ_A, and the UTS are required to calculate the limiting values of the allowable stresses. For preliminary design purposes, the effect of σ_s may be neglected, and Eq. (2.38) simplified to

$$[\tau_a/(\sigma_A/\sqrt{3})]^2 + \sigma_{Ms}/(\sigma_u/2) = 1 \tag{2.39}$$

A similar theory to that of McDiarmid has been proposed by Brown and Miller [*41*], but is more physically based in that it considers the geometry of the cracks that form by relating the planes of maximum shear to the orientation of the free surface. It recognizes the importance of the maximum shear strain (rather than stress) and of the tensile strain generated across this plane in assisting decohesion and crack propagation. Again using a crack propagation approach, Tomkins [*42*] has shown that fatigue data for cylinders in the limited life region can be correlated with simple push-pull or torsion fatigue data. These approaches are discussed more fully in Section 10.9. Here, we may note that a fracture mechanics approach to fatigue design, in which the concern is with the propagation of cracks from possible pre-existing defects, is the more appropriate one to take when dealing with pressure vessels, as the incidence of fatigue failures in these is (fortunately) exceedingly small, and the total number of cycles to which such a component is likely to be subjected during its life is relatively small (e.g., 175 000 cycles for hourly pressurization over a twenty-year life).

The discussion so far has been confined to in-phase multiaxial stresses, and the more general case where the stresses are out of phase has received relatively little attention until recently. There is some confusion in the literature over the importance of combined stresses being out of phase, probably as a result of the paper by Nishihara and Kawamoto [*43*] which is cited on a number of occasions as indicating that, in their studies of combined twisting and bending, the combined stress fatigue strengths with the torsion and bending out of phase were never less than the corresponding in-phase case. However, Little [*44*] has pointed out that this statement is misleading and that when the fatigue limit data of Nishihara and Kawamoto are stated in terms of the *true* shear stress amplitudes instead of the in-phase shear stress amplitudes, the fatigue limit actually decreases as the phase difference increases: thus design on the basis of combined stresses being in phase is not necessarily conservative when applied to a situation where phase differences exist.

Several attempts have been made recently to extend analytical procedures to deal with out of phase multiaxial stresses. For example, McDiarmid and Kumar [*45*] have applied McDiarmid's earlier analysis [*40*] to cover this situation, and have produced plots of allowable principal stress amplitudes as a function of their ratio and out of phase angle, ϕ, for both reversed and repeated biaxial stresses. Again, the basis of the approach is that the shear stress amplitude, modified for the effects of any normal stresses occurring on the plane of maximum shear stress, is the major parameter. When compared with experimental data, the method provided close agreement for values of ϕ of 60° or greater, although the agreement for $\phi < 60°$ was poor [*45*]. Miller [*46*] has extended his earlier model [*41*] to cover out of phase stresses, this depending on the growth of cracks having different orientations with respect to the surface, and he predicts a possible interference between the three possible crack systems, leading to a decrease in crack growth rate and an increase in endurance. For the case of $|\sigma_1| = |\sigma_2| \neq \sigma_3 = 0$, a maximum endurance is predicted at $\phi = 60°$ for sinusoidal loading or 90° for a triangular waveform, in conformity with the experimental data quoted by McDiarmid and Kumar [*45*]. Additional supporting evidence is provided by experiments reported by Grubisic and Simburger [*47*].

REFERENCES

1. CALLADINE, C. R., *Engineering Plasticity*, Pergamon Press, Oxford (1969).
2. TAYLOR, G. I., and QUINNEY, H., *Phil Trans. Roy. Soc.*, **A230**, 323, (1931).
3. TRESCA, H., *Comp. rend. Savants étrangers*, Paris (1865, 1868 and 1870).
4. VON MISES, R., *Nachr. Ges. Wiss. Göttingen, Math.-physik. Klasse* (1913).
5. NADAI, A., *Theory of Flow and Fracture of Solids*, Vol. 1, 2nd Edition, McGraw-Hill, New York (1950).
6. SINES, G., *Elasticity and Strength*, Allyn and Bacon, Boston (1969).
7. FORD, H., *Advanced Mechanics of Materials*, Longmans, Green, London (1963).
8. LODE, W., *Z. Physik*, **36**, 9 (1926).
9. WATTS, A. B., and FORD, H., *Proc. I. Mech. E.*, **169**, 1141 (1955).
10. BACKOFEN, W. A., HOSFORD, W. F., Jr., and BURKE, J. J., *Trans. ASM*, **55**, 264 (1962).
11. BABEL, H. W., EITMAN, D., and MCIVER, R., *Trans. ASME, J. Basic Eng.*, **89**, 13 (1967).
12. BACKOFEN, W. A., *Deformation Processing*, Addison-Wesley, Reading, Mass. (1972).
13. HILL, R., *Proc. Roy. Soc.*, **A193**, 281 (1948).
14. VON KÁRMÁN, T., *Z. VDI*, **55**, 1749 (1911).
15. BRIDGMAN, P. W., *Studies in Large Plastic Flow and Fracture*, McGraw-Hill, New York (1952).
16. COULOMB, C. A., *Memoires de Mathematique et de Physique*, Academie Royal des Sciences, Paris (1776).
17. MARIN, J., *Mechanical Behavior of Engineering Materials*, Prentice-Hall, London (1962).
18. PAUL, B., *Trans. ASME, J. Appl. Mech.*, **28**, 259 (1961).
19. GRIFFITH, A. A., in *Proc. of the First Int. Cong. for Applied Mechanics*, p. 55, Delft (1924).
20. FISHER, J. C., *ASTM Bulletin*, 74, April (1952).
21. BABEL, H. W., and SINES, G., *Trans. ASME, J. Basic Eng.*, **90**, 285 (1968).
22. FRISHMUTH, R. E., and MCLAUGHLIN, P. V., *Trans. ASME, J. Eng. Matls. and Technology*, **98**, 69 (1976).
23. INGLIS, C. E., *Trans. Inst. Naval Architects*, **55**, 219 (1913).
24. BROUTMAN, L. J., KRISHNAKUMAR, S. M., and MALLICK, P. K., *J. Am. Ceramic Soc.*, **53**, 649 (1970).

25. SINES, G., and ADAMS, M., in *Proc. of the 1971 Int. Conf. on Mechanical Behavior of Materials*, Vol. V, p. 295, The Society of Materials Science, Japan (1972).
26. McCLINTOCK, F. A., and ARGON, A. S., *Mechanical Behavior of Materials*, Addison-Wesley, Reading, Mass. (1966).
27. NEUBER, H., *Kerbspannungslehre*, Springer, Berlin (1937).
28. SINES, G. and ADAMS, M., ASME Paper No. 75-DE-23 (1975).
29. SINES, G., *Bull. of the Japan Soc. for Mechanical Engineers,* **4**, 443 (1961).
30. GOUGH, H. J., *Some Experiments on the Resistance of Metals to Fatigue Under Combined Stresses*, Great Britain Aero. Res. Council, Ministry of Supply, R & M 2552, London (1951).
31. SAWERT, W., *Z. VDI,* **87**, 609 (1943).
32. FROST, N. E., MARSH, K. J., and POOK, L. P., *Metal Fatigue*, Clarendon Press, Oxford (1974).
33. SINES, G., *Failure of Materials under Combined Repeated Stresses with Superimposed Static Stresses*, NACA Tech. Note 3495 (1955).
34. BOOTH, S. E., M.Sc. Thesis, University of London (1970).
35. KREMPL, E., in *Specialists Meeting of Low Cycle High Temperature Fatigue*, p. 5-1, AGARD Conf. Proc. No. 155 (1974).
36. MARIN, J., and SHELSON, W., NACA Tech. Note 1889 (1949).
37. BUNDY, R. W. and MARIN, J., *Proc. ASTM,* **54**, 755 (1954).
38. MORIKAWA, G. K. and GRIFFIS, L., *Weld. J.,* **24**, 167s (1945).
39. TOMKINS, B., in *Proc. of the 2nd Int. Conf. on Pressure Vessel Technology*, Part II, p. 835, ASME, New York (1973).
40. McDIARMID, D. L., in *Proc. of the 2nd Int. Conf. on Pressure Vessel Technology*, Part II, p. 851, ASME, New York (1973).
41. BROWN, M. W., and MILLER, K. J., *Proc. Inst. Mech. Engrs.,* **187**, 745 (1973).
42. TOMKINS, B., *Int. J. on Pressure Vessels and Piping,* **1**, 37 (1973).
43. NISHIHARA, T., and KAWAMOTO, M., *Memoirs, Coll. Eng., Kyoto Imperial University,* **11**, 85 (1945).
44. LITTLE, R. E., *The Aeronautical Quarterly,* **20**, 57 (1969).
45. McDIARMID, D. L., and KUMAR, V. M., Research Memo. ML79, The City University, London (1975).
46. MILLER, K. J., in *Fatigue Testing and Design*, Vol. 1, p. 13.1 (R. G. Bathgate, ed.), Soc. of Environmental Engrs., London (1976).
47. GRUBISIC, V., and SIMBURGER, A., in *Fatigue Testing and Design*, Vol. 2, p. 27.1 (R. G. Bathgate, ed.), Soc. of Environmental Engrs., London (1976).

3

TENSILE DEFORMATION AND DUCTILITY

3.1 TENSILE TESTING

Although in many ways a crude test of an engineering material's mechanical properties, the tensile test is an extremely useful one, and can provide considerable insight into a material's serviceability, and information concerning its strength and ductility. Hence, it is proposed to discuss the test in some detail. It may be noted at the outset that the tensile test is probably the test most commonly used to characterize materials, apart, possibly, from hardness measurements, and the latter do not indicate to any great extent both strength and ductility values for a material.

The engineering stress-strain curve for a ductile metal was shown in Fig. 1.3: such a curve is shown again in Fig. 3.1, and compared with the stress-strain curves for an ideally brittle solid, and for a material having very limited ability to deform plastically. The engineering stress-strain curve may be constructed from measurements of load and extension with engineering stress, σ, and conventional (or engineering) strain, e, derived from these.

3.1.1 Engineering Properties

The engineering properties that may be determined from the engineering stress-strain curve include the elastic modulus, yield strength (or yield point), tensile strength, ductility (as determined by the uniform strain, fracture strain,

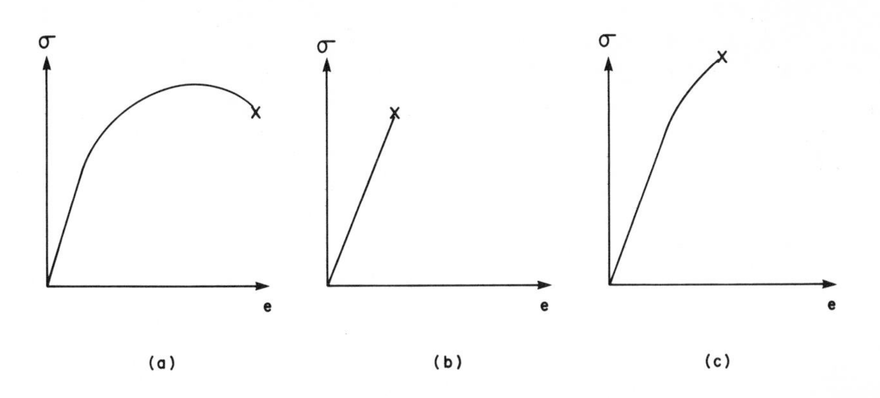

Fig. 3.1 Engineering stress-strain curves: (a) ductile metal; (b) ideally brittle solid; (c) solid having very limited plastic deformation

zero gage length strain, or reduction of area), toughness, and Poisson's ratio. These are all defined below for completeness[1], although several have been discussed previously in Chapter 1. It should be noted that all tensile testing and evaluation of these various properties must be done in accordance with a standard test procedure, commonly the ASTM "Standard Methods of Tension Testing of Metallic Materials" (ASTM E8-78) in North America.

Elastic modulus (Young's modulus)[2], *E*

As defined in Chapter 1, this is the slope of the initial linear part of the stress-strain curve. It is a measure of the stiffness of a material, or its resistance to elastic strain. Its value depends on the forces binding the atoms of the material together, and it is relatively unaffected by changes in structure, such as cold work, alloying, or heat treatment.

The modulus of elasticity for elevated temperature conditions may be determined in a dynamic manner [*1*], owing to the tendency of a metal to creep under such conditions, giving rise to a shallower slope the lower the strain rate applied.

In materials having a non-linear elastic region, such as concrete or cast iron, the modulus depends on the strain, and either tangent, secant, or chord modulus may be determined as required (see Fig. 3.2 and ASTM E111-62).

---

1 Definitions of terms relating to mechanical testing are given in ASTM E6-76, "Standard Definitions of Terms Relating to Methods of Mechanical Testing".

---

2 See ASTM E231-69, "Standard Method for Static Determination of Young's Modulus of Metals at Low and Elevated Temperatures".

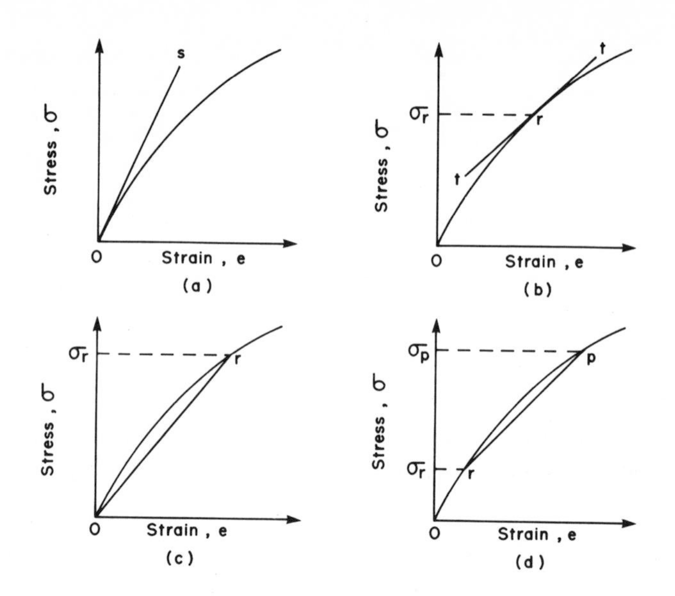

Fig. 3.2 Stress-strain curves showing (a) initial tangent modulus, the slope *Os*; (b) tangent modulus at specified stress, σ_r, being the slope of the tangent *trt* at σ_r; (c) secant modulus, being the slope *Or* between the origin and a point on the curve at specified stress, σ_r; (d) chord modulus, the slope of the line *rp* between specified levels, σ_r and σ_p

Yield strength (σ_Y)

The yield strength (or yield point) refers to the engineering stress at which a small specified plastic strain takes place. In the case of an idealized material having a sharp and well-defined yield point, this can be simply defined (see Fig. 3.3) as the stress at which plastic deformation commences; however, for the general case, it is necessary to specify the plastic strain (usually 0.2% offset), and to draw a line parallel to the linear portion of the stress-strain curve and offset by the specified amount to determine the intercept with this curve, as shown in Fig. 1.3. It should be noted that the stress is the engineering stress defined in terms of the original cross-sectional area.

The yield strength is extremely sensitive to the structure and prior history of a material.

Tensile strength (σ_u)

This is also referred to as the ultimate tensile strength (UTS), and is the maximum tensile stress which the material is capable of carrying. It is calculated from the maximum tensile load occurring during a tensile test, divided by the original cross-sectional area of the specimen.

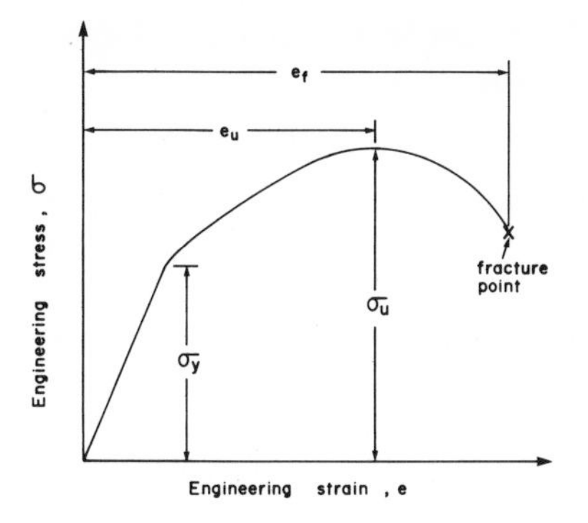

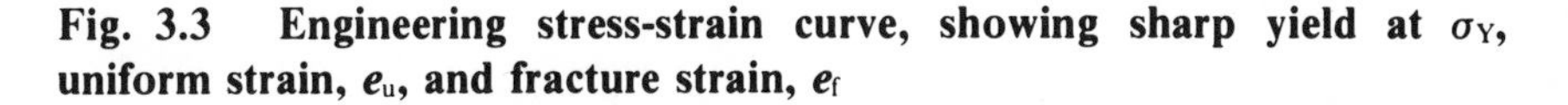

Fig. 3.3 Engineering stress-strain curve, showing sharp yield at σ_Y, uniform strain, e_u, and fracture strain, e_f

As with the yield strength, tensile strength is sensitive to the structure and processing of a material, but to a much lesser degree. Correspondence between the tensile strength and other properties on a rational and mechanistic basis is not good, however there are useful empirical correlations between it, hardness and fatigue strength [*2, 3*], particularly for steels, and it is a useful quantity in the quality control of materials.

Ductility

Ductility is the ability of a material to deform plastically before fracturing[3]. It is an extremely important property in that in the absence of sufficient ductility for some required local adjustment to take place in an engineering member, this will fracture. For the case of a crack in a material, the resistance of the material to crack propagation is dependent on its having sufficient ductility to relieve local stresses at the tip of the crack or notch. Hence, ductility and *toughness* are seen to be closely related.

Ductility can be specified in many different ways. This is discussed further in Sections 3.1.2 and 3.2, definitions here being confined to the engineering stress-strain curve.

Commonly, ductility is measured on the basis of the *percentage elongation at fracture*, the fracture strain e_f, being expressed as

$$e_f = (l_f - l_o)/l_o \tag{3.1}$$

where l_f is the length between gage marks at fracture, and l_o is the original gage length (commonly 50.8 mm). Expressed on a percentage basis, e_f gives the percentage elongation at fracture.

[3] This definition is according to ASTM E6-76.

Alternatively, ductility may be defined on the basis of the *reduction of area* at fracture, q, (or RA), as

$$q = (A_o - A_f)/A_o \tag{3.2}$$

where A_o and A_f are respectively the cross-sectional area of the original specimen and the minimum cross-sectional area after fracture.

During deformation in a tensile test, the strain is uniform up to the point of maximum load, after which localized necking may begin in ductile materials[4]. The contribution which this localized necking makes to the fracture strain, e_f, will depend on the gage length. The shorter this is, the greater is the contribution of necking and the greater the percentage elongation. Hence the need for standardized specimens, and for the reporting of gage length used in connection with elongation measurements.

Hence, ductility may, in some cases, be recorded in terms of the *uniform strain*, e_u (see Fig. 3.3) or the *zero gage length strain*, e_o. The former is an important parameter in the formability of sheet metal, as its magnitude specifies the allowable strain, beyond which local thinning will occur.

To define and determine zero gage length strain, e_o, we note first that plastic deformation takes place under essentially constant volume conditions, i.e., $Al = A_o l_o$, where A and l are respectively the instantaneous cross-sectional area and length between gage marks. Hence

$$l/l_o = A_o/A = 1/(1 - q)$$

from Eq. (3.2). Thus

$$e_o = (l - l_o)/l_o = (A_o/A) - 1 = 1/(1 - q) - 1 \tag{3.3}$$

or

$$e_o = q/(1 - q) \tag{3.4}$$

Thus, the zero gage length strain can be determined from the reduction of area at fracture. This quantity is of importance in forming operations where the gage length is very short.

To illustrate the variation in the values of ductility when measured in these different ways, Table 3.1 provides some examples for two aluminum alloys in two different conditions.

The importance of using standard specimens has already been noted, but this is not always possible, particularly when sub-standard specimens only can be cut from a component for testing. In such cases, it is usual to use a fixed

[4] In some ductile metals, yielding may begin at the point of maximum load, with localized thinning or shear failure taking place, e.g., in some textured zirconium alloys; hence the uniform strain may be virtually zero while the material still deforms extensively by plastic deformation before failure. In brittle solids, fracture of an essentially brittle nature will take place without any necking.

Table 3.1 Tensile Ductility of Two Aluminum Alloys on the Basis of Several Different Indices [4]

Alloy	e_o	$e_{2.0}$	e_u
24S-0 (2024-0)	1.22	0.18	0.16
24S-T (2024-T)	0.64	0.18	0.15
75S-0 (7075-0)	1.56	0.16	0.11
75S-T (7075-T)	0.44	0.11	0.09

Note:
e_o = zero gage length strain
$e_{2.0}$ = fracture strain on a 2-in. (50.8 mm) gage length
e_u = uniform strain

ratio of gage length to diameter or to the square root of the cross-sectional area, although this ratio has different values in different countries with differing sets of standards (see Ref. [3]). The use of different specimens having a common ratio of l/D (length/diameter) makes the effect of the necked region constant as shown below.

At fracture, the extension of a specimen is given by

$$l_f - l_o = \alpha + e_u\, l_o \tag{3.5}$$

where α is the local necking extension and the other terms are as defined previously. Thus

$$e_f = \alpha/l_o + e_u \tag{3.6}$$

Barba's law [5] states that geometrically similar specimens form geometrically similar necks. Hence the local necking extension is proportional to the square root of the cross-sectional area, or $\alpha = \beta\sqrt{A_o}$. Thus

$$e_f = \beta\sqrt{A_o}/l_o + e_u \tag{3.7}$$

Eq. (3.7) indicates that e_f will be unaffected as long as the ratio $l_o/\sqrt{A_o}$ or l/D is kept constant.

Toughness

The toughness of a material refers to its resistance to crack propagation, or to its ability to absorb energy in the plastic range of deformation. The definition is a loose one, and the information obtainable from a tensile test is *qualitative only*. For more quantitative determination of the *fracture toughness* of a material, see Chapter 9 and ASTM E399-78, "Standard Test Method for Plane-Strain Fracture Toughness of Metallic Materials".

A qualitative picture of the relative toughness of materials may be obtained from considering the total energy absorbed during a tension test, that is, the area under a stress-strain curve with subtraction of the elastic stored energy at fracture.

Poisson's ratio (ν)

This is the ratio between the transverse strain and the corresponding axial strain at stresses below the proportional limit of the material. It requires accurate measurement of strain in the transverse direction, and specimens of rectangular section are normally used.

3.1.2 True Stress-Strain Testing

A more correct picture of the stress-strain characteristics is given by a plot of true stress versus true strain, the resulting curve being a *flow curve* for the material. This arises because the engineering stress-strain curve is based on the original dimensions of the test specimen, which change as deformation takes place: also, necking occurs in a ductile material, and the load (and engineering stress) fall after this point, although the material continues to strain harden, and the true stress to cause plastic deformation continues to increase.

The method of determining a true stress-strain curve is to record the diameter of the test specimen simultaneously with load measurement. This may be done either by means of a round-nosed micrometer or from a clip gage attached to the specimen. After the maximum load point is reached and a neck begins to form, it is necessary to correct the value of true stress to take into account the triaxial state of stress occurring in this region. Dieter [2] has discussed this matter fully, and provides correction factors to be applied to the average true stress to compensate for the effect of transverse stresses at the neck. Table 3.2 provides values of this Bridgman correction factor [6] for various values of a/R, where a is the radius of the minimum cross-section at the neck, and R is the radius of curvature of the neck, which is readily determined by projecting the contour of the neck onto a screen.

As previously noted in Chapter 1,

$$\sigma = P/A_i \tag{3.8}$$

where P is the applied force and A_i the instantaneous cross-sectional area, while true strain is defined as

$$\epsilon = \ln(l/l_o) = \ln(A_o/A) \tag{3.9}$$

where l is the current length of the specimen originally of gage length l_o.

Figure 3.4 illustrates the true stress-strain curve for a metal, comparing it with the conventional stress-strain curve, and also shows the form of the correction for necking. It is found that the true stress and true plastic strain, ϵ_p, can be related approximately by a power law,

$$\sigma = K\epsilon_p^n \tag{3.10}$$

where K is termed the *strength coefficient*, and n the *strain hardening exponent*. It may be seen from Eq. (3.10) that K is the value of true stress to cause a true plastic strain of unity.

Table 3.2 Bridgman Correction Factors to be Applied to Average True Stress to Compensate for Transverse Stresses at the Neck of a Tensile Specimen [2]

a/R	Factor
0	1.000
1/3	0.927
1/2	0.897
1	0.823
2	0.722
3	0.656
4	0.606

Note: a is the radius of the minimum cross-section at the neck
R is the radius of curvature of the necked region

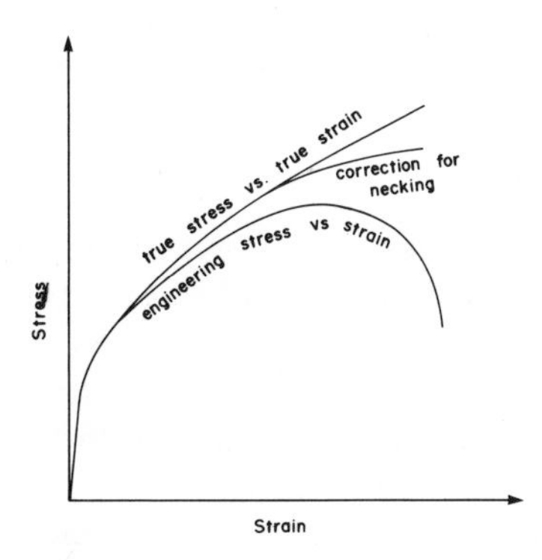

Fig. 3.4 True stress-strain curve for a metal, showing correction for necking, and also comparison with the conventional stress-strain curve

At small strains, the true and engineering values of the elastic modulus and yield strength are nominally equal, but several other strength values may be defined from true stress-strain data, in addition to K and n. They are discussed below.

True fracture strength (σ_f)

This is the true tensile stress to cause fracture, = P_f/A_f, with correction made for triaxiality due to necking.

True fracture ductility (ϵ_f)

This is the true plastic strain to cause fracture, = $\ln(A_o/A_f)$, and can be related to the reduction of area, q, by

$$\epsilon_f = \ln(A_o/A_f) = \ln[1/(1-q)] \tag{3.11}$$

True uniform strain (ϵ_u)

This is the true strain up to a maximum load, at which point necking commences, and is of relevance to the formability of metals.

$$\epsilon_u = \ln(A_o/A_u) \tag{3.12}$$

where A_u is the cross-sectional area at maximum load.

True local necking strain (ϵ_n)

This is the true strain required to deform a specimen from maximum load to the point of fracture, and is given by

$$\epsilon_n = \ln(A_u/A_f) \tag{3.13}$$

The values of K and n in Eq. (3.10) can be determined from a log-log plot as shown in Fig. 3.5. Such a plot must be corrected for triaxiality. From Fig. 3.5, we see that

$$n = d(\log\sigma)/d(\log\epsilon_p) = d(\ln\sigma)/d(\ln\epsilon_p) = (\epsilon/\sigma)\, d\sigma/d\epsilon \tag{3.14}$$

3.1.3 Criterion for Necking

In a ductile material, necking commences at the maximum load. If no strain hardening took place, it would start as soon as yielding began, but in normal strain hardening alloys, necking is delayed until a point is reached where the increase in true stress due to reduction in cross-sectional area as the specimen elongates is more than sufficient to compensate for the increase in load carrying capacity due to strain hardening.

At maximum load, the current yield stress (σ_u), is reached, the

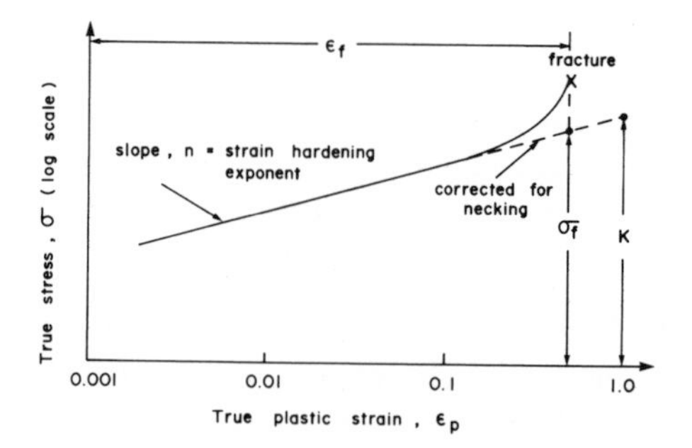

Fig. 3.5 True stress versus true plastic strain (log - log plot)

corresponding true stress being σ_m. Thus at this point

$$\sigma_u = P_{max}/A_o,\ \sigma_m = P_{max}/A_u,\ \epsilon_u = \ln(A_o/A_u) \tag{3.15}$$

Hence

$$\sigma_u = \sigma_m \exp(-\epsilon_u) \tag{3.16}$$

This equation provides a relationship between the UTS and the true stress at maximum load.

The condition for instability, or necking, is that $dP = 0$. Now, at any time

$$P = \sigma A$$

where σ is the true stress. Hence, for instability

$$dP = \sigma\, dA + A d\sigma = 0 \tag{3.17}$$

From constancy of volume

$$dl/l = -dA/A \tag{3.18}$$

Thus

$$-dA/A = dl/l = d\sigma/\sigma = d\epsilon = de/(1+e) \tag{3.19}$$

considering the relationship between true and engineering strain [Eq. (1.54)]. Therefore

$$d\sigma/d\epsilon = \sigma \tag{3.20}$$

or

$$d\sigma/de = \sigma/(1+e) \tag{3.21}$$

From Eq. (3.20) it may be seen that necking will begin at a strain at which the slope of the true stress-strain curve equals the true stress.

Considère's construction [7], based on Eq. (3.21), is shown in Fig. 3.6, and,

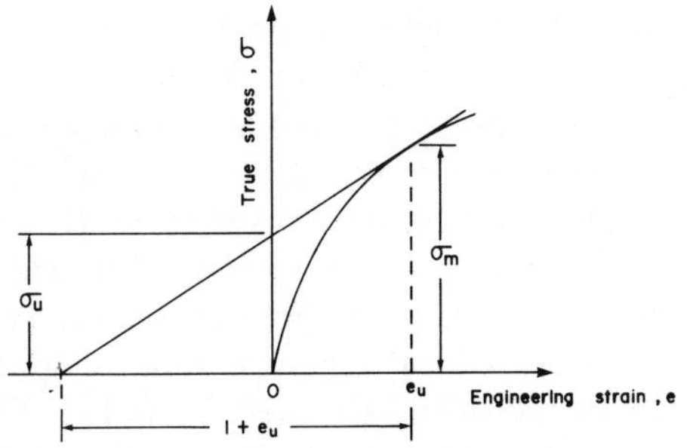

Fig. 3.6 Considère's construction to determine point corresponding to initiation of necking

using this plot of true stress against engineering strain, the point of maximum load (or value of true stress at maximum load) may be determined geometrically by drawing a tangent to the curve from a true strain of -1.

Now the true stress at maximum load and the UTS, σ_u, are related by

$$\sigma_u/\sigma_m = A/A_o = l_o/l = l_o/l_o(1 + e) = 1/(1 + e) \tag{3.22}$$

Hence

$$\sigma_u = \sigma_m/(1 + e) \tag{3.23}$$

From Fig. 3.6, we see that the intersection of the tangent with the ordinate axis corresponds to the UTS.

If the flow curve corresponds to Eq. (3.10), then, at instability, from Eq. (3.20),

$$d\sigma/d\epsilon = \sigma = K\epsilon_p^n = nK\epsilon_p^{n-1} \tag{3.24}$$

Hence

$$\epsilon_u = n \tag{3.25}$$

or the strain at which necking begins is numerically equal to the strain hardening exponent. This rule is obeyed in practice by most metals, although exceptions do occur, as not all metals obey the power law flow curve [(Eq. (3.10)]. For example, austenitic stainless steels which undergo a strain-induced transformation to martensite during straining can have true stress-strain curves which are *concave upwards* for the earlier part of the deformation, before reverting to a more normal parabolic shaped curve [*8*]. This is due to the extremely effective strain hardening induced by the transformation of the austenite, which occurs more extensively in the earlier stages of deformation.

Criticism of the classical analysis of tensile instability may be made on the basis that real materials do not have constancy of volume during deformation, i.e., the Poisson's ratio, ν, <0.5. This is certainly correct, as ν will vary from a constant value (~ 0.3) during elastic deformation, to a value approaching, but not reaching, 0.5 during plastic deformation. In order to account for this, the classical theory has been modified [*9*], but further calculations have shown that we can take $\nu = 0.5$ at the point of instability for all practical purposes [*10*].

So far our discussion has been confined to *diffuse* necking, which involves a length at least equal to the width of the test specimen. However, another type of necking instability which may arise is *local* necking, which may be formed in sheet metal, particularly after it has been cold rolled [*11*].

In this latter instability, a local region is thinned over a narrow band inclined at angle ϕ to the specimen axis, as shown in Fig. 3.7. The orientation of the band is such as to allow plane strain deformation within it, as there can be no plastic flow along its edges in the direction x_2', since the neighboring region of the specimen has not yielded.

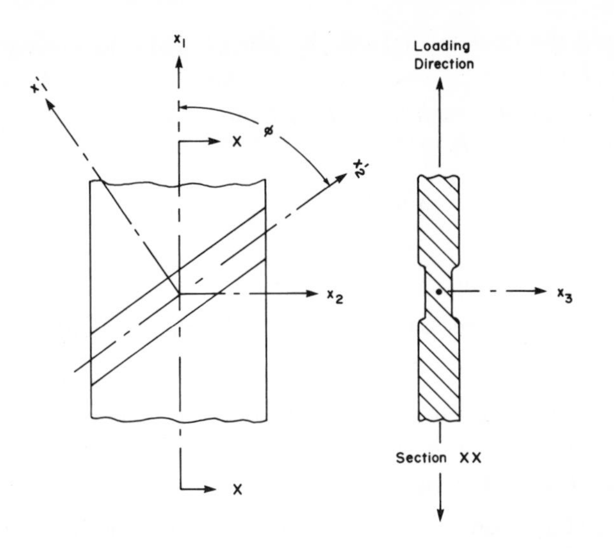

Fig. 3.7 Local necking in a sheet specimen under tensile loading

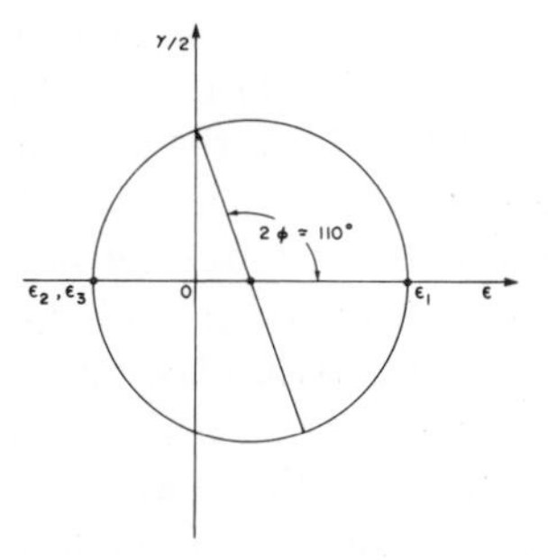

Fig. 3.8 Mohr's circle for deforming metal in the locally-necked band of Fig. 3.7

Consider the Mohr circle for strain, as shown in Fig. 3.8. From constancy of volume, $\epsilon_2 = \epsilon_3 = -\epsilon_1/2$, and the line for zero normal strain is at an angle $2\phi \simeq 110°$. Thus $\phi \simeq \pm 55°$. This corresponds to the observed angle for the formation of Lüder's bands in mild steel, these being regions of local necking formed without strain hardening, and which propagate sideways along the specimen.

The condition for local necking can be set down similarly to that for diffuse necking, namely that strain hardening is balanced by reduction in load bearing capacity from reduction in section. Here, however, the width (w) is constant and the thickness (t) only is reduced. Thus, by analogy with Eqs. (3.17) and (3.19),

$$-dA/A = -wdt/wt = d\sigma/\sigma = -d\epsilon_3 = d\epsilon_1/2 \tag{3.26}$$

Hence

$$d\sigma/d\epsilon_1 = \sigma/2 \tag{3.27}$$

or

$$d\sigma/de_1 = \sigma/[2\ (1 + e_1)] \tag{3.28}$$

If we substitute the power law relationship, we obtain the uniform strain ϵ_{1u} as

$$\epsilon_{1u} = 2n \tag{3.29}$$

From the foregoing, we see that local necking occurs at uniform strains twice those for diffuse necking, local necking having the restriction that $\epsilon_2' = 0$.

3.1.4 Notch Tensile Testing

Tensile testing using circumferentially notched specimens is a useful method of assessing a metal's *notch sensitivity*. Figure 3.9 shows the geometry of such a circumferentially notched tensile specimen, and the important parameters are the notch sharpness (a/r) and notch depth. Commonly, a 60° notch is used with root radius of 0.025 mm, and a depth such that the cross-sectional area at the root is half that of the unnotched section.

The effect of the geometry is to cause the state of stress to be biaxial at the surface of the notch and triaxial in the interior. A material which is notch sensitive or *notch brittle* will have a lower net section stress at the notch at failure than the tensile strength of the smooth specimen, while in a *notch ductile* material the reverse is true, as the plastic constraint at the notch tip will increase the local yield and tensile strengths. The normal practice is to determine the net section stress at failure, and the reduction in area at this section, both quantities being indicative of notch sensitivity when compared with unnotched data.

The *notch strength ratio* (NSR) is defined as the ratio of the notch strength to the ultimate tensile strength, and is the most common index used to evaluate notch sensitivity, especially for lower temperature conditions. The factors influencing this ratio are discussed by Weiss [*12*], but it may be noted here that for high strength levels and low ductility, a reduction in notch sharpness, which reduces the elastic stress concentration factor but does not affect the triaxiality of stress significantly, will cause an increase in NSR. At lower strength levels there is no effect on increasing the notch radius by a considerable amount. Changes in notch depth affect triaxiality greatly but elastic stress concentration factor very little. Hence, at low strength levels NSR is a linear function of notch depth, while at higher strength levels it is dependent on the notch ductility, which generally decreases to very low values for high tensile strengths [*2*].

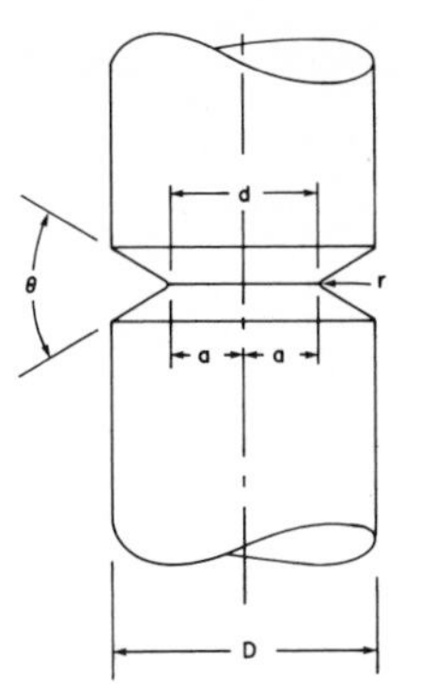

Fig. 3.9 Geometry of circumferentially notched tensile specimen.

Notch tensile testing is of importance in testing materials which have metallurgical structures that may lead to low local ductility, and also where environmental factors (e.g., the presence of hydrogen, causing embrittlement) may cause similar effects. Notched tests are also of value in elevated temperature studies, where the ratio of rupture strength (for a given time to rupture in a creep-rupture test) in a notched specimen to that in an unnotched specimen will give a measure of notch strengthening or notch weakening, this being a function of notch acuity [*13*].

3.1.5 Effects of Strain Rate

ASTM E8-78, "Standard Methods of Tension Testing of Metallic Materials", sets limits on the strain rate to be used when conventional tensile properties are to be determined on the basis that the loads and strain recorded should be accurately indicated and not affected by the strain rate used. Conventionally, strain rates of around 10^{-3} s^{-1} are used, but for most engineering materials the room temperature stress-strain curve does not vary greatly for strain rates an order of magnitude different from this. However, under very high rates of deformation, for example in cold rolling or wire drawing where engineering strain rates of up to 10^{3} s^{-1} may be employed, the tensile and yield strengths may be increased substantially. The effect is accentuated by testing at elevated temperature.

Figure 3.10 shows the load-strain curves produced when testing a steel at deformation rates corresponding to road speeds of 0.5 km/h and 80 km/h, [5] and shows the displacement of the entire curve, although for steels the yield strength is usually affected to a greater extent than is the UTS [*2*]. The

[5] The data were derived to determine the effect of a vehicle impacting at different road speeds.

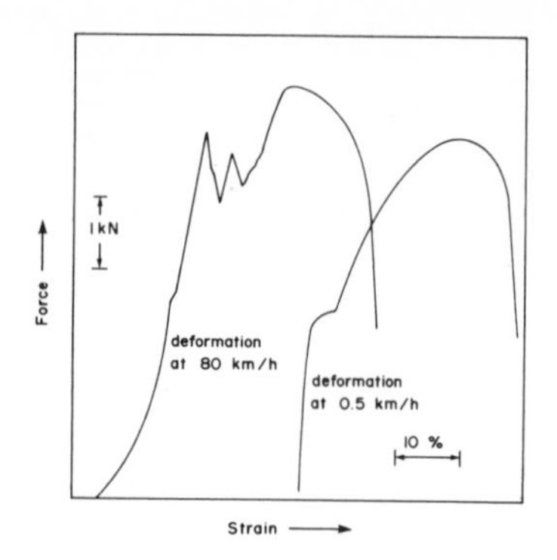

Fig. 3.10 Load-strain curves for a hot-rolled 0.32 w/o C steel deformed at widely different strain rates. (After Davies and Magee[*14*])

oscillations in the high strain rate curve of Fig. 3.10 are caused by reflection of elastic stress waves in the specimen and machine. Such oscillations make it difficult to obtain reliable yield stress data for strain rates in excess of ~80 s^{-1}. This is of particular relevance in testing low carbon steels treated to eliminate the yield point prior to forming, as the yield point returns under rapid straining.

The flow stress for a given true strain and temperature as a function of true strain rate, $\dot{\epsilon}$, is generally written in the form of a power law [*15*] as

$$\sigma_{\epsilon,T} = C_1 (\dot{\epsilon})^m \tag{3.30}$$

where m is the *strain rate sensitivity*. We may obtain the value of m (assuming Eq. (3.30) to hold) by rapid alteration of the strain rate during a test at some predetermined strain, and recording the flow stresses before and after the change (σ_1 and σ_2). Thus m is given by

$$m = \log(\sigma_2/\sigma_1)/\log(\dot{\epsilon}_2/\dot{\epsilon}_1) \tag{3.31}$$

Davies and Magee [*14*] have made an extensive study of strain rate on the tensile behavior of a wide range of engineering materials at room temperature, representative results for two steels being shown in Fig. 3.11. It was found that, for each material, the dynamic factor, R, defined as the ratio of the flow stress or UTS at the high strain rate ($\dot{\epsilon}_2$) to that for quasi-static conditions ($\dot{\epsilon}_1$) obeyed the equation

$$R = 1 + K_r \log(\dot{\epsilon}_2/\dot{\epsilon}_1) \tag{3.32}$$

where K_r is a constant (the slope of the straight line when R is plotted against $\dot{\epsilon}$). Values of R, together with best values of K_r, are given in Table 3.3 for all materials tested, for $\dot{\epsilon}_2$ = 833 s^{-1} and $\dot{\epsilon}_1$ = 0.016 s^{-1}. From this the dynamic factors for other strain rates can be obtained.

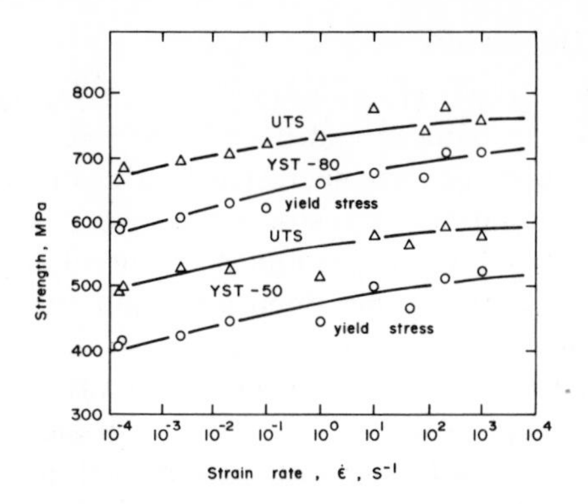

Fig. 3.11 Yield stress and ultimate tensile strength of two high strength, low alloy steels as a function of strain rate. (After Davies and Magee [*14*])

Table 3.3 Dynamic Factor for Increase in Tensile Strength with Increase in Strain Rate [*14*]

Material	Dynamic Factor, R, $= (UTS_{\dot{\epsilon}=833\ s^{-1}})/(UTS_{\dot{\epsilon}=0.016\ s^{-1}})$	K_r
Hot-rolled steel		
0.10% C	1.37	0.0870
0.19% C	1.32	0.0622
0.32% C	1.26	0.0549
0.47% C	1.23	0.0456
0.58% C	1.20	0.0415
0.70% C	1.19	0.0384
0.82% C	1.18	0.0363
Hard-rolled 1010 steel	1.19	0.0413
Q & T 4340 steel	1.07	0.0110
Fine pearlitic steel	1.12	0.0264
HSLA YST-50	1.12	0.0291
YST-80	1.10	0.0191
Stainless steel 302	1.00	0.0
310	1.12	0.0146
Al alloys 6061	1.00	0.0
7075	1.03	0.0038
Glass reinforced plastic		
fibreglass	1.55	0.1178
shell molding compound	1.43	0.0925

3.1.6 Effects of Temperature

The stress-strain curve is generally lowered as temperature is increased, strain hardening being reduced and ductility at fracture raised. However, exceptions to this rule occur when microstructural changes take place, such as strain aging, precipitation, transformations or recrystallization. As temperature is increased, creep phenomena (continuous deformation with time under steady load) become important, and the stress-strain curve becomes very sensitive to rate of straining.

In order to compare the stress-strain behavior of various metals at different temperatures, it is appropriate to consider the homologous temperature, which is the ratio of the test temperature to the melting point, both on the absolute scale, and metals at the same value of homologous temperature have generally similar behavior for similar crystal structure and absence of complicating structural effects. Also, in comparing the yield or flow stresses of a metal at different temperatures, it is more satisfactory to correct for the effect of temperature on modulus by comparing ratios of σ/E rather than simple ratios of yield stress.

Face centered cubic (FCC) metals, which have a gradual yielding process, have a small temperature dependence of yield strength and a large variation in strain hardening exponent, as seen from Figs. 3.12 and 3.13. They do not have any transition to brittle behavior at low temperatures, except in some specific alloys. Hexagonal close packed (HCP) metals have a generally similar temperature dependence of yield strength, but their tensile behavior is very sensitive to the presence of interstitial impurities and their deformation mode (see Chapter 4): they also undergo a transition to brittle fracture at low temperature in almost all cases. Body centered cubic (BCC) metals, which have a distinct discontinuous yield point, have a strong temperature dependence of their yield stress (see Fig. 3.12), while the strain hardening exponent is affected only slightly.

One exception to the general rule for BCC metals should be noted. For mild steel (and some other BCC alloys), the strength increases as the temperature is raised from room temperature, maximum strength, together with a minimum in ductility, occurring at around 200° C. This is caused by strain aging[6], and is commonly referred to as "blue brittleness" in steels on account of the color of the surfaces from slight oxidation.

The flow stress at given strain and strain rate can generally be expressed as a function of temperature by the equation

$$\sigma_{\epsilon,\dot{\epsilon}} = C_2 \exp(Q/RT) \qquad (3.33)$$

[6] Strain aging, caused by the pinning of dislocations by mobile impurity atoms (interstitial C or N in the case of steel) is described in Section 4.4.3.

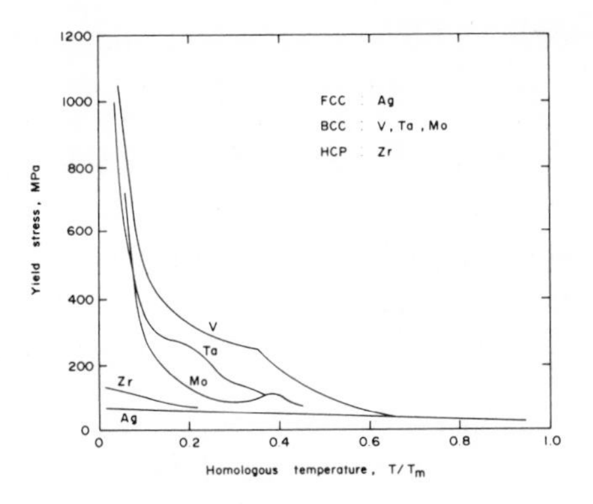

Fig. 3.12 Temperature dependence of yield stress for various polycrystalline metals. (After Tegart [*15*])

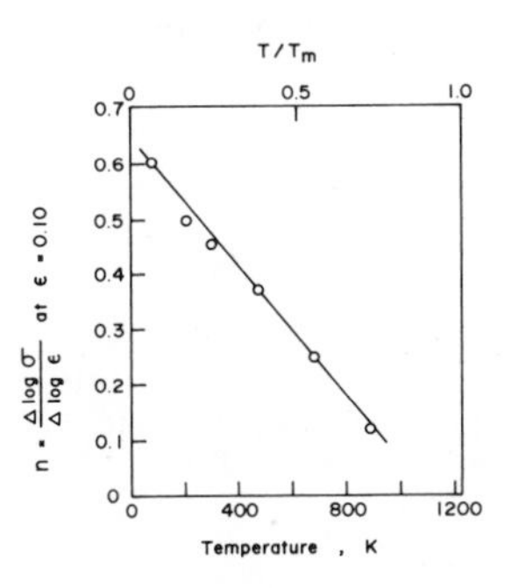

Fig. 3.13 Temperature dependence of strain hardening exponent as a function of temperature for polycrystalline pure silver. (After Carreker [*16*])

where Q is an activation energy for plastic flow, R the universal gas constant, and T the absolute temperature of testing. If this relationship is obeyed, a plot of ln $\sigma_{\epsilon,\dot{\epsilon}}$ versus $1/T$ will give a straight line, from the slope of which Q can be determined. Eq. (3.33) has been found to be applicable to a wide range of alloys, but it is not satisfactory for BCC metals at low temperatures, where a relation of the form

$$\sigma_{\epsilon,\dot{\epsilon}} = \text{const.}/T \tag{3.34}$$

is found to be applicable.

3.1.7 Combined Effects of Strain Rate and Temperature

The effect of strain rate on tensile properties is, as noted already, greatly affected by temperature, and a commonly used parameter to allow for the flow stress dependence on both variables is the following [*17*]:

$$\sigma_\epsilon = f(Z) \tag{3.35}$$

where Z is termed the Zener-Hollomon parameter, defined as

$$Z = \dot{\epsilon} \exp(\Delta H / RT) \tag{3.36}$$

where ΔH is an activation energy which is related to Q in Eq. (3.33) by $Q = m\Delta H$, where m is the strain rate sensitivity [*15*].

If Eq. (3.35) holds, a plot of $\ln\dot{\epsilon}$ versus $1/T$ for given values σ and ϵ should give a straight line, and it has been found to hold for a number of alloys including mild steel, copper, aluminum and molybdenum [*2*].

As discussed by Tegart [*15*], the validity of the Zener-Hollomon parameter requires that Q vary in some fashion with temperature, such that Q/m is constant, since ΔH is a constant which must be independent of temperature and strain rate, and m is temperature dependent. Thus it appears that, where the relationship is obeyed, the ranges of temperatures and strain rate are such as to maintain Q/m approximately constant.

3.2 DUCTILITY INDICES

Ductility is a most important property concerning a material's resistance to fracture. It may be defined as the ability of the material to flow and redistribute stress without generating a fracture instability under a given loading [*18*][7]. This definition applies at low temperature and also at high temperature under creep conditions, where ductility exhaustion may be expected in time, either at a defect present in the material, or at a geometric discontinuity in the member. Accordingly, many indices of ductility have been proposed both for low temperature and elevated temperature application. Some of these have been described earlier in this chapter, and the various indices are reviewed by Manjoine [*18*] in a paper having particular relevance to elevated temperature conditions.

Table 3.4 lists the various ductility indices of relevance to a designer. Indices 1c to 5c and 7c to 12c are equivalent to the indices for low temperature conditions, but refer to creep-rupture test conditions. Detailed discussion of all the indices is provided by Manjoine [*18*], and further discussion here will be confined to Index 21, the theoretical ductility ratio. It should be noted that the index appropriate to a particular design situation is determined by the type of service loading, manufacturing specifications, and the reliability required of the component.

[7] Compare with the ASTM E6-76 definition given in Section 3.1.1.

3.2.1 Elevated Temperature Damage

Under conditions of elevated temperature we can consider two types of damage as occurring, attributed phenomenologically to either stress or strain. Damage represents the adverse effect of deformation and stress on a material's strength or ductility, while ductility represents the material's ability to redistribute stress concentrations without accumulating damage giving rise to fracture instability. The two damage modes are [*13, 18*]:

(a) damage accumulated by the history of the maximum principal stress, this being relevant to a low ductility material where crack initiation at grain-boundaries is the dominant effect, and
(b) damage accumulated by the shear stresses until ductility is exhausted or a strain limit is reached. This mode is relevant to high ductility materials.

Combinations of the two modes can also be expected, these interacting together.

Hence the damage for a given temperature, metallurgical state, and environment, may be represented by the sum of the two components, and a linear damage rule has been proposed as

$$D_{T,\phi,\ \mathrm{Env}} = A\Sigma(t_i/t_r)_{\sigma_1} + B\Sigma(e_i/e_L)_{\dot{e},\ \tau_{ij}} \tag{3.37}$$

where

$D_{T,\phi,\mathrm{Env.}}$ is the damage for given temperature T, metallurgical state ϕ, and environment (Env.)
t_i is the time increment for the maximum principal stress $(\sigma_1)_i$
t_r is the rupture time at the maximum principal stress $(\sigma_1)_i$
e_i is the strain increment at a strain rate $\dot{e}_i$ under a stress state τ_{ij}
e_L is the strain limit for a strain rate $\dot{e}_i$ under a stress state τ_{ij}
A and B are interaction coefficients.

3.2.2 Index for Multiaxial and Elevated Temperature Conditions

The amount by which a material can deform is limited by the constraints imposed upon it, and the geometry of the component or the applied loads may produce multiaxial stresses which will reduce the material's ductility. Davis and Connelly [*19*] suggested that the ductility will be influenced by the normal stress and the shear stress on the octahedral plane[8], and introduced a triaxiality factor, T.F., as

$$\mathrm{T.F.} = (\sigma_1 + \sigma_2 + \sigma_3)/(1/\sqrt{2})[(\sigma_1 - \sigma_2)^2 + (\sigma_2 - \sigma_3)^2 + (\sigma_3 - \sigma_1)^2]^{1/2} \tag{3.38}$$

This is the ratio of the normal stress to the shear stress on the octahedral plane,

[8] See Section 1.2.2.

Table 3.4 Ductility Indices for Low and Elevated Temperatures [*18*]

Index		Strain	Deformation
No.	Name	Measurement	Type
1	Unform elongation	$e_u = (l_u - l_o)/l_o = (A_o - A_u)/A_u$	Plane stress in tension (TT)
2	Elongation	$e_f = (l_f - l_o)/l_o$	TT
3	True fracture strain	$\epsilon_f = \ln (A_o/A_f)$	TT
4	Reduction in area	$\mathrm{RA} = q = (A_o\ A_f)/A_o$	TT
5	Zero gage length strain	$e_o = (A_o/A_f)/A_f = q/(1 - q)$	TT
1c to 5c	Same as above for tensile creep-rupture.		
6	Secondary creep strain	$e_{2c} = \acute{e}_m\ t_r$ $\acute{e}_m$ is minimum creep rate; t_r rupture time	Creep-rupture
7	Strain instability factor	$\alpha = e_o/e_u$	TT
8	Necking factor	$\beta = A_u/A_f$	TT
9	Void factor	$\nu = e_f/e_o$	TT
10	Maximum shear strain	$\gamma_{max} = r_o\ \theta$	Torsion test
11	Notched true strain	$N\epsilon = \ln A_o/A_f$	Notched tension test
12	Notched RA	$\mathrm{NRA} = (A_o - A_f)/A_o$	Notched tension test
7c to 12c	Same as above for creep-rupture		

Table 3.4 continued

Index		Strain	Deformation
No.	Name	Measurement	Type
13	True stress ratio	TSR = True stress at fracture/UTS	TT
14	Notch strength ratio	NSR = Notched strength/unnotched strength	Static or dynamic tension
15	Notch rupture strength ratio	NRSR = (NSR) $t_f = c$	Tensile creep-rupture for a given time, t_f
16	Rupture time ratio	$\tau\sigma$ = Notched/rupture time/Unnotched rupture time	Tensile creep-rupture at a given stress
17	Notch strength analysis	Ultimate strength reduced by plastic stress concentration factor	Plane stress
18	Propagation-Initiation ratio	Fracture propagation time/Fracture initiation time	General
19	Hot working	Absence of cracking	Process simulation
20	Fracture ductility	Fracture appearance: Brittle — cleavage or intergranular Ductile — shear or transgranular	General
21	Theoretical ductility ratio	Fracture strain/Theoretical strain limit	Multiaxial stress

normalized to unity for simple tension. The denominator is the *effective stress* or Mises stress, σ_{eff} [see Eq. (1.60)].

When damage is primarily due to the maximum principal stress, corresponding to a low ductility material, we may represent the damage criterion for failure as

$$\Sigma(t_i/t_r)_{\sigma_1} = 1 \tag{3.39}$$

Under creep conditions, the strain rate can reasonably be represented by a power function of the stress (see Section 11.3.1),

$$\dot{e}_{eff} = B\,\sigma_{eff}^{m} \tag{3.40}$$

where [cf. Eq. (1.61)]

$$\dot{e}_{eff} = (\sqrt{2}/3)\,[(\dot{e}_1 - \dot{e}_2)^2 + (\dot{e}_2 - \dot{e}_3)^2 + (\dot{e}_3 - \dot{e}_1)^2]^{1/2} \tag{3.41}$$

and m is the strain rate sensitivity.

The theoretical failure strain for multiaxial stress, e_{f3}, will be the accumulated strain based on the history of the effective strain rate and effective stress, and may be written as

$$e_{f3} = \Sigma(\dot{e}_{eff}\,t_i)_{T_{ij}} \tag{3.42}$$

while the strain in simple tension for the same strain rate, and stress = σ_1, is

$$e_f = (\dot{e}_{eff}\,t_r) \tag{3.43}$$

Figure 3.14 shows the ratio of the theoretical failure strain for multiaxial stress to the failure strain for uniaxial stressing at the same strain rate, using the assumptions noted above, and for different values of the strain rate sensitivity, m, plotted as a function of the triaxiality factor. This shows that the failure strain may be decreased to a very great extent for high values of m, and this can account for cracking in welds at low strains in some materials during cooling, or during stress relaxation.

For a material of high ductility, with damage due primarily to accumulation of strain until ductility is exhausted or a limiting value of strain is reached, the limiting damage criterion is

$$\Sigma(e_1/e_L) = 1 \tag{3.44}$$

Manjoine [*13, 18*] proposed that a theoretical reduction of ductility with triaxiality of stress is the strain limit for simple tension divided by the triaxiality factor, and the data derived for annealed Type 304 stainless steel at 593°C are plotted in Fig. 3.15. In this, uniform strain (Indices 1 and 1c) is plotted as a function of T.F.

Where the two damage modes are operative, Eq. (3.37) may be utilized to combine them, although values for the interaction coefficients should be estimated from experimental data before quantitative application of the theory can be made. It should be noted in any event that, while the theory

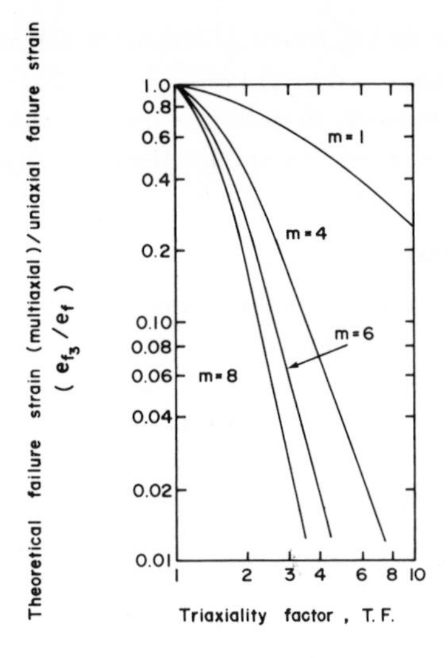

Fig. 3.14 Relation between theoretical failure strain and triaxiality factor. (After Manjoine [*13*])

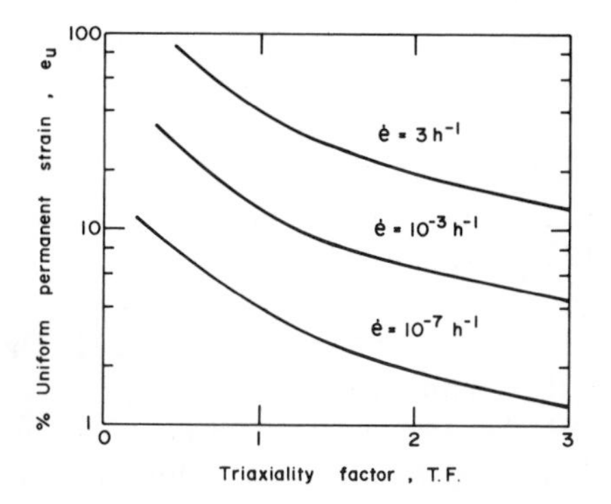

Fig. 3.15 Uniform strain as a function of triaxiality factor for annealed Type 304 stainless steel at 593° C. (After Manjoine [*13*])

proposed by Manjoine and discussed above is supported *qualitatively* by observations, there is a need for good *quantitative* data derived from multiaxial testing to support it before it can be used with confidence by the designer. Nevertheless, it does represent a first and useful approach towards solution of an important and practical problem.

Referring back to Table 3.4, Index 21, the theoretical ductility ratio, can be determined from the measured strain at failure under specified conditions of multiaxial stress, temperature, metallurgical state, and environment, and the theoretical strain at failure determined for the same multiaxial condition and taking into account the mode of damage.

3.2.3 Concluding Remarks

The foregoing remarks have been intended to emphasize the importance of the concept and measurement of ductility, but a detailed coverage of the subject or its very extensive literature has not been attempted. Many conferences have been concerned with this important property, two being of particular relevance [*20, 21*]. In addition, the important question of ductility limits for elevated temperature service has been discussed at a conference on elevated temperature design [*22*]. Another area where ductility is of key importance is fatigue, particularly with the now widespread use of testing methods and design bases utilizing cyclic strains rather than stresses, this being considered further in Chapter 10.

3.3 SUPERPLASTIC DEFORMATION

Some materials, notably glass and polymers at elevated temperature, are able to withstand very large deformations in tension without necking. Under the proper conditions it has been found that some metals can also exhibit large uniaxial ductility (to engineering strains $> 1000\%$ in some cases), because localized necking does not occur, but rather necking is diffuse. In recent years considerable interest has developed in producing engineering alloys which behave in this way [*23*], as the processing of these *superplastic* alloys can be done without the normal restrictions on processing metals of limited uniform ductility which require that they be squeezed to shape. Thus such metals can be processed using the techniques of polymer processing, and extensive biaxial stretching, either in free space or into a cavity by means of a small pressure differential, can be accomplished without fear of local thinning. In the extreme case, one can envisage "blowing" superplastic alloys in a manner similar to glass.

In the following section we shall first look at the criteria for the avoidance of localized necking, and then discuss some of the factors which are necessary for superplastic deformation, identifying the operative mechanisms of deformation.

3.3.1 Phenomenology of Resistance to Necking

As seen in Section 3.1.3., necking begins when the slope of the true stress-true strain curve equals the true stress. Thus, to avoid necking starting, or an incipient neck from growing, the following condition must be maintained:

$$d\sigma/d\epsilon > \sigma$$

where σ is the true stress. This implies that superplastic materials should have large values of $d\sigma/d\epsilon$.

In order to understand the phenomenon better, we may note that the true stress required to deform a material is a function of a number of factors such as strain, strain rate, temperature, and surface energy (γ). Thus we may write

$$\sigma = \sigma(\epsilon, \dot{\epsilon}, T, \gamma, \ldots\ldots) \tag{3.45}$$

And the strain hardening characteristics may be defined by [24]

$$d\sigma/d\epsilon = (\partial\sigma/\partial\epsilon) + (\partial\sigma/\partial\dot{\epsilon})(d\dot{\epsilon}/d\epsilon) + (\partial\sigma/\partial T)(dT/d\epsilon) + (\partial\sigma/\partial\gamma)(d\gamma/d\epsilon) + \tag{3.46}$$

In conventional engineering alloys most of the strain hardening is contributed by the first term on the right hand side, but this falls in magnitude as strain is increased, leading to instability and necking, usually at engineering strains of less than 50%.

In superplastic materials it has been found that there is a large sensitivity of flow stress to strain rate, and the resistance to necking is contributed by the second term on the right hand side of Eq. (3.46). Both parts of this term are normally positive, $\partial\sigma/\partial\dot{\epsilon}$ since stress normally increases with strain rate, and $d\dot{\epsilon}/d\epsilon$ because of changes in geometry as a neck starts to form. Even though the crosshead speed, v, in a testing machine does not change, the effective gage length is shortened to the length of the incipient neck, l_n, and the local strain rate, $\dot{\epsilon} = v/l_n$, is increased.

We can express the relationship between stress and strain rate in the form of Eq. (3.30) as

$$\sigma = C_1 (\dot{\epsilon})^m = P/A \tag{3.47}$$

where m may or may not be a function of $\dot{\epsilon}$, and the pulling force P acts on cross-sectional area A. For a rod length l being extended under conditions of constant volume,

$$\dot{\epsilon} = (1/l)(dl/dt) = -(1/A)(dA/dt) \tag{3.48}$$

The rate of shrinkage of the cross-section is $-dA/dt$, and substituting $\dot{\epsilon} = (P/C_1A)^{1/m}$ from Eq. (3.47) in Eq. (3.48), we obtain the shrinkage rate as

$$-dA/dt = (P/C_1)^{1/m}(1/A^{(1-m)/m}) \tag{3.49}$$

Thus it may be seen that the rate of shrinkage of cross-sectional area is very dependent on the value of m, and this is illustrated in Fig. 3.16 for $P = C_1$.

As $m \to 1$, the reduction in area becomes more independent of cross-sectional area, and at m=1 the flow becomes Newtonian viscous and any irregularities in cross-section which were present at the beginning will be preserved, and no necking will develop.

In hot glass $m = 1$, while in hot polymers it is in the range 0.3 - 1.0 depending

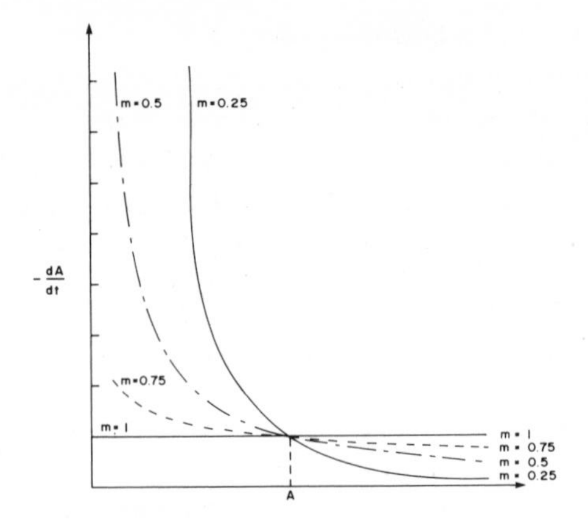

Fig. 3.16 Variation in the rate of shrinkage of cross-sectional area with cross-sectional size as a function of the strain rate exponent, *m*

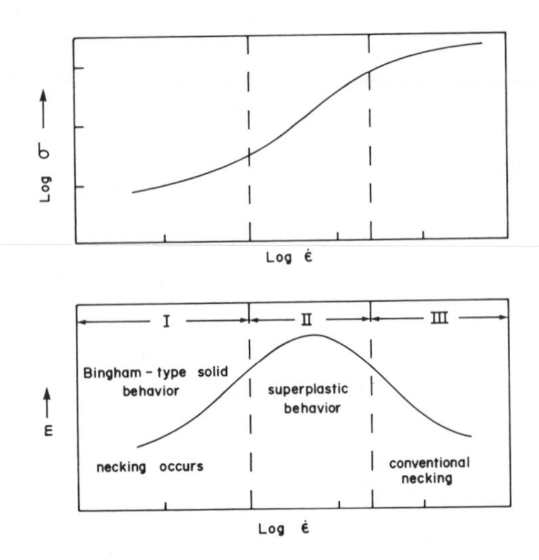

Fig. 3.17 Schematic representation of strain rate dependence of flow stress and strain rate exponent, *m*, on strain rate for a superplastic material

on temperature: in most metals m varies from 0.02 - 0.20 even at temperatures up to $0.9T_m$. For superplastic behavior in metals to occur, values of $m \geqslant 0.3$ are generally found to be necessary, and this is found to require that T/T_m be >0.4 [25]. At temperatures in this range the effects of deformation history will be small, and σ approximates to a single-valued function of strain rate, temperature, and initial microstructure [24].

It is found that m is principally a function of ϵ, T, and grain size, d, the last mentioned being conveniently defined in terms of the metallographic mean-free path [24]. The schematic variation of m with strain rate for a superplastic alloy is shown in Fig. 3.17, and it should be noted that increasing d or decreasing T will shift the curves to the left. High values of m, sufficient for extensive superplastic deformation, occur over an intermediate range of strain rates (region II) having an upper bound of ~ 0.1 - $0.2\ s^{-1}$, and grain sizes in the micron range are required [11]. At low stresses or strain rates, m decreases, with a loss of superplasticity, although it may be noted that region I has only been observed in any detail in a few studies, because of the very low strain rates involved.

Before discussing the mechanisms governing superplastic behavior, it may be noted that long uniform elongation and resistance to necking can also be promoted by application of appropriate conditions to provide positive contributions from the third and fourth terms of the right hand side of Eq. (3.46). Normally the third term is negative as $\partial\sigma/\partial T$ is generally negative. Thus, to make the term positive we require $dT/d\epsilon$ to be negative. This can be done by applying external sources of heat and cooling to a moving rod under tension as shown in Fig. 3.18. A neck is developed as the rod moves past the heat source but on reaching the cooled region ($dT/d\epsilon$ -ve) the strength increases ($\partial\sigma/\partial T$ -ve), and further deformation ceases. Such a process of *die-less drawing* in effect propagates a neck along the rod in the opposite direction to that of the rod's travel. The process is potentially attractive, although difficult to control to provide the required shape consistency.

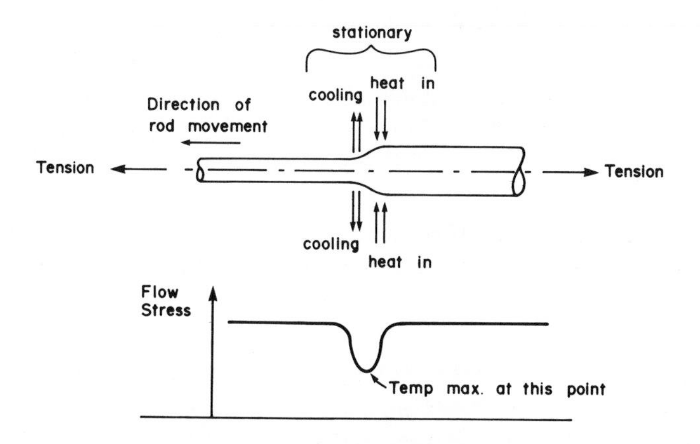

Fig. 3.18 Die-less drawing by means of alternate heating and cooling zones through which a rod is passed

The fourth term of Eq. (3.46) is of importance in describing the stability of soap bubbles and foams. The term $\partial\sigma/\partial\gamma$ is positive and $d\gamma/d\epsilon$ is likewise positive, so the film will resist any incipient thinning. In deforming solids, however, the value of surface energy is too small for it to make any significant contribution to their stability.

3.3.2 Mechanisms of Superplasticity

From the studies which have been made of many superplastic alloys, it is clear that a small and stable grain size (d) is a primary requirement, as m is very dependent on d, and grain growth during deformation would soon produce conditions under which superplastic deformation could not continue. It has also been found from metallographic examinations that the imposed strain may be at least an order of magnitude greater than the elongation of individual grains, and that both grain boundary shear and migration occur [*24*]. Small grains are promoted by refinement through phase transformation and by the hot working of two-phase (mainly eutectic) alloys. The presence of multiple phases helps to prevent grain growth in any one and to preserve the superplastic nature of deformation.

Several models for superplastic deformation have been put forward, based on microstructural observations, but no single model satisfies all observations concerning mechanical behavior and microstructure. However, it is generally agreed that three mechanisms operate to a greater or lesser extent during superplastic deformation. These are:

(a) grain boundary sliding;
(b) diffusional creep;
(c) dislocation creep/dynamic recovery.

Grain boundary sliding has been observed by optical microscopy during superplastic deformation [*24*] and has also been observed directly in a high voltage (1 MV) electron microscope [*26*]. However, it cannot occur alone, or grain boundary compatibility would be lost and pores develop. Hence, an accommodation process such as diffusional creep or dislocation creep must be associated with it.

Diffusional creep can be of two types, Nabarro-Herring [*27, 28*], or Coble creep [*29*]. In the former, stress-induced diffusion takes place in a transgranular fashion as shown in Fig. 3.19, while in the latter the diffusion path is intergranular.

In Nabarro-Herring deformation

$$\sigma = \eta_{\text{N-H}}\,\dot{\epsilon} \tag{3.50}$$

where $\eta_{\text{N-H}}$ is a "viscosity", given by [*23*]

$$\eta_{\text{N-H}} \simeq d^2kT/\alpha v D_{\text{L}}$$

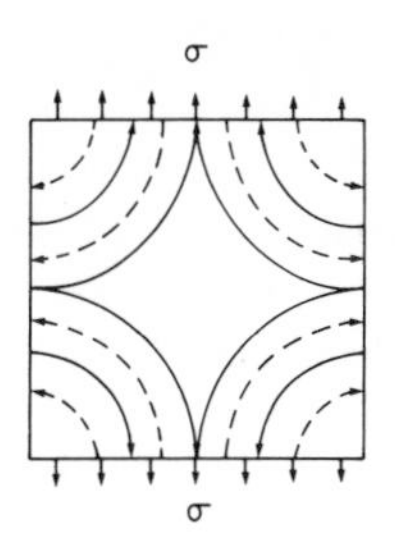

Fig. 3.19 Schematic representation of Nabarro-Herring creep, under tensile stress, σ. The solid lines represent the flow of atoms; the dotted lines represent the flow of vacancies

In this d is the mean intercept grain size, k Boltzmann's constant, T the absolute temperature, v the atomic volume, D_L the lattice diffusion coefficient, and α a constant. For intergranular diffusion,

$$\sigma = \eta_C \dot{\epsilon} \tag{3.51}$$

and

$$\eta_C \simeq d^3 kT / \beta v w D_{gb}$$

where β is a constant $\simeq$ 150, w the grain boundary width and D_{gb} the grain boundary diffusion coefficient.

Diffusional creep predicts a linear relationship between stress and strain rate, which is not normally observed, and it has been pointed out by Lifshitz [*30*] that grain boundary sliding in conjunction with it is still inevitable as otherwise voids would be formed at the boundaries where vacancies deposit.

Dislocation creep in common with *dynamic recovery* requires that the activation energy of the process be that for lattice diffusion, if it is to be the controlling process. This is certainly not always the case, and the mechanism is not a sufficient one by itself.

The evidence available leads one to conclude that superplastic deformation occurs by some combination of the above mechanisms, although the major theories developed do not account fully for all the observed phenomena. The various creep mechanisms are discussed more fully in Chapter 11 after having reviewed basic deformation processes in Chapter 4.

REFERENCES

1. ANDREWS, C. W., *Metal Progress*, **58**, 85 (1950).
2. DIETER, G. E., *Mechanical Metallurgy*, McGraw-Hill, New York (1961).

3. *Metals Handbook*, Vol. 1, *Properties and Selection of Materials*, ASM, Metals Park, Ohio (1961).
4. LOW, J. R., and PRATER, T. A., in *Symposium on Deformation of Metals as Related to Forming and Service*, STP 87, ASTM, Philadelphia (1948).
5. BARBA, M. J., *Mem. Soc. Ing. Civils*, Part I, 682 (1880).
6. BRIDGMAN, P. W., *Trans. ASM*, **32**, 553 (1944).
7. CONSIDÈRE, A., *Ann. ponts et chaussées*, **9**, ser. 6, 574 (1885).
8. POWELL, G. W., MARSHALL, E. R., and BACKOFEN, W. A., *Trans. ASM*, **50**, 487 (1958).
9. BERT, C. W., MILLS, E. J., and HYLER, W. S., *Trans. ASME, J. Basic Eng.*, **89**, 35 (1967).
10. LAL, K. M., *Trans. ASME, J. Eng. Matls. and Technology*, **97**, 284 (1975).
11. BACKOFEN, W. A., *Deformation Processing*, Addison-Wesley, Reading, Mass. (1972).
12. WEISS, V., in *Fracture*, Vol. III, p. 227 (H. Liebowitz, ed.) Academic Press, New York (1971).
13. MANJOINE, M. J., in *Proc. of the Symposium on Mechanical Behavior of Materials*, Kyoto, Japan (1974).
14. DAVIES, R. G., and MAGEE, C. L., *Trans. ASME, J. Eng. Matls. and Technology*, **97**, 151 (1975).
15. TEGART, W. J. McG., *Elements of Mechanical Metallurgy*, Macmillan, New York (1966).
16. CARREKER, R. P., *Trans. AIME*, **209**, 112 (1957).
17. ZENER, C., and HOLLOMON, J. H., *J. Appl. Phys.*, **15**, 22 (1944).
18. MANJOINE, M. J., *Trans. ASME, J. Eng. Matls. and Technology*, **97**, 156 (1975).
19. DAVIS, E. A., and CONNELLY, F. M., *Trans. ASME, J. Appl. Mech.*, **81**, 25 (1959).
20. *Ductility*, ASM, Metals Park, Ohio (1968).
21. *Toward Improved Ductility and Toughness*, Iron and Steel Inst. of Japan, Japan Inst. of Metals, and Climax Molybdenum Development Co. of Japan (1971).
22. *Proc. of the Int. Conf. on Creep and Fatigue in Elevated Temperature Applications*, I. Mech. E., London (1975).
23. JOHNSON, R. H., *Metall. Rev.*, **15**, No. 146, 115 (1970).
24. BACKOFEN, W. A., AZZARTO, F. J., MURTY, G. S., and ZEHR, S. W., in *Ductility*, p. 279, ASM, Metals Park, Ohio (1968).
25. BACKOFEN, W. A., TURNER, I. R., and AVERY, D. H., *Trans. ASM*, **57**, 980 (1964).
26. NAZIRI, H., PEARCE, R., HENDERSON BROWN, M., and HALE, K. F., *Journal of Microscopy*, **97**, 229 (1973).
27. NABARRO, F. R. N., in *Proc. Conf. on Strength of Solids*, p. 75, Physical Society, London (1948).
28. HERRING, C., *J. Appl. Phys.*, **21**, 437 (1950).
29. COBLE, R. L., *J. Appl. Phys.*, **34**, 1679 (1963).
30. LIFSHITZ, I. M., *Soviet Phys. JETP*, **17**, 909 (1963).

4

DEFORMATION MECHANISMS IN METALS

4.1 INTRODUCTION

The discussion of deformation and fracture in the preceding chapters has been based largely on the premise that engineering alloys are homogeneous and isotropic, except where some effects of texture in modifying the yield locus have been examined briefly, and that they behave in a fairly idealized manner. This is, of course, not true of real materials, although it is a satisfactory basis on which to base the design of many components and structures, using the theory of elasticity and assuming ideal elastic-plastic stress-strain relations to hold, and it is necessary to discuss the mechanisms of deformation and effects of microstructure in order to obtain a clear understanding of the strengthening mechanisms which may be utilized to improve the load-bearing capacity of simple materials, and of the mechanisms whereby structural components fail in service at stresses far below those required to cause failure under simple tensile loading.

Accordingly, the deformation mechanisms occurring in metals are first reviewed in this chapter, following which some aspects of dislocations and their interaction are examined. The stress-strain behavior of single crystals and polycrystalline aggregates is then reviewed briefly in the light of the discussion on dislocations, leading into a short discussion of deformation textures. The effects of temperature on the behavior of polycrystalline materials are not dealt with specifically, although mention is made of mechanisms of deformation at high temperature: rather this matter is discussed in Chapter 11 when examining creep and creep design procedures. Although the discussion is confined largely to metals, many of the principles apply to non-metallics, both polymeric and ceramic, but any detailed consideration of these materials has been avoided in the interests of space.

The material presented in this chapter is deliberately brief, and the interested reader is advised to consult the various reference texts listed [*1-5*] for more detailed discussion. It is hoped, however, that the points covered are relevant to and sufficiently detailed for an appreciation of the later discussions on strengthening mechanisms and of the quantitative relationships between strength and microstructure presented in Chapter 6.

To many engineers, detailed discussions concerning dislocations, their stress fields and interactions, may appear to be of little direct value, as the equations derived are not apparently of use in that they cannot be extrapolated to practical engineering situations. To some extent this is true, but an understanding of basic dislocation mechanisms is important in explaining the "why" of many aspects of materials behavior. In addition, a considerable effort has been made recently, and continues to be made, to develop constitutive relations for the plastic flow and high temperature creep of bulk materials on the basis of dislocation models [*6, 7*], and to explain fully and quantitatively the processes and laws of crack growth and fracture from consideration of the movements of dislocations [*8-10*]. Hence, it appears probable that engineers involved in aspects of mechanical metallurgy will require more detailed study of these topics in the future, in order to provide a good physical basis for the various constitutive and other relations developed for bulk materials, and this approach is employed in discussing creep theories in Chapter 11.

4.2 BASIC MECHANISMS

Plastic deformation of crystalline solids takes place by means of the processes of *slip* and *twinning*, the importance of each being dependent on various factors, including the chemical bonding and bond energy, temperature, and the presence of defects in the crystal structure. At elevated temperature ($T > 0.4\ T_m$[1]) or in the presence of radiation, other processes may become important in allowing deformation to take place readily, these being dependent on diffusion of atoms within the crystal lattice. In non-crystalline or partially crystalline solid materials (such as polymers), additional

[1] T_m is the homologous temperature, that is the actual temperature divided by the melting point of the material, both on the absolute scale.

mechanisms may become important, such as the viscous flow of the polymer molecules, these being discussed fully elsewhere [*11, 12*].

In this section, discussions are confined to crystalline solids, and it is demonstrated that, while slip and twinning allow us to explain shape changes in real materials deformed under stress, the concept of the *dislocation* is necessary to explain the difference between theoretical and measured flow stresses.

4.2.1 Slip and Slip Systems

The principal means by which plastic deformation occurs in crystalline solids is slip, in which movement of one part of the crystal takes place with respect to another, the sliding taking place along *slip planes*. At the surface of a crystal, a step is produced, this being a *slip band* as shown in Fig. 4.1. Such slip bands may be seen on the surface of deformed crystals which have been polished previously (Fig. 4.2), and demonstrate the inhomogeneity of the slip process. Individual slip bands or lines are made up of many discrete lamellae which may be resolved using scanning or transmission electron microscopy, and this is shown schematically in Fig. 4.3.

Slip does not occur with equal ease on all crystallographic planes: rather, it is anisotropic and takes place more easily on some crystal planes and in certain specific directions. The directions of slip have been found to be the directions of closest packing and the slip planes are those which are most densely packed, these being spaced furthest apart. The combination of a slip plane and slip direction is termed a *slip system*.

Face-centered cubic (FCC) metals slip along <110> on {111}: thus they have 12 possible slip systems. Body-centered cubic (BCC) metals slip along <111>, and on {110}, {112},and {123}, all these planes having a high atomic density. Accordingly, it may be shown that BCC metals have 48 possible slip systems, although not all are operative for specific alloys and specific temperature conditions. In the case of hexagonal close-packed (HCP) metals, slip almost invariably takes place along <11$\bar{2}$0> and generally on {0001},

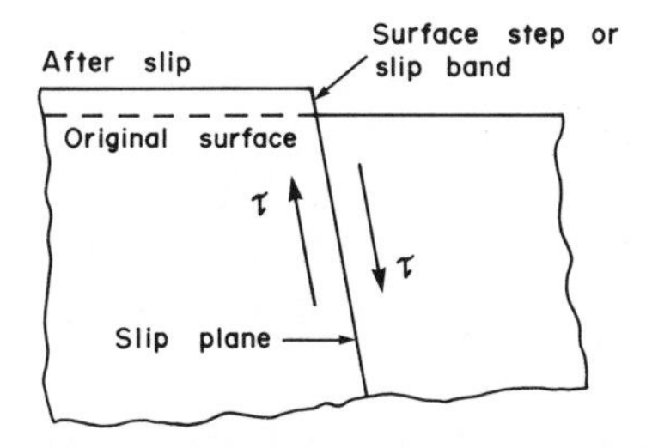

Fig. 4.1 Schematic view of the classical model for slip

which are the only close-packed planes under conditions of ideal close-packing of the atoms in this structure. Thus HCP metals may possess three slip systems only, in which case their deformation will be strongly dependent on the direction of the applied stress with respect to the slip planes, and this leads to their ductility being highly dependent on orientation. However, the situation is complicated by the fact that the c/a ratio[2] of HCP metals deviates considerably from the ideal value of 1.633 for ideally close packing. The

Fig. 4.2 Slip bands on polycrystalline α-brass deformed at room temperature (X 800). (Courtesy of W. E. White, University of Calgary)

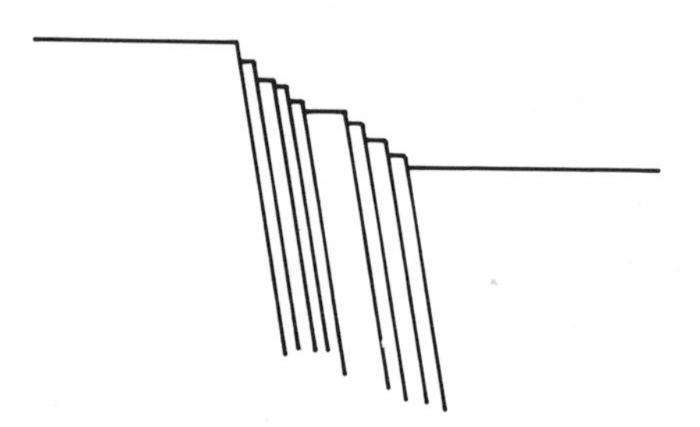

Fig. 4.3 Schematic drawing of section through slip band lamellae

[2] The ratio of the height of the unit cell, c, to the lattice parameter in the basal plane, a — see Fig. 4.4. The ratio may vary from 1.886 for cadmium to 1.567 for beryllium, with even lower ratios for alloy solid solutions.

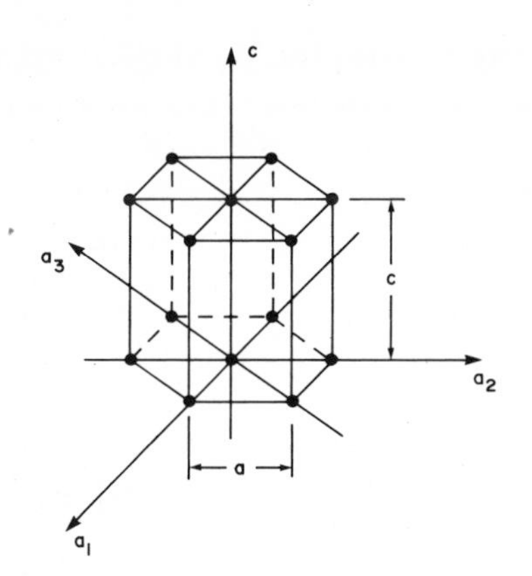

Fig. 4.4 The unit cell in the hexagonal close-packed lattice

smaller this ratio becomes, the less ideally close-packed do the basal planes become, and the likelihood of deformation on prism $\{10\bar{1}0\}$ or pyramidal planes $\{1\bar{1}01\}$ becomes greater. In almost all cases the slip direction remains $\langle 11\bar{2}0 \rangle$, and this means that elongation or thinning along the *c*-axis cannot take place as a result of slip. Hence, to allow for such shape changes or deformation of each grain in a polycrystalline aggregate, additional deformation modes are required. This matter is referred to further in Sections 4.2.5 and 4.4.2.

The slip systems for the three common crystal structures discussed above are listed in Table 4.1, together with a listing of some specific metals for each slip mode.

Additional slip systems over and above those discussed for room temperature may be important in some materials deformed at elevated temperature.

4.2.2 Critical Resolved Shear Stress

When the shear stress acting on a slip plane along a slip direction reaches a critical value, slip occurs, this first being recognized by Schmid [*21*]. If a single crystal is stressed in tension in different directions, the tensile stress at which slip begins will vary; however computation of the resolved shear stresses will produce a unique value, the *critical resolved shear stress* (CRSS), in accordance with Schmid's law. This has been demonstrated for many metallic crystals.

If the axis along which a tensile stress is applied to a single crystal makes an angle λ with the slip direction and an angle ϕ with the normal to the slip plane, the resolved shear stress is given by [*2*]

$$\tau_R = \sigma \cos\phi \cos\lambda \qquad (4.1)$$

Table 4.1 Slip Systems for Metallic Crystal Structures

Structure	Examples	Slip Systems (slip plane and slip direction)	Ref.
FCC	Ag, Au, Cu, Ni, Al	$\{111\}\ \langle 110\rangle$	[*13*]
	Al at elevated temperatures	$\{100\}\ \langle 10\bar{1}\rangle$	[*14*]
BCC	α-Fe	$\{100\}\ \langle 11\bar{1}\rangle$ $\{112\}\ \langle 11\bar{1}\rangle$ $\{123\}\ \langle 11\bar{1}\rangle$	[*13*]
	Mo, Na, Nb	$\{110\}\ \langle 111\rangle$	[*13*]
	Na, K at 0.8 T_m	$\{123\}\ \langle 11\bar{1}\rangle$	[*13*]
HCP	Ti, Zr	$\{0001\}\ \langle 11\bar{2}0\rangle$ $\{10\bar{1}0\}\ \langle 11\bar{2}0\rangle$ $\{10\bar{1}1\}\ \langle 11\bar{2}0\rangle$	[*15, 16*]
	Be	$\{0001\}\ \langle 11\bar{2}0\rangle$ $\{10\bar{1}0\}\ \langle 11\bar{2}0\rangle$	[*17, 18*]
	Zn	$\{0001\}\ \langle 11\bar{2}0\rangle$ $\{11\bar{2}2\}\ \langle 11\bar{2}\bar{3}\rangle$	[*19*]
	Zn at high temperature	$\{1\bar{1}00\}\ \langle 11\bar{2}0\rangle$	[*20*]
	Cd	$\{0001\}\ \langle 11\bar{2}0\rangle$ $\{10\bar{1}1\}\ \langle 11\bar{2}0\rangle$	[*19*]
	Cd at high temperature	$\{10\bar{1}0\}\ \langle 11\bar{2}0\rangle$	

For HCP metals, where the number of slip systems is limited, the variation in yield stress, σ_Y, can be large: however, in FCC metals, where the number of equivalent slip systems is large and the possible variation in orientation between slip direction and applied stress axis is relatively small, the maximum variation in yield stress is only by about a factor of 2. In BCC metals, the variation is smaller still because of the even larger number of slip systems.

When the tensile stress is such that slip may take place with equal ease on more than one slip system, *double glide* or *multiple glide* may take place. A good example of a material in which double glide occurs is rock salt, in which the slip plane is $\{110\}$ and the slip direction $\langle 110\rangle$. For a cubic lattice $[hkl]$ is perpendicular to (hkl), and hence the slip direction in one slip plane is also normal to another slip plane, and the orientation factor $\cos\phi\cos\lambda$ is constant for the two conjugate slip systems.

4.2.3 Theoretical Stress for Slip

The theoretical stress for slip to occur in a perfect lattice can be estimated as follows [*2*], the stress concerned being the theoretical value of the CRSS for the material.

The sliding of one plane of atoms over another is illustrated in Fig. 4.5, and as a first approximation we may assume that the force required to cause movement is sinusoidal in nature, with a period b. When atom A has been moved from its original position to an equivalent position B, clearly no external force is required: also, at the mid-position it will be in unstable equilibrium and require no force for reasons of symmetry. Hence, the shear stress, τ, may be expressed as

$$\tau = \tau_{max} \sin(2\pi x/b) \tag{4.2}$$

where τ_{max} is the amplitude of the sine wave. For small displacements, Hooke's law will hold, and

$$\tau = G\gamma = Gx/a \tag{4.3}$$

Also, for x small with respect to b, we may rewrite Eq. (4.2) as

$$\tau \simeq \tau_{max}\, 2\pi x/b$$

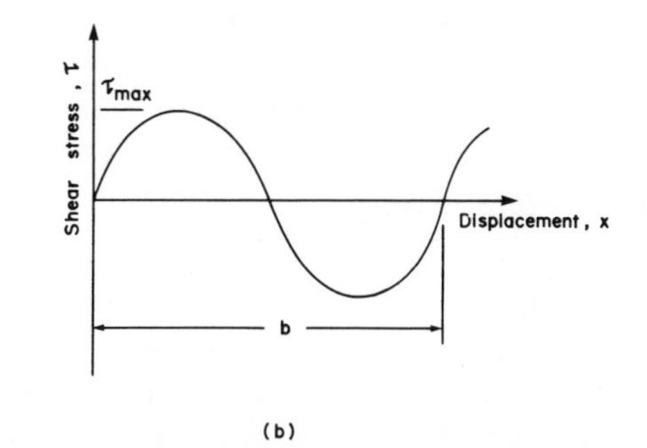

Fig. 4.5 (a) Idealized view of one plane of atoms sliding over another; (b) variation of shear stress with displacement

and combining this with Eq. (4.3), we obtain

$$\tau_{max} = Gb/2\pi a \tag{4.4}$$

As an approximation, we may take $b = a$, and hence the theoretical shear stress for slip in a perfect crystal is given approximately by

$$\tau_{max} \simeq G/2\pi \tag{4.5}$$

Eq. (4.4) may be evaluated more accurately for different crystal structures. For example in FCC metals, $b = a_o/\sqrt{6}$ and $a = a_o/\sqrt{3}$, where a_o is the lattice parameter. Hence

$$\tau_{max} \simeq G/9 \tag{4.6}$$

For a layered structure such as graphite, $b/a = 0.4$, for b equal to the slip vector of a partial dislocation (see Section 4.3), and

$$\tau_{max} \simeq G/15 \tag{4.7}$$

The values of CRSS obtained from Eqs. (4.5-4.7) are several orders of magnitude greater than the observed values for single crystals or polycrystalline materials, and this may be illustrated for the most commonly used engineering alloy, namely steel. The shear modulus G of steel is approximately 83 000 MPa, hence τ_{max} is approximately 13 000 MPa, or the uniaxial tensile yield stress is predicted to be approximately 26 000 MPa, whereas commonly used structural steels have a tensile yield stress in the region of 200-400 MPa, and pure iron yields at stresses of the order of 30 MPa [*13*].

In order to reduce the discrepancy between the real and theoretical stresses for slip in crystalline solids, the simple sine wave expression used in Eq. (4.3) can be replaced by a more exact function taking into account that the critical shear strain should be less than $b/4$ when a more realistic law for the forces between atoms is used [*2*], and that there may be possible positions of mechanical stability in the lattice while slip takes place [*22*]. Taking these considerations into account, the theoretical shear stress in Eq. (4.5) may be reduced by a factor of about 5, i.e., to around $G/30$, but this is still very much greater than the observed strength of single crystals of soft materials.

4.2.4 Role of Dislocations in Slip

The concept of line defects in crystalline solids was proposed to explain the discrepancy between the actual and theoretical values of stress for slip in materials described in the preceding section. Instead of the entire plane of atoms moving together, we may consider local movement only, with slip starting at some position and gradually spreading across the slip plane under the action of the applied stress. At any time from the initiation of the slip until the whole plane of atoms has slipped, there will be a discontinuity between the slipped and unslipped regions: this is termed a *dislocation*. This localized movement of atoms will require a much lower force than simultaneous movement of all on the slip plane; hence the shear stress required will be much less than the theoretical values predicted as in Section 4.2.3. This concept of slip taking place by the movement of dislocations was proposed independently in 1934 by Taylor [*23*], Polanyi [*24*], and Orowan [*25*], long before such defects could be observed in materials.

Figure 4.6 illustrates a crystal lattice containing a dislocation, and it may readily be visualized that the stress required to be applied across the slip plane to move the dislocation will be small, its magnitude being determined by the way in which the energy of misfit between the faces of the slip plane changes as the dislocation is moved across the plane. Solution of this problem is complicated because it depends on the atomic arrangements at the center of the dislocation, but has been done through the efforts of Peierls [*26*] and Nabarro [*27*], excellent summaries being presented by Cottrell [*2*] and Friedel [*1*]. As a result, the small force required to move a dislocation through a crystal lattice is termed the *Peierls-Nabarro force* and it is substantially smaller than the observed stress for the plastic flow of engineering materials or indeed for slip in single crystals of pure metals. This is accounted for by the strengthening conveyed by impurities or alloying elements present in these materials even in minute quantities, and by other effects of structure. Nonetheless, the concept of the Peierls-Nabarro force or stress is an important one, and it is necessary to provide means to raise the stress required to move dislocations over substantial distances in the crystal lattice, so preventing measurable plastic deformation, in order to produce useful engineering alloys. These matters are discussed further in Chapter 6.

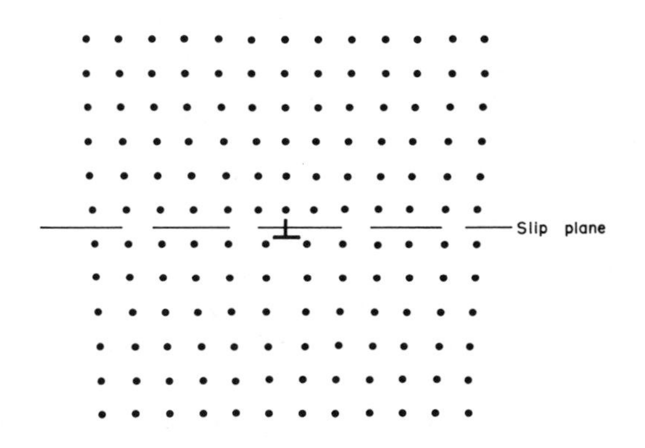

Fig. 4.6 Schematic drawing of an edge dislocation in a simple cubic lattice

4.2.5 Twinning

The second important mechanism of deformation in crystalline solids is twinning, which involves the simultaneous movement of atoms on one side of a twinning plane, so that they take up positions constituting a mirror image of the atoms on the other side of this plane. Figure 4.7 shows the classical picture of twinning, and it may be seen that the essential difference between this and slip is that, in the latter, atoms on one side of the slip plane all move through an equal distance, while in twinning the distance moved is proportional to the atom's distance from the twinning plane.

Mechanical twinning can take place very rapidly (in microseconds) and is of particular importance in deformation at very high strain rates. (Slip band formation may take several milliseconds). Mechanical twins can be formed in FCC, BCC, or HCP crystal lattices (to deal only with the common metallic structures), although they do not occur readily in FCC metals except at very low temperatures. In BCC metals, twinning can occur at moderately high to high strain rates, particularly in shock loading or explosive forming, their formation again being favored by low temperature. In HCP metals, twinning is a regularly occurring phonomenon, because of their limited number of slip systems and the absence of a slip vector in the *c*-direction: however, the actual amount of deformation which twinning can produce is small, and its importance is more that it brings parts of the grain (or crystal) into a different orientation, so placing new slip systems in an operative position with respect to the stress axis.

Twinning takes place on specific planes in specific directions, and these may

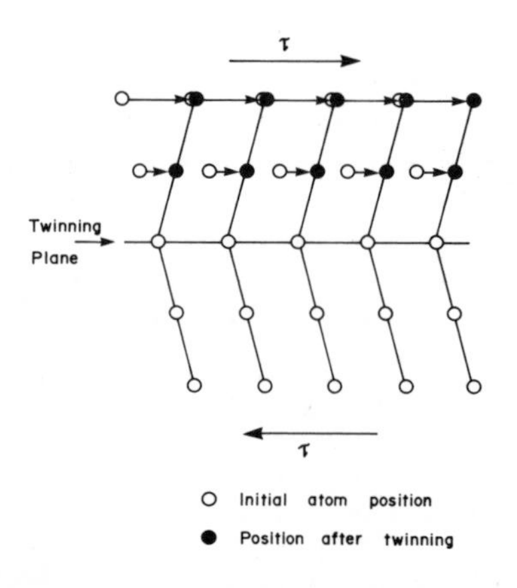

Fig. 4.7 Twinning in a crystal lattice

Table 4.2 Common Twinning Planes and Directions [*14*]

Structure	Twinning Planes	Twinning Directions	Comments
FCC	$\{111\}$	$\langle 11\bar{2} \rangle$	
BCC	$\{112\}$	$\langle 11\bar{1} \rangle$	
HCP	$\{11\bar{2}3\}$	$\langle 11\bar{2}6 \rangle$	Tensile Twinning
	$\{10\bar{1}2\}$	$\langle 10\bar{1}1 \rangle$	Tensile Twinning
	$\{11\bar{2}2\}$	$\langle 11\bar{2}3 \rangle$	Compressive Twinning

vary between twins formed in tension and those formed in compression. Table 4.2 lists some of the common planes and directions. As with slip, it appears that some critical resolved shear stress is required to initiate twinning, its value being dependent on the particular twinning plane and direction in the alloy of interest, as well as on the temperature.

4.2.6 Effects of Temperature

The critical resolved shear stress for slip increases as temperature is decreased [*13*], and it appears that that required for twinning does not increase to the same extent. Hence mechanical twinning becomes relatively easier at low temperatures. This is particularly noticeable in FCC metals where twins are not normally formed at ambient temperature but in which ready twinning can be induced at temperatures down to 4 K [*13, 14*].

Two additional mechanisms of deformation become important at high temperature, for example during creep deformation. These are grain boundary sliding and stress-induced diffusion such as Nabarro-Herring or Coble creep. These were referred to previously when discussing superplastic behavior in Chapter 3.

Grain boundary sliding becomes important at temperatures greater than approximately $0.4T_m$. In effect the grain boundaries become more easily deformed than the grains themselves, and hence the old concept of an amorphous cement arose, although it is now clear that grain boundaries are very narrow, with regions of fit and occasional small regions of misfit [*28*]. When such sliding occurs, there must also be deformation within the grains to preserve compatibility, although the opening up of grain boundary cracks is an important aspect of creep fracture (see Chapter 11).

Nabarro-Herring and Coble creep are important at temperatures greater than $0.8T_m$, and for small grain sizes. As seen in Chapter 3 (see Fig. 3.19) they involve the movement of vacancies produced at grain boundary sources under tension to sinks at grain boundaries under compression. Further discussion is given in Chapter 11.

It should also be restated here that deformation at elevated temperature may cause additional slip systems to operate than do so at normal ambient temperature, aluminum, for example, slipping on {110}. The general character of slip changes with increasing temperature because of the greater ease of cross-slip (see Section 4.3.6) and the operation of multiple slip systems, so that a metal which deforms at room temperature with the formation of long planar slip lines, may deform at high temperature ($T > 0.4\ T_m$) with a wavy slip mode.

4.3 DISLOCATIONS

As pointed out in Section 4.2.4, the concept of line defects or dislocations occurring in crystalline solids was introduced to explain the discrepancy between theoretical and observed values of yield stress, and subsequent developments in electron microscopy have enabled these defects to be observed directly and studied extensively [*29-31*]. Apart from the major reviews of this subject by Friedel [*1*], Cottrell [*2*] and Hirth and Lothe [*5*], the interested reader is referred to the excellent short texts by Weertman and Weertman [*3*] and Hull [*4*] for more detailed consideration of the subject.

4.3.1 Dislocation Types and Strength

Dislocations are of two distinct types, *edge* and *screw*. They are defined in terms of their *Burgers vector*, this vector being obtained by making a *Burgers circuit* around the dislocation in the crystal lattice. This is illustrated for both types of dislocation in Fig. 4.8, the closure vector representing the Burgers vector, **b**, of the dislocation, and it may be seen that a circuit made around several dislocations will give a closure vector equal to the sum of their separate Burgers vectors.

In the general case, dislocations are part edge, part screw, and occur in the form of curves or loops forming a 3-dimensional network (see Section 4.3.9). Figure 4.9 illustrates a dislocation loop with Burgers vector **b**, and it may be seen that the dislocation is pure screw at point *A*, where **b** is parallel to the dislocation line, and pure edge at point *B*, where **b** is perpendicular to the dislocation line. As already noted, a dislocation represents the boundary between slipped and unslipped regions of a crystal, and this is indicated in Fig. 4.9. Hence it either closes in the form of a loop, ends at a surface (either an external surface or a grain boundary), or terminates at a *node* where several dislocations meet. At such a point, the sum of the Burgers vectors is zero, i.e., for three dislocations with Burgers vectors $\mathbf{b}_1$, $\mathbf{b}_2$ and $\mathbf{b}_3$, $\mathbf{b}_3$ is given as the negative sum (or resultant) of the vectors $\mathbf{b}_1$ and $\mathbf{b}_2$ (see Section 4.3.6).

A dislocation is said to be of unit strength when its Burgers vector equals the lattice parameter. In general, larger dislocations dissociate from energy considerations, the strain energy being proportional to the square of the Burgers vector (see Section 4.3.2).

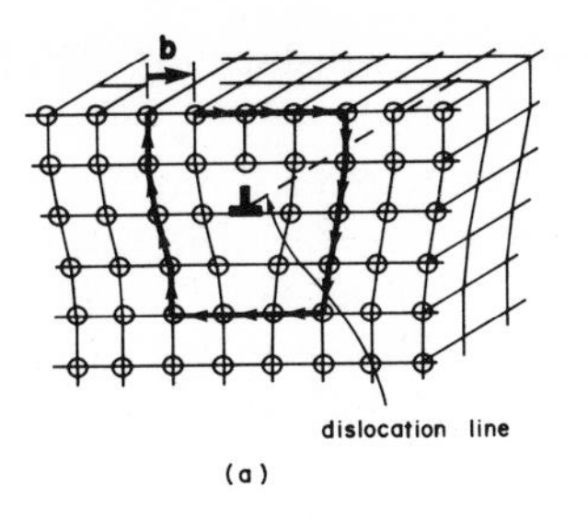

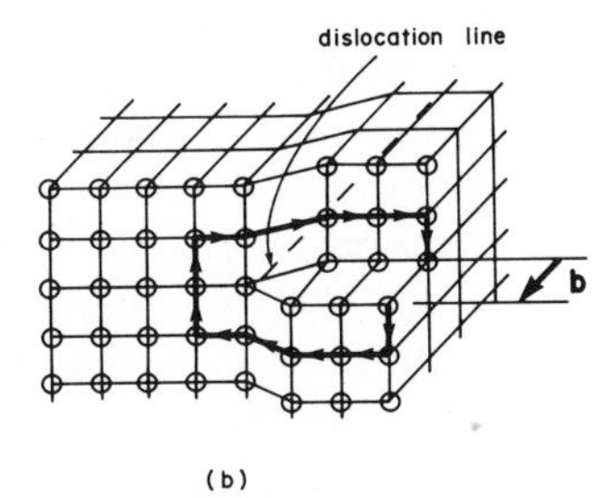

Fig. 4.8 (a) Edge dislocation; (b) screw dislocation. The Burgers circuit and Burgers vector *b* are shown for each case.

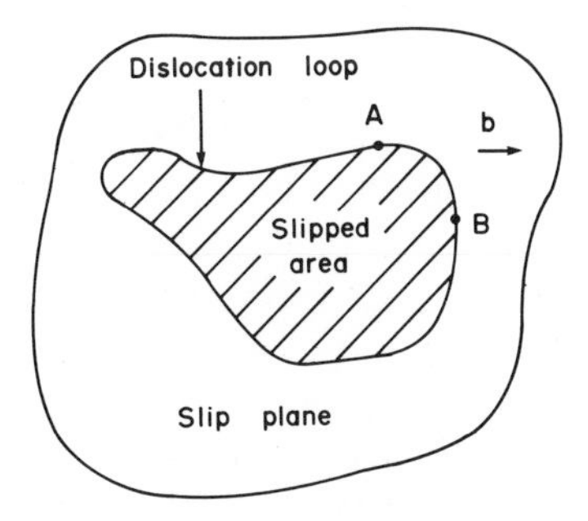

Fig. 4.9 A dislocation loop on a slip plane, showing the pure screw and pure edge components at *A* and *B*, respectively

In close-packed structures, dislocations of a strength less than unity are possible, and Burgers vectors are specified by giving the components along the axes of the crystallographic structure cell: for example, in a cubic lattice, for slip from a cube corner to the center of one face, the Burgers vector has the components $a/2$, $a/2$, 0, and is given as $[a/2, a/2, 0]$ or $\mathbf{b} = (a/2)\,[110]$ as is shown in Fig. 4.10. The strength of a dislocation with Burgers vector $a[uvw]$ is $|b| = a[u^2 + v^2 + w^2]^{1/2}$.

When a unit dislocation translated by one Burgers vector produces an identity translation, it is termed a *perfect dislocation*. This occurs when **b** is parallel to a direction of closest packing in the crystal lattice. Such a dislocation can dissociate to form *imperfect dislocations* when a translation by one Burgers vector does not produce an identity translation. Between two such imperfect dislocations formed in this way, the stacking sequence of the atoms will be disturbed and a *stacking fault* will exist. This is stable only if the decrease in energy due to dissociation of the perfect dislocation is greater than the increase in interfacial energy of the faulted region.

Such stacking faults occur readily in FCC and HCP metals on the close-packed planes, {111} and (0001) respectively. In the former the {111} stacking sequence is *ABCABCA*..., and atoms in, say, the *C* positions may move to other *C* positions as shown in Fig. 4.11, giving rise to a perfect dislocation, or they may move in two steps *via* the intermediate *A* positions, giving rise to a stacking sequence *ABABCA*..., and effectively a layer of HCP material, which lattice has the normal stacking sequence on (0001) of *ABABA*.... Similarly, in the HCP lattice slip may occur in two stages, producing an intermediate layer of what is effectively FCC lattice sandwiched between the (0001) planes.

The perfect dislocation in the FCC lattice is given by $a/2\,[\bar{1}01]$, and the two partials into which it may dissociate are $a/6\,[\bar{2}11]$ and $a/6\,[\bar{1}\bar{1}2]$: the resultant stacking fault on {111} is shown in Fig. 4.12, and the dissociation reaction may be written as

$$a/2\,[\bar{1}01] \rightarrow a/6\,[\bar{2}11] + a/6\,[\bar{1}\bar{1}2] \tag{4.8}$$

The important role that stacking faults play in deformation is discussed further in Section 4.3.6.

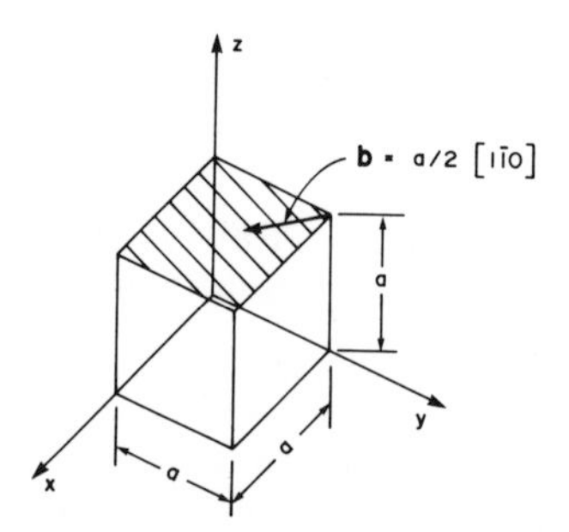

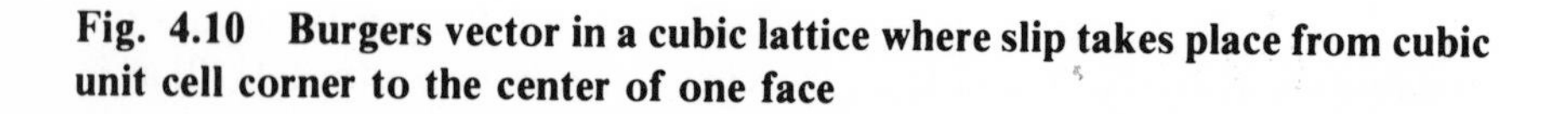
Fig. 4.10 Burgers vector in a cubic lattice where slip takes place from cubic unit cell corner to the center of one face

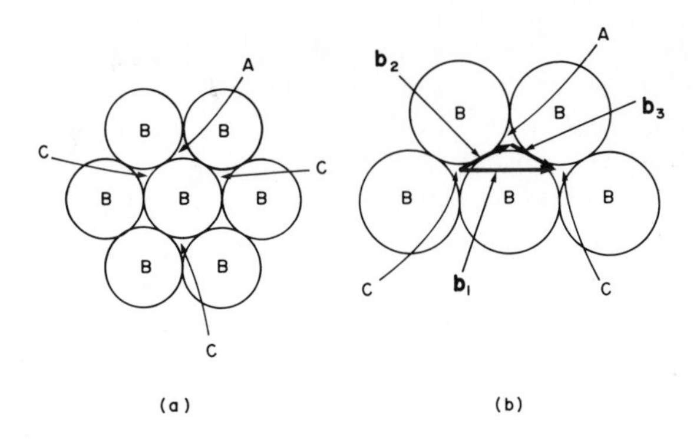

Fig. 4.11 Slip in the FCC lattice on a close-packed plane

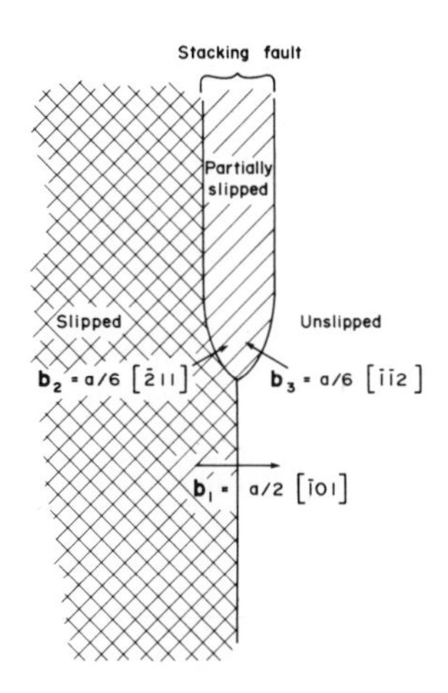

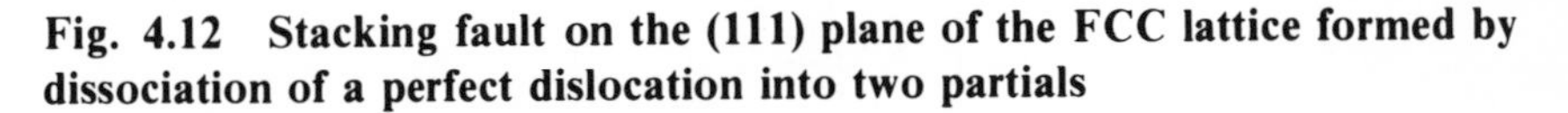

Fig. 4.12 Stacking fault on the (111) plane of the FCC lattice formed by dissociation of a perfect dislocation into two partials

4.3.2 Elastic Properties of Dislocations

Dislocations are not thermodynamically stable and, as a result of the lattice distortion around them and the interatomic forces involved, they possess an elastic strain energy. This property enables us to explain many physical observations including the following:

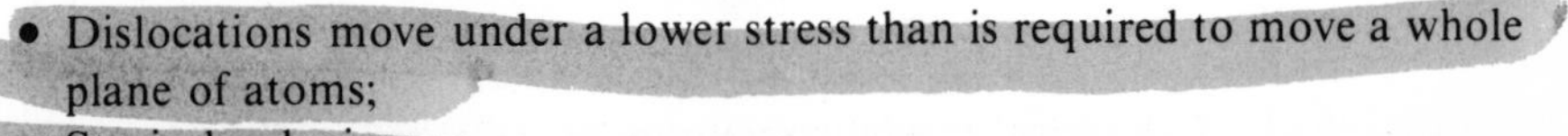

- Dislocations move under a lower stress than is required to move a whole plane of atoms;
- Strain hardening;

- Softening by means of recovery and recrystallization;
- The stability of low angle grain boundaries;
- Strengthening by means of substructure or precipitates;
- The splitting of dislocations to form partials;
- The identification of dislocations by means of etch pits.

Rather than discussing each of these items in turn, it is proposed first to derive the expression for elastic strain energy, and then to discuss the forces required to move a dislocation and those between interacting dislocations and between dislocations and other defects. From these basic considerations all else follows.

Consider a cylindrical crystal of length L with a screw dislocation of Burgers vector **b** along its axis. The elastic shear strain (γ) in a thin annular section of radius r and thickness dr is given by $\gamma = b/2\pi r$ (see Fig. 4.13) where $b = |\mathbf{b}|$.

The strain energy per unit volume, dU/dV, of the annular region is $dU/dV = \tau\gamma/2 = G\gamma^2/2 = G(b/2\pi r)^2/2$, where G is the shear modulus. The volume of the annular ring is $dV = 2\pi rL\,dr$. Hence

$$dU = G(b/2\pi r)^2 \pi r L dr = (GLb^2/4\pi)dr/r$$

And the total strain energy is given by

$$U = \int_{r=r_o}^{r=R} (GLb^2/4\pi)(1/r)dr$$

Hence

$$U = (GLb^2/4\pi)\ln(R/r_o) \qquad (4.9)$$

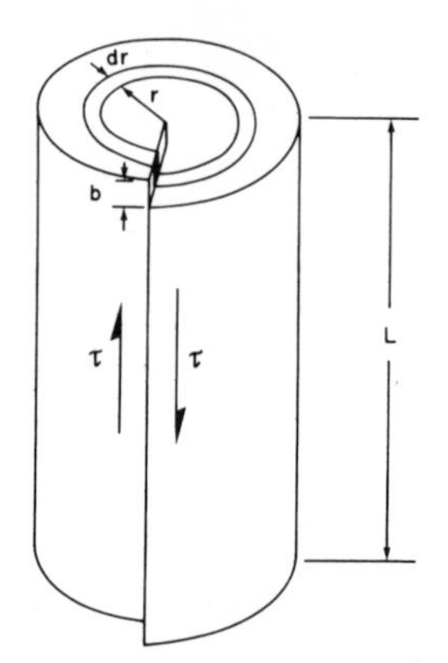

Fig. 4.13 Cylindrical crystal containing an axial screw dislocation

Hooke's law does not hold for large strains at the dislocation core, and U would tend to infinity as r tended to zero. Also R cannot be infinity as again U would tend to infinity, and the strain field will, in any event, be cancelled out by interaction with those of other dislocations. From a physical point of view, r_o cannot be smaller than one atomic spacing as the material is crystalline rather than a continuum, and R cannot be larger than half the grain or subgrain considered. In real crystals the dislocations form a network (see Section 4.3.9) and the maximum value of R will be half the spacing of the network dislocations. A realistic value of r_o is considered to be $\sim 5b$ [3], and taking appropriate values of R, [3] $\ln(R/r_o)$ may be estimated as being close to 4π, hence $U \simeq GLb^2$ [2]. However, more detailed analysis suggests that the energy is closer to half this value [32], and this is the value normally used in analyses of strength increments which depend on dislocation interaction. Hence, we shall make use of this relationship in the form

$$U \simeq GLb^2/2 \tag{4.10}$$

For an edge dislocation, the analysis is more complex because of the absence of radial symmetry. However, it may readily be shown that the value of U in this case is given by [2]

$$U \simeq [1/(1-\nu)](GLb^2/4\pi)\ln(R/r_o) \tag{4.11a}$$

or

$$U \simeq GLb^2/2(1-\nu) \tag{4.11b}$$

where ν is Poisson's ratio. If we take $\nu = 1/3$, it is seen that the strain energy per unit length of an edge dislocation is approximately 1.5 times greater than that of an equivalent screw dislocation. In most cases, we can consider the values as being equal for all practical purposes.

More importantly, we can see that strain energy is proportional to b^2, and hence the most stable dislocations are those whose Burgers vectors lie in close-packed directions (hence the smallest values of b).

Because of the proportionality of strain energy to dislocation length, it is convenient to consider the energy along a dislocation line as a line tension, in a similar manner to the way in which we consider surface energy as being equivalent to a surface tension. Hence a curved dislocation will tend to move to minimize its length as a straight line, although interactions with other dislocations or obstacles with which there is an interaction energy may prevent this being accomplished (see Chapter 6).

Consider a curved dislocation as shown in Fig. 4.14. It will have a line tension (T) lying along it such that

$$T = \partial U/\partial L \simeq Gb^2/2 \tag{4.12}$$

[3] Note that the value of U is *relatively* insensitive to R, because of the logarithmic term.

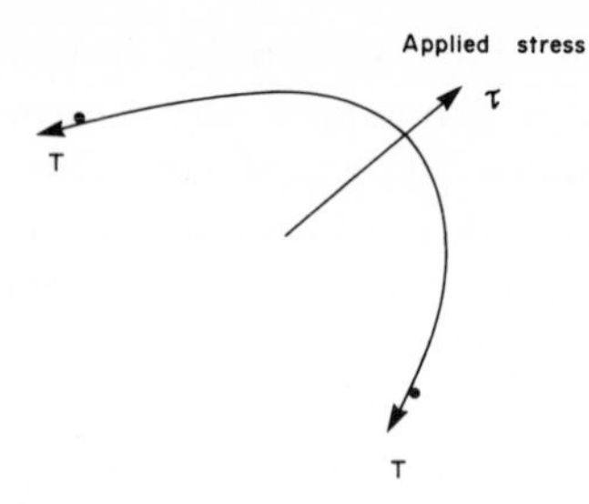

Fig. 4.14 Line tension acting on a dislocation being bowed under the action of an applied stress

This line tension is balanced by the applied stress, τ, acting across the length of the dislocation.

4.3.3 Forces on Dislocations

Consider a crystal containing an edge dislocation of Burgers vector, **b**. When it is moved across the crystal, the force on the dislocation, f, moves through a distance equal to the length, l, of the crystal (see Fig. 4.15), while the applied shear stress, τ, moves through a distance b. The work done by the applied force must equal the work done in moving the dislocation. Hence

$$fl = \tau A b$$

where A is the area of the slip plane over which movement takes place (= l per unit crystal width or per unit length of dislocation). Thus, $f = \tau b$, per unit

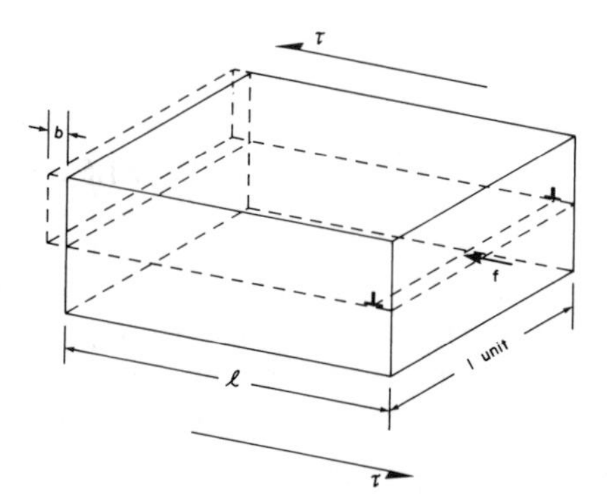

Fig. 4.15 Edge dislocation being moved through a crystal to produce a surface step, *b*

length of dislocation, or, for a dislocation of length, L, the force on the dislocation, F, is given by

$$F = \tau b L$$

4.3.4 Dislocations in Motion

When a dislocation is moved through a crystal, it may come up against obstacles to its motion. Consider a dislocation restrained by a pair of obstacles (precipitates, other sessile dislocations, etc.) as shown in Fig. 4.16. It is bowed out by the applied stress, τ. Hence, for equilibrium, the force on the dislocation, $\tau b L$, is balanced by the components of the line tension, $T\sin\theta$, or

$$\tau b L = 2T\sin\theta$$

Substituting for T from Eq. (4.12),

$$\tau = (Gb/L)\sin\theta \tag{4.13}$$

and the stress required to bow the dislocation out between the obstacles ($\theta \rightarrow 90°$) is

$$\tau_{max} = Gb/L \tag{4.14}$$

The mutual interaction of dislocations when they move through a crystal lattice, as well as their interaction with other defects, will be discussed further after examination of the forces acting between dislocations and the stress fields surrounding them.

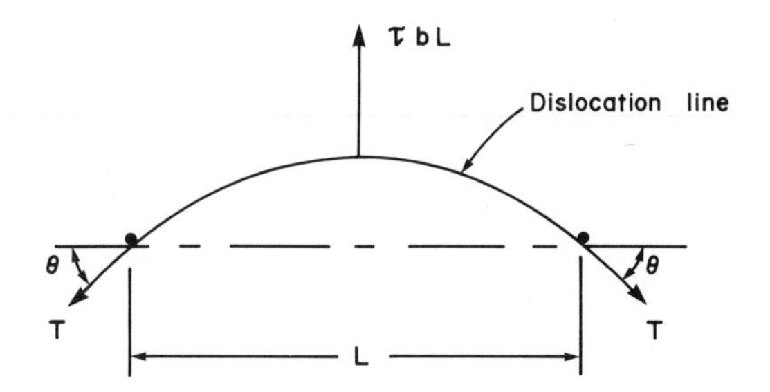

Fig. 4.16 Balance of forces on a dislocation held up at obstacles

4.3.5 Forces Between Dislocations

It is convenient to consider the stress fields surrounding dislocations when discussing their mutual interaction. The general form of these stress fields is as shown in Fig. 4.17 for both edge and screw dislocations.

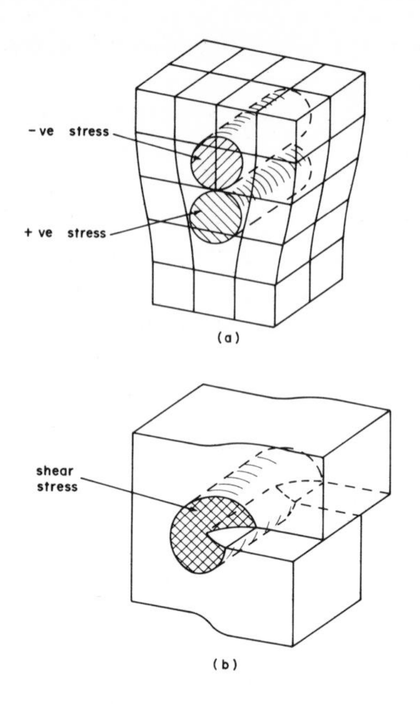

Fig. 4.17 General form of stress fields around dislocations: (a) edge dislocation; (b) screw dislocation

It is evident that dislocations which are of the opposite sign and lie on the same slip plane will tend to run together and annihilate each other, whereas dislocations of the same sign will repel each other and attempt to move further apart. This rearrangement of dislocations is what takes place during the process of *recovery* (in addition to movement and annihilation of vacancies) and may also lead to the formation of polygonized structures containing many subgrains with a small degree of misorientation.

The forces operating between two dislocations may be evaluated most simply for two parallel screw dislocations. For the screw dislocation, there is radial symmetry and the stress is a shear stress only. Hence τ in Fig. 4.13 is given by

$$\tau = G\gamma = Gb/2\pi r \tag{4.15}$$

The force between two parallel screw dislocations acts at right angles to the dislocation lines and has a magnitude

$$F = Gb_1b_2/2\pi r, \text{ per unit length of line} \tag{4.16}$$

where b_1 and b_2 are the magnitudes of the Burgers vectors of the two dislocations. This force is attractive for unlike dislocations and repulsive for like ones. The cases for edge dislocations and for edge-screw interaction forces are more complex, and the interested reader is referred to more detailed treatments elsewhere[4].

A closely related matter is the force between a dislocation and a free surface. Clearly, a dislocation will be attracted to a free surface as it will then move out of the crystal, so decreasing the lattice strain energy.

The force on a dislocation arising from a free surface is determined by the *method of images* [*1*]. Figure 4.18 shows a screw dislocation, A, lying parallel to a free surface, and an "image" dislocation, B, this being of opposite sign. If each has the same magnitude, then, at the equidistant points lying along the surface, the stresses will cancel each other out, assuming the medium to be infinite on both sides of the surface. The medium can then be cut at the surface and the portion containing the image dislocation removed without affecting the forces on the real dislocation. Hence the attractive force on the dislocation, pulling it towards the surface, is

$$F = -Gb^2/4\pi l \qquad (4.17)$$

Where the different medium is not vacuum but another substance of different modulus G', as in a boundary between two phases or for two anisotropic crystals of the same phase in different orientations, it may be shown [*33*] (see also discussion in Friedel [*1*], p. 45) that attraction of the dislocation to the surface occurs if $G' < G$, and repulsion occurs if $G' > G$, and the interaction force varies as $1/l$.

For a region of modulus G' of finite thickness which is in contact with a vacuum as in Fig. 4.19, the screw dislocation at A will be drawn to the free surface if $G' < G$; while if $G' > G$, it will take up an equilbrium position at $l = h$ [*1*].

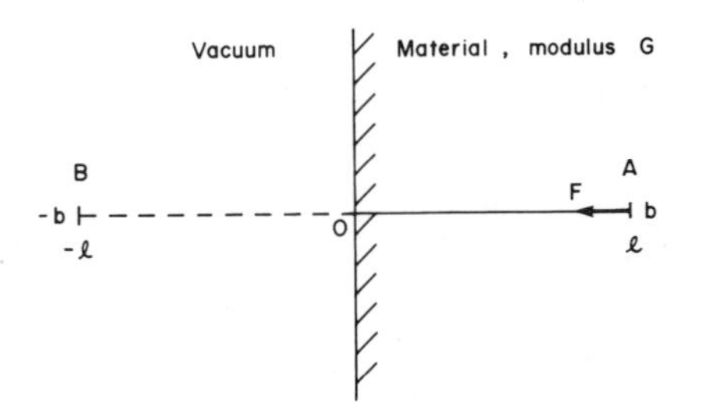

Fig. 4.18 Image force on a dislocation at a free surface

[4] See, for example, Friedel [*1*] pp. 34-44 or Weertman and Weertman [*3*] pp. 65-72.

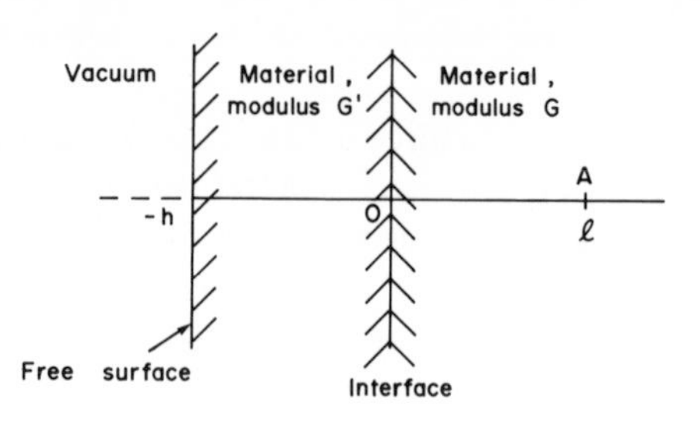

Fig. 4.19 Dislocation lying below a surface layer of dissimilar material

One implication of the foregoing is that a plated film or an oxide layer on a metal's surface, for which $G' > G$, will hinder the egress of dislocations from the surface of a material, increasing apparent work hardening or reducing creep rate. Removal of such a layer can indeed cause additional deformation [*34*] or an increase in creep rate [*35*].

4.3.6 Dislocation Reactions and Sessile Dislocations

As noted in Section 4.3.1, when dislocations meet in a crystal lattice, they do so at a node, at which point the sum of the Burgers vectors of the dislocations whose positive directions point towards the node is equal to the sum of the Burgers vectors of those whose positive directions point away from the node. Some examples of dislocation nodes are shown in Fig. 4.20, and it may be emphasized again that a dislocation in a network can terminate only at a node or at the crystal boundary.

It is now proposed to discuss dissociation reactions of dislocations in the three crystal structures of primary concern here.

(a) FCC lattice

Dissociation of any dislocation of Burgers vector $\mathbf{b}_3$ into two dislocations $\mathbf{b}_1$ and $\mathbf{b}_2$ may occur if energetically favorable. Since the strain energy is proportional to b^2, dissociation will be favored if $b_3^2 > b_1^2 + b_2^2$. An example of dissociation was shown in Fig. 4.12, which showed the formation of a stacking fault in an FCC crystal lattice. In this structure, a perfect dislocation, $a/2$ $[\bar{1}01]$, will always tend to dissociate to two partials of magnitude $(a/6)$. The separation of the two parallel partial dislocations will depend primarily on the *stacking fault energy* (SFE). This quantity varies considerably for different metals and alloys, and some typical values are listed in Table 4.3.

SFE is an important quantity in deformation, as one of the major mechanisms by means of which dislocations bypass obstacles is cross-slip, and the ease of cross-slip is determined largely by the SFE: the higher the SFE, the narrower the stacking fault, and the greater the ease of cross-slip. As a result of

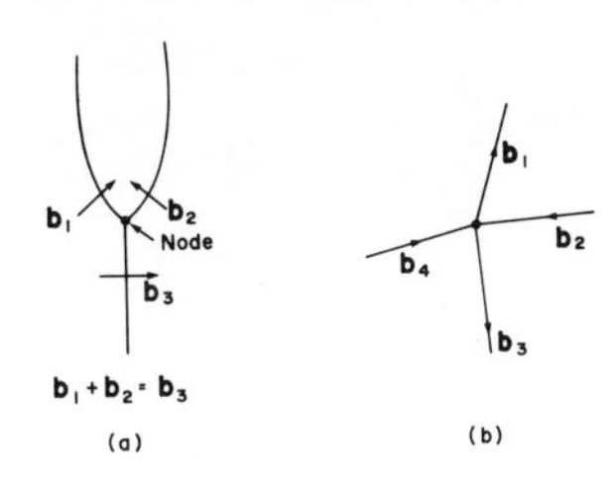

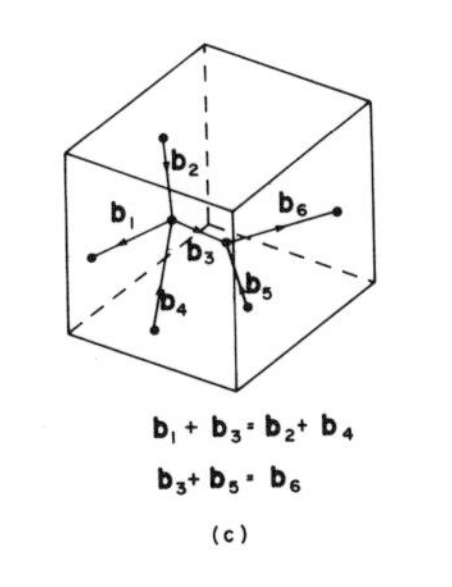

Fig. 4.20 Nodes in dislocation networks

Table 4.3 Typical Values of Stacking Fault Energy [36]

Metal	Stacking Fault Energy ($mJ.m^{-2}$)
Ag	21 ± 7
Al	280 ± 50
Au	52 ± 15
Cu	85 ± 30
Ni	450 ± 90

extensive cross-slip during deformation at elevated temperature, a polygonized structure may develop.

Figure 4.21 shows cross-slip in an FCC lattice where a *screw* dislocation moves from its original slip plane, $(\bar{1}11)$, onto an intersecting $(1\bar{1}1)$ plane.

Let us now consider the interaction between two dislocations on intersecting {111} planes in the FCC lattice as shown in Fig. 4.22. The two dislocations react to form a new dislocation according to the reaction

$$a/2[101] + a/2[01\bar{1}] \rightarrow a/2[110] \quad (4.18)$$

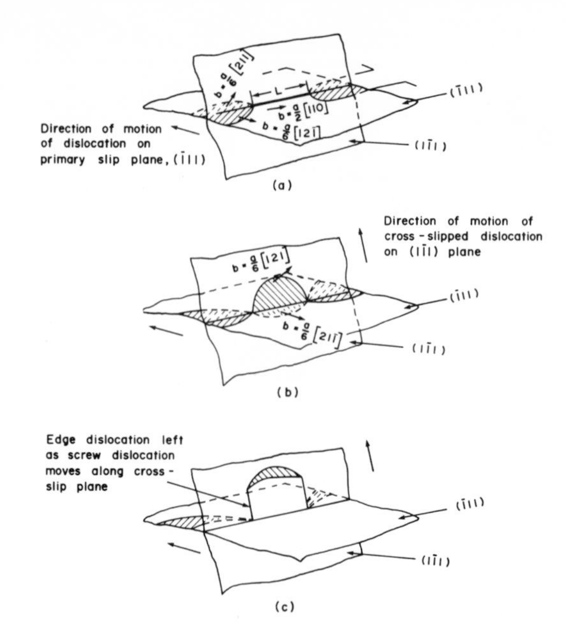

Fig. 4.21 Cross-slip of a screw dislocation in the FCC lattice

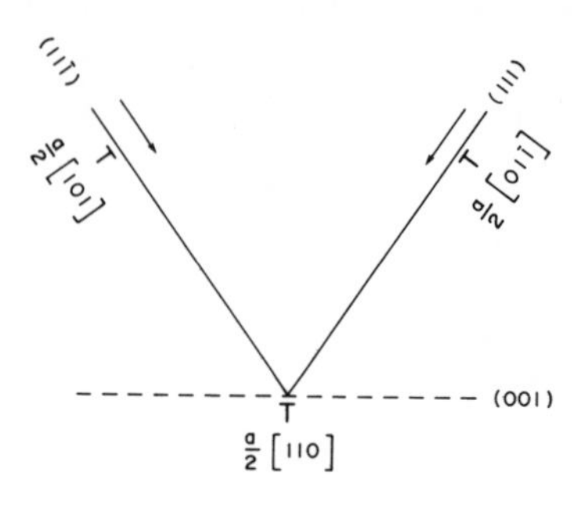

Fig. 4.22 A Lomer lock

The resulting edge dislocation, $a/2[110]$, lies on the (001) plane, which is not a slip plane. Hence the dislocation is unable to move and is sessile. A dislocation of this type is a *Lomer lock*, and its importance lies in the fact that it impedes movement of further dislocations on either side of the {111} planes.

A sessile partial dislocation can be created in the FCC lattice by the collapse of a disc of condensed vacancies on a (111) plane as shown in Fig. 4.23. The

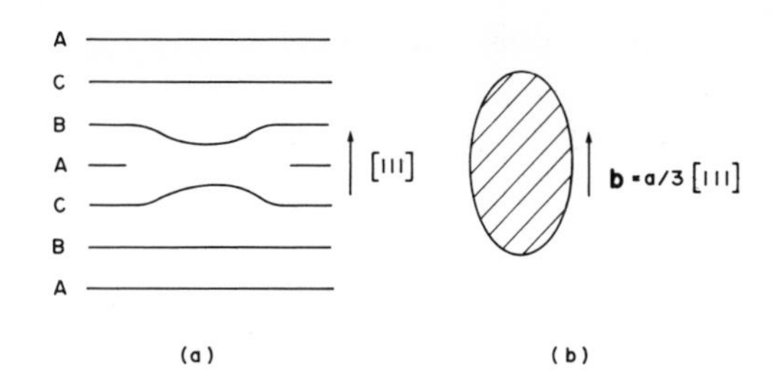

Fig. 4.23 Collapse of a disc of condensed vacancies to produce a Frank sessile dislocation; (b) shows a plan view of the stacking fault bounded by the dislocation line

Burgers vector of a *Frank sessile dislocation*, as this dislocation is termed, is normal to the (111) plane and has a magnitude $a/\sqrt{3}$. As the dislocation lies on a plane normal to the (111) slip plane, there is little possibility of its moving.

As shown previously (see Fig. 4.12), a perfect dislocation in the FCC lattice can split into two partials with the dissociation reaction

$$a/2\,[\bar{1}01] \rightarrow a/6\,[\bar{2}11] + a/6\,[\bar{1}\bar{1}2] \qquad (4.8)$$

and the resulting partials are *Shockley partial dislocations*. Suppose two pairs of such partials meet on intersecting {111} planes as shown in Fig. 4.24. The reaction may be written as

$$a/6\,[\bar{1}2\bar{1}] + a/6\,[2\bar{1}1] \rightarrow a/6\,[110] \qquad (4.19)$$

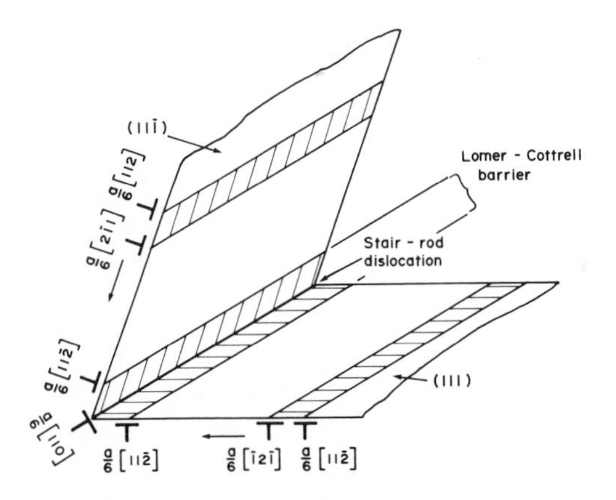

Fig. 4.24 A Lomer-Cottrell barrier formed by the meeting of Shockley partial dislocations on intersecting {111} planes

and the resulting dislocation, termed a *stair-rod dislocation*, is sessile. The combination of the three partials is known as a *Lomer-Cottrell barrier*, and the reaction is an important one as it provides a strong barrier to slip.

(b) HCP lattice

As discussed in Section 4.2.1, for ideal packing in the HCP lattice slip is on the close-packed (0001) planes along $<\bar{1}\bar{1}20>$, and stacking faults can occur, giving rise to local FCC structure between these planes. For this situation the perfect dislocation has the notation

$$b = a/3\ [\bar{1}\bar{1}20]$$

and it may split into two Shockley partials in the reaction

$$a/3\ [\bar{1}\bar{1}20] \rightarrow a/3\ [\bar{1}010] + a/3\ [0\bar{1}10] \quad (4.20)$$

The resulting extended dislocation moves only in the (0001) plane.

For non-ideal packing, slip may take place on the prism or pyramidal planes as noted earlier, both of these containing the smallest possible Burgers vector of the type $a/3\ [\bar{1}\bar{1}20]$.

(c) BCC lattice

In the BCC lattice, slip occurs in the $<111>$ directions with a Burgers vector of $a/2\,[111]$. Several dissociation reactions have been postulated, and stacking faults have been observed in BCC metals [*1*]; however, a particularly interesting dislocation reaction was suggested by Cottrell [*2*]. This is as

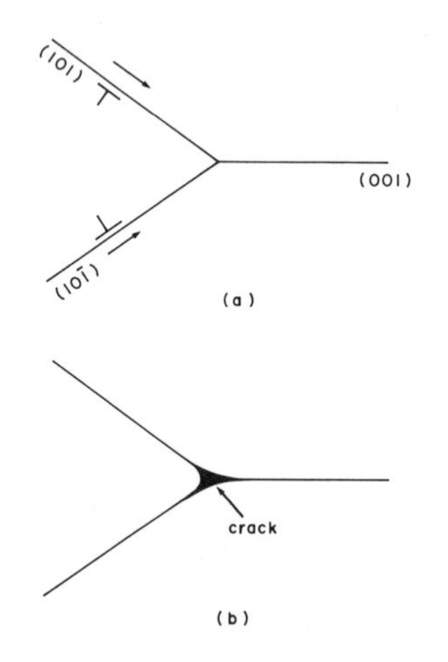

Fig. 4.25 Crack nucleation in BCC lattice by interaction between dislocations to produce immobile dislocation

follows:

$$a/2\,[\bar{1}\bar{1}1] + a/2\,[111] \rightarrow a\,[001] \quad (4.21)$$

The resulting dislocation is pure edge, lying on the (001) plane. It is immobile since (001) is not a slip plane, but this is the plane on which cleavage occurs in BCC metals. Hence, we may envisage a crack nucleating at the intersection of {110} planes on which dislocations are moving together, and growing as more such dislocations move down to the intersection (see Fig. 4.25).

4.3.7 Dislocation Generation and Multiplication

To account for the large plastic deformations which can take place in metals, a very large number of dislocations are required to move within the crystal lattice; indeed many more than could be accommodated physically in the material. Also, it has been observed from transmission electron microscopy and X-ray diffraction studies, that the number of dislocations in a crystal increases with increasing deformation, rather than dislocations being used up in the production of surface steps, and the conclusion reached is that dislocations are generated during plastic deformation. For example, in an annealed polycrystalline metal, the observed density[5] may be as low as $10^5/\mathrm{mm}^2$, while it may rise to $10^{10}/\mathrm{mm}^2$ with severe cold working [*37*]. Approximate densities of dislocations in metals are listed in Table 4.4.

Considerable controversy took place in the past concerning the mechanisms by which dislocation multiplication occurs; however, it has been established by direct observation of thin foils in the transmission electron microscope (TEM) that the mechanism proposed independently by Frank and Read [*38*] in 1950 is the operative one. This is illustrated in Fig. 4.26, where a pinned segment of dislocation expands under the influence of a shear stress to produce a series of concentric loops. The dislocation segment may be

Table 4.4 Approximate Dislocation Densities in Various Conditions [*37*]

Material	Dislocation Density, lines/mm^2
Carefully grown high-purity single crystals	0 to 10^1
Ordinary single crystals, annealed or unstrained	10^3 to 10^4
Polycrystalline specimens, annealed	10^5 to 10^6
Severely cold-worked specimens	10^9 to 10^{10}

[5] Being line defects, the density of dislocations is given per unit area, or intersecting lines per unit area.

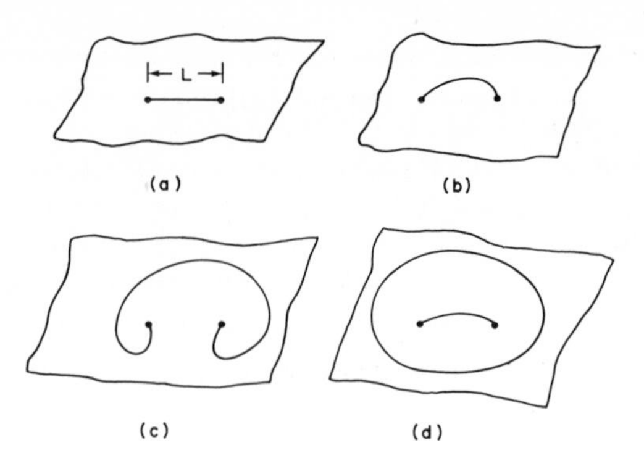

Fig. 4.26 Generation of dislocations at a Frank-Read source

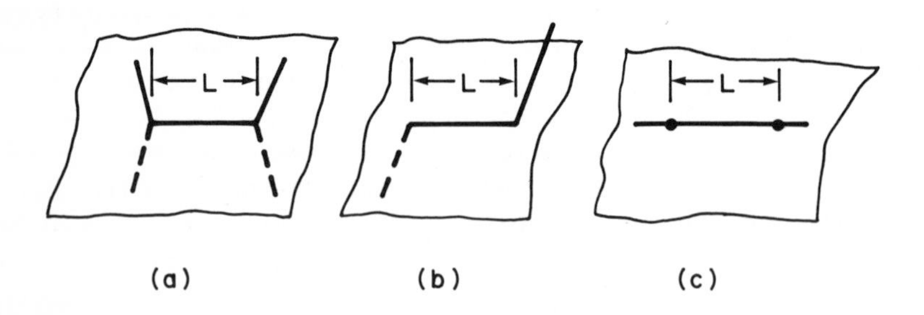

Fig. 4.27 Pinning of a dislocation segment (a) by other dislocations; (b) where it leaves the slip plane; (c) by impurity atoms or precipitates

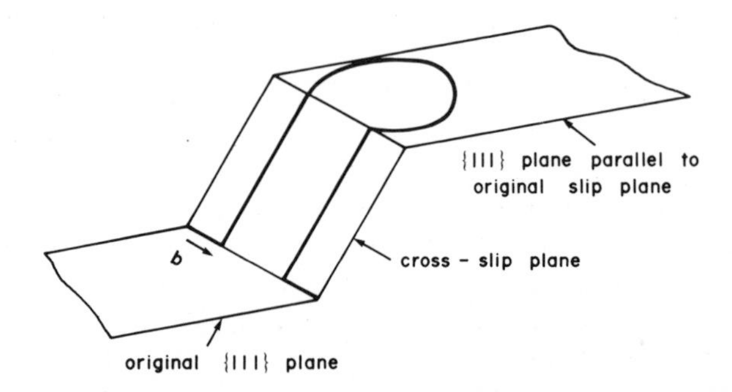

Fig. 4.28 Dislocation generation on another plane after cross-slip has taken place

pinned at its ends in several ways, for example on intersection with other dislocations on different planes [Fig. 4.27(a)], where the dislocation leaves the slip plane [Fig. 4.27(b)] or where it is pinned by impurity atoms or precipitates [Fig. 4.27(c)]. A *Frank-Read source* may also be produced when a portion of a screw dislocation cross-slips, as shown in Fig. 4.28.

The stress required to generate a dislocation is a function of the distance between pinning points, L, as given by Eq. 4.14. When L is large, as in a metal with low dislocation density, the stress is small: as deformation increases and there are more dislocations present, there will be increased dislocation interaction and a consequent reduction in L. From this, *strain hardening* occurs, and the effect is increased as dislocations pile up at barriers as discussed in Section 4.3.10. These piled-up dislocations also set up a *back stress* making further deformation increasingly difficult.

4.3.8 Interactions With Point Defects

Point defects, such as vacancies, interstitials and solute atoms, play a major role in assisting or in restraining the movement of dislocations. We shall deal with these separate point defects in turn.

(a) Vacancies

Unlike dislocations, vacancies are thermodynamically stable, their equilibrium number in metals depending on temperature according to the relation,

$$n/N = A \exp(-E_f/kT) \qquad (4.22)$$

where n is the number of vacancies, N is the number of atom sites in the lattice, A is an entropy term (usually taken as unity), E_f is the energy required to form one defect by removing an atom from the lattice and depositing it in a site at the crystal surface, k is Boltzmann's constant, and T is the absolute temperature[6]. Vacancies are, therefore, present in increased numbers at high temperature, and also have greater mobility with increased temperature. Hence, one important mechanism they facilitate is the *climb* of edge dislocations past obstacles at elevated temperature, as in creep deformation. Vacancies are attracted to the stress fields of dislocations, and Fig. 4.29 illustrates climb as vacancies diffuse to an edge dislocation. This figure also shows the formation of a *jog* by diffusion of vacancies to a dislocation. Although this is not a point defect, its effects may conveniently be touched upon here, in view of its close relationship with vacancy diffusion.

A jog in an edge dislocation (produced, for example, as discussed above) is generally mobile so will not impede dislocation mobility. However, a jog in a

---

[6] For a derivation of this relation, see, for example, Smallman [*39*].

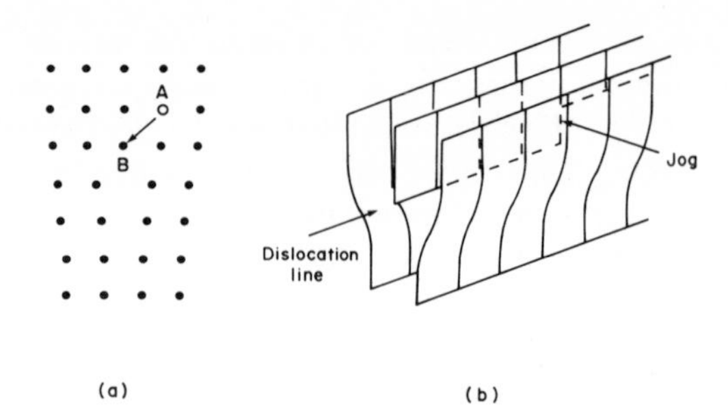

Fig. 4.29 (a) Edge dislocation climb by vacancies diffusing to it; (b) jog formed by vacancies diffusing to part of an edge dislocation

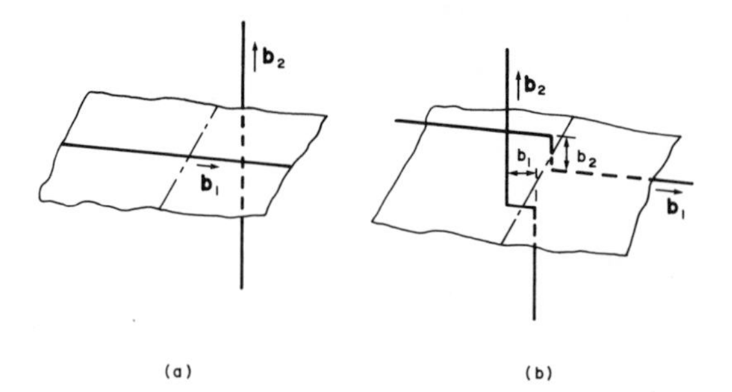

Fig. 4.30 Intersection of two screw dislocations to produce a jog

screw dislocation, which may, for example, be produced by the intersection of two screw dislocations as shown in Fig. 4.30, is of an edge nature and, to be moved, must be dragged through the lattice creating a row of point defects, as its Burgers vector does not lie in the plane in which it is constrained to move with continued movement of the screw dislocation. This mechanism constitutes another major component in the work hardening of metal crystals (see Section 4.4.1), of particular importance at elevated temperature.

(b) Interstitials

Interstitial atoms can congregate at a dislocation, so reducing the total strain energy in the lattice and making it more difficult to move the dislocation. Such an "atmosphere", termed a *Cottrell atmosphere*, is responsible for the discontinuous yield point observed in carbon steels and certain other alloys. Once dislocations are freed from such groups of interstitial atoms, they can move under reduced stress, so producing distinct upper and lower yield points (see Section 4.4.3).

(c) Substitutionals

In a similar manner to interstitials, substitutional atoms tend to congregate at dislocations to reduce the total strain energy of the system. Again a discontinuous yield point may be observed. The interaction of substitutional atoms with dislocations is considered further in Chapter 6 in dealing with strengthening mechanisms.

4.3.9 Dislocation Arrays

When a crystal contains many dislocations, these seek to take up positions of minimum energy, and it is observed that a network, known as a Frank network (Fig. 4.31), is set up by the residual dislocations after prolonged holding at high temperature. When a large number of additional dislocations of like sign are produced by cold working, these may form a fine network superimposed on the Frank network, and forming parallel arrays on its faces (Fig. 4.32).

During the processes of hot working, creep, fatigue, or recovery after cold work, a polygonized structure may be formed [*1, 13*], and Fig. 4.33 shows a

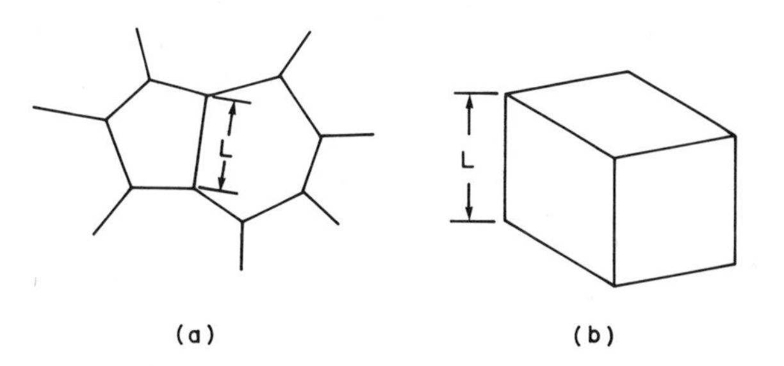

Fig. 4.31 The Frank dislocation network in a crystal. The cube shown in (b) schematically represents one block in the network

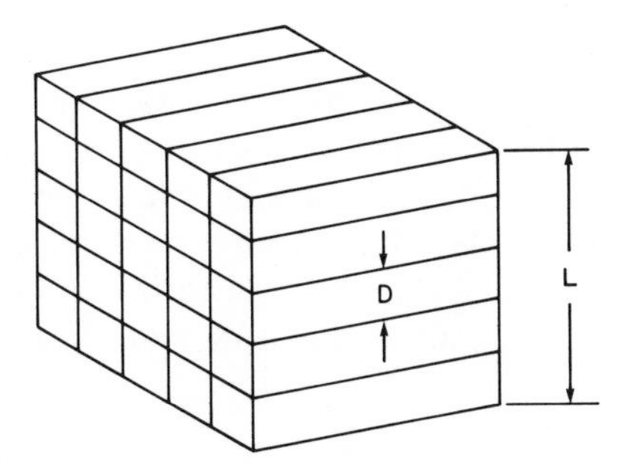

Fig. 4.32 Polygonization on the surface of a block in the Frank network

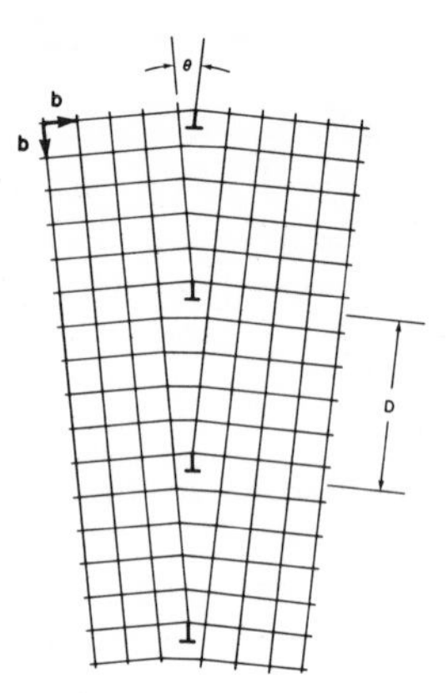

Fig. 4.33 Tilt boundary formed of edge dislocations

section through the face of the "block" depicted in Fig. 4.31(b). Here we see that the misorientation of the lattice across this face can be stated in terms of the spacing of the edge dislocations on the *tilt boundary*, as

$$\tan \theta = b/D \tag{4.23}$$

where b is the Burgers vector magnitude and D the spacing between dislocations. In terms of the stress fields illustrated in Fig. 4.17(a), it is clear that the dislocations have taken up positions of minimum energy[7].

Screw dislocations, in the form of planar arrays, give rise to *twist boundaries*, as illustrated in Fig. 4.34. These, together with tilt boundaries, are termed *low-angle grain boundaries*, the misorientation being generally less than 10°. At misorientations greater than about 10 to 15°, it has been suggested that there can no longer be a simple dislocation structure to a grain boundary, and that such *high-angle grain boundaries* are essentially matching surfaces between two practically undisturbed grains. A few atoms may belong to both grains; others may belong to neither, as illustrated in Fig. 4.35.

However, recent studies indicate that most grain boundaries, whether of low or high angle, can be represented in terms of dislocation models [*40*], being either a superposition of a uniform atomic structure and a low-angle dislocation boundary or of sets of lattice and grain boundary dislocations, the latter being shown in Fig. 4.36. For a historical review of the theories concerning the structure of grain boundaries, and illustration of the classic

[7] Obviously a still lower energy is achieved when all dislocations anneal out of the structure, or when recrystallization takes place.

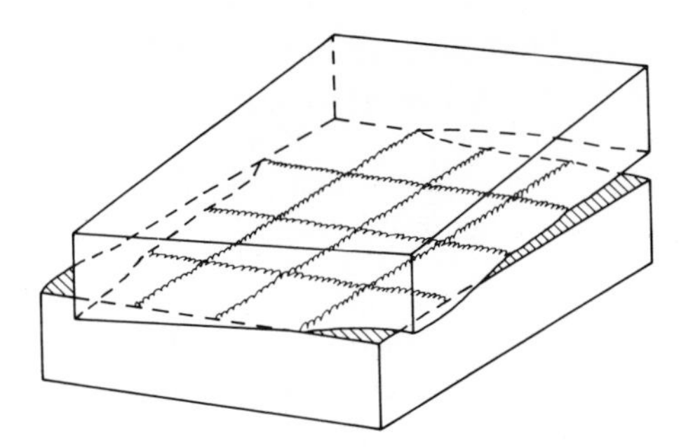

Fig. 4.34 Twist boundary formed by intersecting screw dislocations

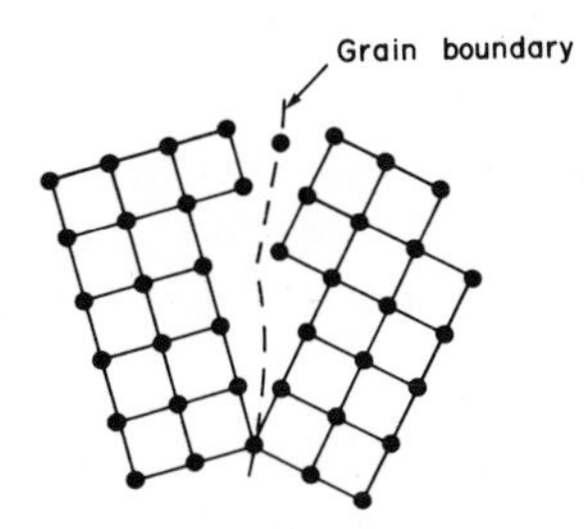

Fig. 4.35 High-angle grain boundary

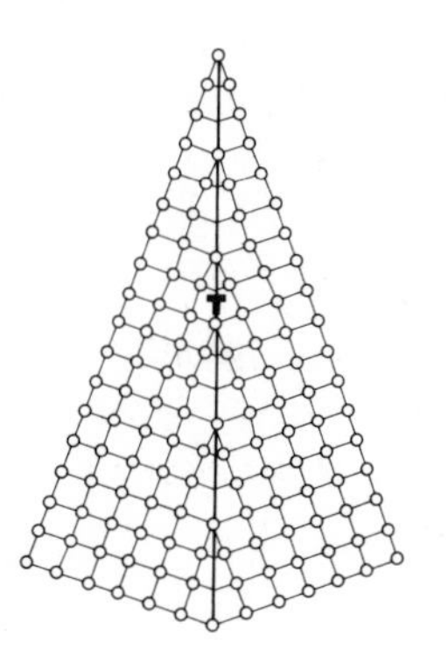

Fig. 4.36 Dislocation model of a grain boundary. After Hirth [*40*]

"bubble raft" models of boundaries of Lomer and Nye [*41*], the reader should consult McLean's detailed monograph [*28*]. It may be noted, however, that the old idea of amorphous grain boundaries can still be a useful one in explaining engineering properties of materials [*42*], although it must not be taken too seriously as a structural model.

4.3.10 Dislocation Pile-Up

Under the influence of an applied stress, dislocations may pile up at an obstacle as shown in Fig. 4.37, which shows a group of n parallel straight edge dislocations after the first (designated as 0) has met an obstacle which cannot be overcome or bypassed under the influence of the applied stress. For equilibrium, the total force on any dislocation must be zero, and for the ith dislocation this is

$$[Gb^2/2(1-v)] \sum_{\substack{j=0 \\ j\neq i}}^{n} 1/(x_i - x_j) - \tau_S b = 0 \tag{4.24}$$

where τ_S is the resolved shear stress on the slip plane and b is the magnitude of the Burgers vector of each dislocation. This equation was solved by Eshelby, Frank and Nabarro [*43*], and it was determined that the lead dislocation experiences a stress

$$\tau_S^* = n\tau_S \tag{4.25}$$

The force exerted at the head of the pile-up is given by

$$F = nb\tau_S \tag{4.26}$$

This is obtained by considering the n dislocations, each of Burgers vector **b**, to be equivalent to a giant dislocation of Burgers vector **nb**, and, at large distances from the dislocation array, this giant dislocation can be considered to be located at the centre of gravity three-quarters of the distance from the source to the obstacle. Slip is equivalent to this giant dislocation moving through the distance $3l/4$,

The number of dislocations in the pile-up is given approximately by[8]

$$n \simeq l\pi\tau_S/Gb \tag{4.27}$$

and where the obstacle is a grain boundary, with the source located at the center of the grain of diameter d, this becomes

$$n \simeq d\pi\tau_S/4Gb \tag{4.28}$$

The factor 4 is required because there is a back stress on the source arising from dislocations piled up on both sides of the source.

---

[8] A proof of this is given by Weertman and Weertman [*3*], p. 128.

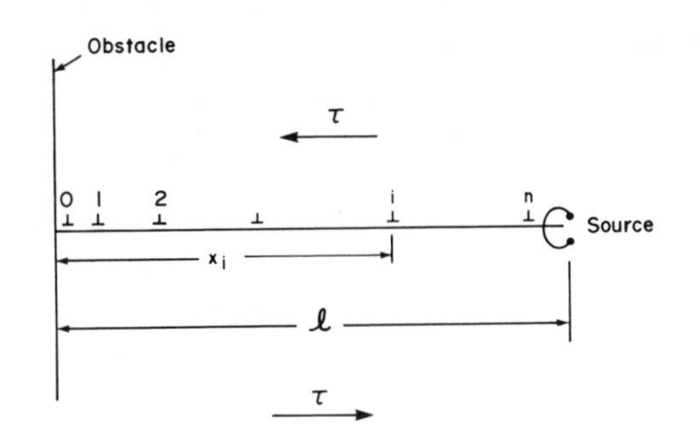

Fig. 4.37 Dislocation pile-up at an obstacle

Yielding occurs when a critical stress, τ_c, is reached at the head of the pile-up, and in the case of the obstacle being a grain boundary, the value of τ_c is assumed to be independent of grain size. Hence, in this case,

$$\tau_c = n\tau_S \simeq d\pi\tau_S^2/4Gb \tag{4.29}$$

The resolved shear stress required to overcome the obstacle, τ_S, may be taken as being equal to the applied stress τ less the internal friction stress required to overcome resistance to dislocation motion, τ_o. Hence

$$\tau_c \simeq d\pi(\tau - \tau_o)^2/4Gb$$

Rearranging,

$$\tau \simeq \tau_o + (\tau_c\, 4Gb/\pi d)^{1/2} \tag{4.30}$$

or

$$\tau = \tau_o + k_S d^{-1/2} \tag{4.31}$$

which is the Hall-Petch relationship [*45, 46*] between yield stress and grain size observed for many metals and alloys. This is discussed further in Chapter 6, when dealing with methods of increasing the yield stress of engineering materials.

4.4 DEFORMATION IN BULK MATERIAL

When bulk material is deformed plastically, many dislocations will be generated, moved and able to interact. The matter is further complicated when grain boundaries are present and deformation of each individual grain must satisfy compatibility relations at these boundaries. It is proposed in this section to review briefly the deformation of single crystals, then of polycrystalline aggregates, with some consideration of the effects of grain rotation which must take place during plastic flow.

4.4.1 Single Crystals

The generalized flow curve for deformation of a single crystal of an FCC metal is shown in Fig. 4.38, and it may be seen that it consists of three stages, labelled I, II, and III, respectively [*46*].

Stage I is that of *easy glide* [*47*], in which little strain hardening occurs, slip takes place on one slip system only, and dislocations move over large distances without meeting obstacles. In Stage II slip takes place on additional slip systems and rapid strain hardening occurs. During Stage III strain hardening falls off with increase in strain.

Cottrell [*2*] has distinguished between two types of plastic flow, terming them *laminar flow* and *turbulent plastic flow*[9], and this is a useful and physically meaningful distinction. The former corresponds to Stage I deformation with a few long slip lines being formed and the amount of strain hardening for purely tensile or shear deformation being small. It commences at a stress greater than the elastic limit, and only a few Frank-Read sources operate where L [see Fig. 4.26(a]) is large, and from regions where the Frank network is especially large. Hence, the distance through which these dislocations move is large and the majority will leave the crystal at the surface with little or no strain hardening taking place. Such strain hardening as occurs will be due to the back stress set up on the Frank-Read sources by dislocations being held up by others in the Frank network. The extent of this easy glide or laminar flow stage will depend on when other slip systems become more favorably oriented than the primary one, and this will depend on the crystal structure and the axis of the applied stress. For HCP crystals, with (primarily) basal slip, or at least very limited slip systems, Stage I can extend over considerable strains, while for cubic materials Stage I is generally small, as, when the crystal rotates to a small extent only, slip can normally take place on one of the many other slip systems. The exception to this is when the tensile axis is in a $\langle 101 \rangle$ direction, when there is appreciably more shear stress on one slip system than others. For a tensile stress oriented close to $\langle 100 \rangle$ or $\langle 111 \rangle$, the shear stresses acting on several slip systems are much the same, and Stage I deformation is very short or non-existent.

The turbulent flow of Stage II corresponds to the operation of slip on several systems, with many short slip lines formed, and the rapid strain hardening is caused by increase in the internal stress from elastic interaction of these dislocations. This hinders the generation of loops from the active sources. Dislocations on intersecting $\{111\}$ planes in an FCC lattice can form

---

[9] On a microscale, slip character is an important characteristic of a material's deformation, and we may distinguish between materials in which the slip mode is *planar* and those in which it is *wavy* [*48*]. The criterion which governs the occurrence of the latter is the ease of initiation of secondary slip systems by the cross-slip of screw dislocations.

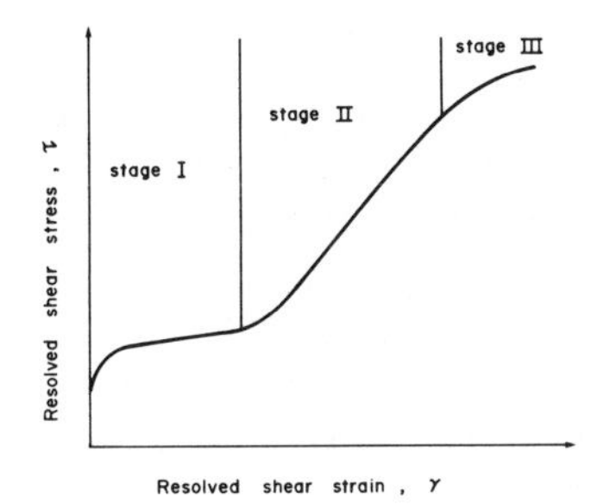

Fig. 4.38 Generalized flow curve for an FCC metallic single crystal

Lomer-Cottrell barriers, as discussed in Section 4.3.6, and Stage II deformation may be thought of as a stage where there is a steady increase in the number of such barriers with increase in strain, and a consequent increase in the stress required to generate additional dislocations. The slope of this curve is found to be largely temperature-independent.

Stage III deformation, with its decreasing strain hardening rate with strain, commences when dislocations are able to bypass the obstacles holding them back. The stress where Stage III begins is strongly dependent on temperature, decreasing with increase in temperature, and the major mechanism whereby the barriers are bypassed is thought to be cross-slip. The fact that strain hardening does continue to take place is attributed to the intersection of the cross-slipped dislocations with the forest of screw dislocations which pierce the active slip planes, producing an increasing number of jogs which cannot move conservatively (see Section 4.3.8).

The ease of cross-slip is increased by increase in stacking fault energy (Section 4.3.6), and this correlates with the observation that in metals of high SFE, Stage II is very short for room temperature deformation.

Two features of interest which may be observed during plastic deformation are *deformation bands* and *kink bands*. The former are thin regions of inhomegeneity in which the lattice is rotated away from the surrounding material, and they form at the end of easy glide. Such bands may form before a second system is oriented for slip, and hence cause the breakdown of Stage I deformation at an early stage, e.g., in aluminum [*2*]. They can form in both FCC and HCP metals. Kink bands are closely related, and may form when an HCP crystal is deformed in an orientation where its basal plane is nearly parallel to the crystal (and stress) axis. They form by a sudden localized rotation of part of the crystal, involving organized slip on many parallel planes [*49*]. Figure 4.39 shows the structure of a kinked crystal, in which the horizontal lines represent basal (slip) planes and a compressive stress has been

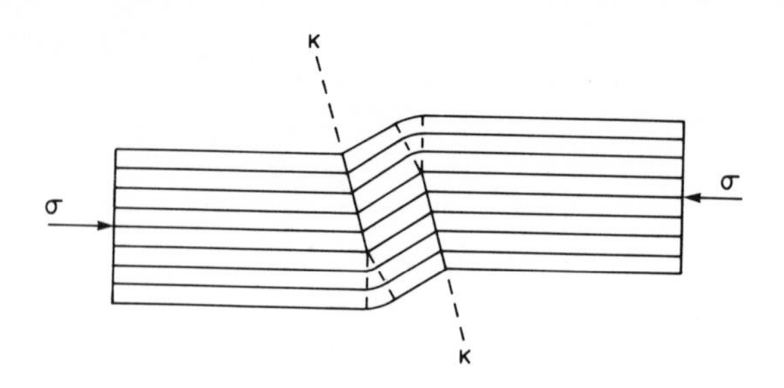

Fig. 4.39 A kink band in the HCP lattice

applied in the direction of the arrows. The planes designated K are the *kink planes* where the orientation changes suddenly.

4.4.2 Polycrystalline Aggregates

The stress-strain curves for polycrystals differ considerably from those for single crystals. Polycrystalline material usually has a higher elastic limit and rate of strain hardening. In addition, no Stage I or easy-glide deformation is possible, the stress-strain curve consisting of Stages II and III only. These effects are due to the presence of the grain boundaries and the differences in orientation between neighboring grains: the grain boundaries act as barriers to the movement of dislocations, these piling up as discussed in Section 4.3.10, while the need for compatibility at their boundaries means that individual grains are not free to deform on any slip system where the resolved shear stress reaches the critical value for single crystals.

Von Mises [*50*] showed that a minimum of five independent slip systems must operate to allow an unrestricted shape change to take place, volume remaining constant, and many attempts have been made to derive the flow curve for polycrystalline material on the basis of single crystal data and consideration of the operative slip systems.

For FCC and BCC metals, the tensile stress σ can be converted to a shear stress τ on the most favored slip system, and the tensile strain e into a shear strain γ in the slip direction. Hence

$$\sigma = m\tau$$

$$\gamma = me$$

or

$$\sigma/\tau = \gamma/e = m \tag{4.32}$$

where m is the orientation factor, $(\cos\phi \cos\lambda)^{-1}$, from Eq. (4.1).

If the grains of the polycrystal are of random orientation and if all the grains have the same stress-strain relation, $\tau = f(\gamma)$, as they would when separated, then the tensile stress-strain curve of the aggregate may be expected to be some average of those for the individual grains. The mean orientation factor $\bar{m}$ for a random aggregate of FCC crystals is approximately 2.24 [51], and Kochendörfer [52] predicted a stress-strain curve of the polycrystal from single-crystal data on the basis that all grains extend equally, in the form

$$\sigma = \bar{m}\,\tau = \bar{m}f(\gamma) = \bar{m}f(\bar{m}e) \tag{4.33}$$

It has been found by experiment that Eq. (4.33) is in good agreement with polycrystalline data [53] when $\bar{m} \simeq 3.1$ rather than 2.2, and Taylor [53] has explained this on the basis that slip must take place on more than the single slip system assumed by Kochendörfer to preserve grain boundary continuity, in line with von Mises' calculations [50]. Taylor then chose, among the 12 possible slip systems in FCC crystals, the combination of five which required the least work: this raises the value of $\bar{m}$ above the average value of 2.2, when it is assumed that only one system operates. Bishop and Hill [54] have taken into account an additional factor, namely that stress discontinuity cannot exist at grain boundaries, and also considered that less work may be required with more than five slip systems. Their studies confirm the value of $\bar{m}$ obtained by Taylor.

That the theories derived for prediction of polycrystalline deformation on the basis of single crystal data are not completely correct is indicated by the fact that grain rotations predicted from these are not completely in agreement with those observed [14]. Hence texture development cannot be properly predicted on the basis of prior orientation and deformation processing. Nonetheless, the value of $\bar{m}$ or *Taylor factor* of 3.1 for FCC materials is well established.

An analysis of constrained deformation in HCP alloys, under conditions of slip and twinning, has been published by Thornburg and Piehler [55]. They utilize the maximum plastic work approach of Bishop and Hill [54] in their analysis.

When external stress is applied to a polycrystal, plastic microstrains of the order of 10^{-5} to 10^{-4} are observed in those grains most favorably oriented for slip, the magnitude of plastic strain increasing with increase in grain size, as this allows the formation of more dislocation loops and the operation of more Frank-Read sources before the back-stress developed at the boundaries inhibits their operation. Subsequently, some parabolic strain hardening may be observed over larger strains to, perhaps, a few percent. This corresponds to the sequential deformation to a very small extent of individual grains: some, with higher internal stresses, deform before others. Under the increasing external stress, slip may propagate from the grains which were most suitably oriented to others, so that the whole specimen deforms plastically. The stress at which this occurs is a function of the grain size, as determined by Hall [44], Petch [45] and others, and decreases with increase in grain size (see Section 6.3.3).

The effect of grain size on the flow stress of HCP metals is greater than for FCC or BCC structures. This is because, in the latter cases, no grain can be very unfavorably oriented for slip, whereas in the former, with a very limited number of slip systems, there may be an orientation difference between grains such that a large increase in applied stress is required to initiate slip in the neighboring grain.

At still larger strains, in order to preserve grain boundary compatibility, simultaneous slip on several systems, or slip in combination with twinning, must take place. In addition, kink bands may be formed as discussed in Section 4.4.1. The effect of slip taking place on several systems (multiple slip) is to produce a high rate of strain hardening. In FCC and BCC metals, this is somewhat greater than for the equivalent single crystals, but for HCP metals the difference in strain hardening rate is very large.

It may be noted that with the widespread use of computers, solution of the problem of deformation in a polycrystalline aggregate is now a more tractable one, and there has been renewed interest in it in recent years [*55, 56*].

4.4.3 Yield Phenomena and Strain Aging

Engineers are generally familiar with the sharp yield point which occurs in annealed low-carbon steels as well as in a number of other engineering alloys. In conducting a tensile test on such materials, the transition from elastic to plastic deformation is sudden, and there is a sharp drop in the load-extension or stress-strain curve as shown in Fig. 4.40(a). The drop from the upper to the lower yield point is accompanied by the formation of a region of localized plastic deformation oriented approximately 55° to the tensile axis, and which then propagates along the test specimen producing plastic extension without increase in load or stress[10]. The band of plastically deformed material is termed a *Lüders band*, and many of these may be formed and give rise to the serrated nature of the deformation curve in Fig. 4.40(a).

In steel, the discontinuous yielding phenomenon is related to the attraction of interstitial carbon or nitrogen atoms to dislocations in the iron lattice, these impurity atoms relieving the local stresses and thus "locking" the dislocations in place. When the applied stress exceeds a critical value (the upper yield point), the dislocations are broken away from their Cottrell atmospheres and move under a reduced stress. In other alloys the locking may be from either interstitial or substitutional atoms. If an annealed specimen of mild steel is deformed as in Fig. 4.40(a), unloaded and retested, there is no drop at the yield point [Fig. 4.40(b)], but an increase in initial yield stress and an accompanying drop can be made to reappear by heating the specimen to, say, 200°C for 1 hour or by holding for several days at room temperature. In these

[10] See discussion of local necking in Section 3.1.3.

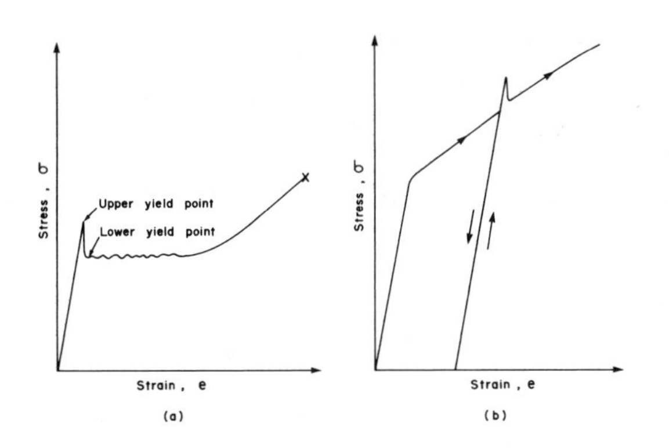

Fig. 4.40 (a) Sharp yield point in steel — schematic; (b) stress-strain curve for the steel specimen previously deformed in (a) to point X, showing absence of drop in stress at the yield point, while restoration of the discontinuity occurs after further unloading and heating to strain-age the material

circumstances, impurity atoms have diffused to the dislocations, reforming the Cottrell atmospheres around them. The phenomenon is known as *strain aging*.

Closely related is the occurrence of a serrated flow curve during tensile testing, as shown in Fig. 4.41, this being found in many alloys (e.g., α-brass and Al-Cu alloys). This behavior is termed the *Portevin-Le Châtelier effect*, and is caused by *dynamic strain aging*, in which solute atoms migrate to dislocations during the deformation process, thus raising the stress under which these would normally move: once they break away, the stress falls again.

In addition to causing a serrated flow curve, dynamic strain aging can

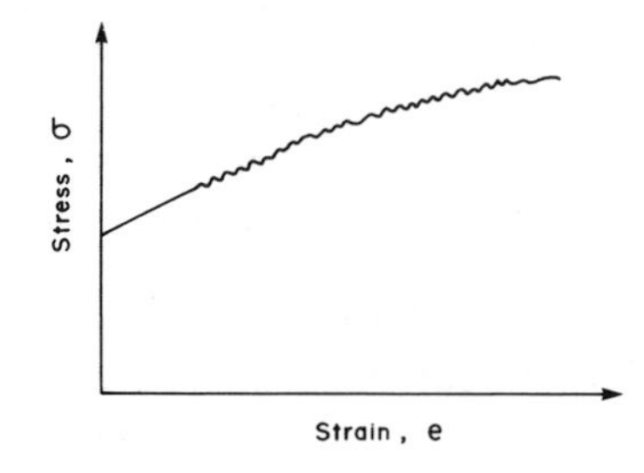

Fig. 4.41 Portevin-Le Châtelier effect

produce several other effects on mechanical properties, including abnormally low or negative strain rate sensitivity of the flow stress, and discontinuities in the variation in flow stress with test temperature. It is proposed to discuss these effects briefly as dynamic strain aging can contribute significantly to the strengthening of alloys, and is an important factor in the resistance of metals to creep deformation at elevated temperature.

In a pure metal, the flow stress generally decreases with increasing temperature and the general relationship is given by curve (a) in Fig. 4.42. However, for alloys in which dynamic strain aging occurs, the stress-deformation temperature curves are as shown by curves (b) - (e) in the figure, these representing alloys in which progressively more dynamic strain aging takes place. It may be noted that the transient in flow stress behavior occurs over a certain range of deformation temperature.

In a "normal" metal or alloy the flow stress increases with increasing strain rate, while the general relationship for a material in which dynamic strain aging occurs is shown in Fig. 4.43, this representing the behavior at specified strain and temperature.

The explanation of these observations is frequently given in terms of the Cottrell concept of dislocations moving at a steady velocity and dragging a solute atmosphere along with them at the same velocity. There is a limiting velocity at which the atmosphere can be dragged and, if the dislocation velocity exceeds this, the dislocation will break away and the flow stress will fall until other solute atoms diffuse to reform the atmospheres. This repetitive process provides an explanation for the serrated flow curve, but a more complete model has been proposed by van den Beukel [57], which permits quantitative prediction of the strain for the initiation of serrated yielding, and leads to curves of stress versus strain rate and stress versus temperature of the form shown in Figs. 4.43 and 4.42, respectively.

Van den Beukel's model recognizes that the movement of a dislocation is, in general, discontinuous, with it being held up periodically at obstacles. During these waiting times solute atoms may diffuse to the dislocations, thus locking them. The solute concentration or atmosphere surrounding a dislocation is dependent on the waiting time and the diffusion coefficient of the solute, and the overall process can be modelled on the basis of dislocations overcoming obstacles by the combined action of effective stress and thermal activation.

4.4.4 The Bauschinger Effect

In the earlier discussion of tensile deformation in Chapter 2, the *Bauschinger effect* [58] was ignored for simplicity. This effect is simply that a lower stress is generally found to be required to initiate plastic deformation in the reverse direction. The effect is observed in all metals and alloys to a greater or lesser extent, and in both single crystals and polycrystalline materials.

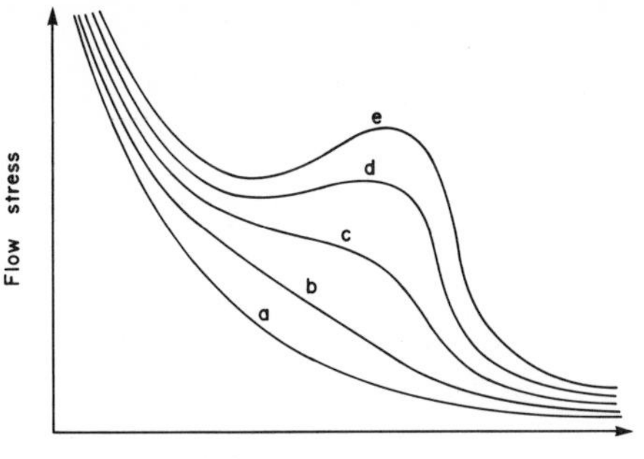

Fig. 4.42 Relationship between flow stress and temperature for a stable metal (a), and for alloys in which progressively more dynamic strain aging occurs (b - e)

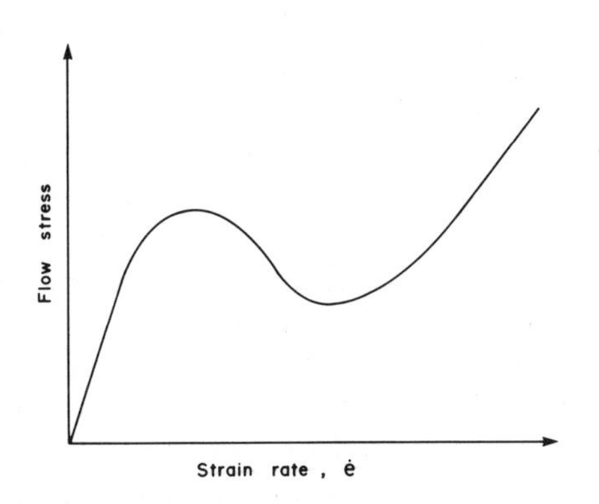

Fig. 4.43 Variation of flow stress with strain rate for material which dynamically strain-ages

Figure 4.44 illustrates the Bauschinger effect schematically, and it is seen that, following tensile deformation, the compressive yield strength is reduced. If the initial strain had been compressive, the tensile yield stress would have been reduced instead.

The reasons for the Bauschinger effect relate to the holding up of dislocations at obstacles and the development of a back stress. When stress is reversed, the back stresses produced will act together with the applied stress, so facilitating dislocations' motion, until these are again held up at obstacles some distance from their initial position.

4.4.5 Development of Texture

As already seen, rotation of a grain takes place when it deforms on one set of

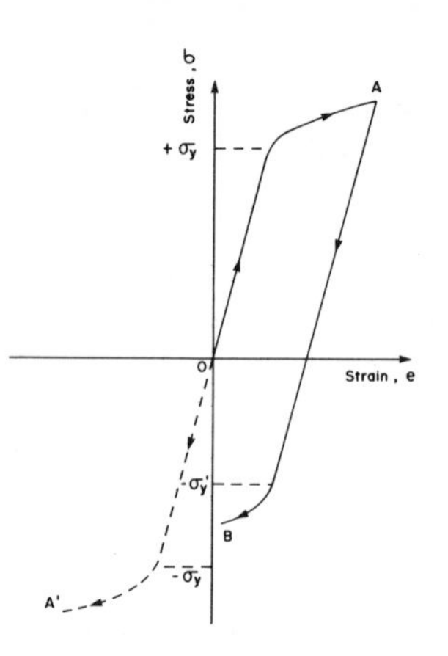

Fig. 4.44 The Bauschinger effect

slip planes, and in polycrystalline material deformed by rolling, forging, deep drawing, or some other process, the grains will rotate from their initial orientations, and a strong degree of preferred orientation or *texture* may be expected after large deformations. The difficulties of predicting the flow curves for tensile deformation of polycrystalline material on the basis of single crystal data were emphasized briefly in Section 4.4.2, and prediction of the degree of preferred orientation arising in a sheet or other shape after processing material of known initial texture is a complex and, as yet, incompletely solved problem. Nonetheless, the general principles are well established and are discussed in detail in the appropriate reference works [*14, 59*].

Crystallographic texture can be measured in several ways, the most convenient and common being the normal and inverse pole figure techniques [*14*]: the former presents information on the intensity of the normals to a specific plane in all directions, while the latter displays the intensities of all planes in a specific direction.[11] Figure 4.45 presents a comparison of the two types of pole figure.

[11] Pole figures representing texture are drawn on the basis of a stereographic projection, which provides a two-dimensional plot in which the angular relationships between planes in the crystal or aggregate are preserved. In a normal pole figure, the center of the projection represents an important direction such as the normal to the sheet in rolling or the axis of a wire, while an inverse pole figure represents the density distribution of poles lying in a specific direction (e.g., the wire axis or the sheet rolling direction) drawn on a stereographic projection of the lattice in a standard orientation. For more detailed discussion of stereographic projections and the construction of pole figures, see, for example, Barrett and Massalski [*14*].

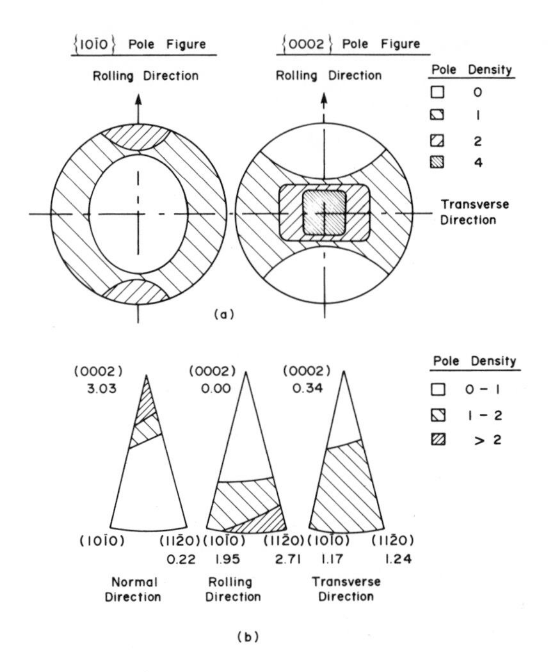

Fig. 4.45 Pole figures for rolled and annealed Zircaloy-4 (HCP). (a) Normal pole figures for {10$\bar{1}$0} and {0002} poles; (b) Inverse pole figures. The pole density values shown are relative values only

The development of texture in a material is important in that the mechanical properties vary with orientation in a particular grain, and hence in heavily textured material they can be anisotropic to a considerable degree. This effect can be used beneficially by arranging that the component is strongest in the direction of highest applied stress (see Section 2.5). Conversely, a processing sequence which develops an undesirable texture with the weakest orientation in the direction in which applied stress will be greatest must be avoided.

Textures in rolled or extruded sheet are frequently described in terms of "ideal orientations" which specify the plane most frequently lying in the plane of the sheet and the direction lying most frequently in the rolling direction. Hence the ideal texture might be described as (*hkl*) [*uvw*], and the scatter in orientations of grains from this ideal, as described by pole figures, ignored for convenience. Similarly, wire or rod "ideal orientations" may be described in terms of a particular orientation which lies along the axis in the majority of grains, these normally having rotational symmetry in the radial direction. Naturally, estimates of anisotropy in mechanical properties based on such simple models are subject to considerable error; however, they represent a good starting point in determining possible deviations from the properties of ideal randomly oriented material.

For a detailed discussion of the particular deformation textures occurring in specific materials, the reader is referred to the appropriate texts [*14, 60, 61*]. However, it may be noted in passing here, that an important factor which causes marked differences in texture between materials of the same crystal structure (e.g., FCC) is SFE, which, as seen previously, controls the ease of cross-slip. It may also be noted that HCP metals, with their limited slip systems, are particularly prone to the development of strong textures and to wide variations in mechanical properties with direction.

Finally, it should be noted that textures are not confined to cold-worked materials: they are also to be found in material which has been subsequently annealed, and such textures may be very strong indeed and may differ markedly from those existing before annealing [*14, 62*]. Of particular interest is the strong "cube" texture, (100) [001], developed in many FCC metals. BCC metals also may form strong annealing textures differing from those of deformation (e.g., the (100)[001] textures produced in transformer-grade silicon steel in which magnetization occurs most easily in the [001] direction), while, in general, the HCP metals retain their deformation textures after annealing [*14*].

REFERENCES

1. FRIEDEL, J., *Dislocations*, Pergamon Press, Oxford (1964).
2. COTTRELL, A. H., *Dislocations and Plastic Flow in Crystals*, Clarendon Press, Oxford (1953).
3. WEERTMAN, J., and WEERTMAN, J. R., *Elementary Dislocation Theory*, Macmillan, New York (1964).
4. HULL, D., *Introduction to Dislocations*, Pergamon Press, Oxford (1965).
5. HIRTH, J. P., and LOTHE, J., *Theory of Dislocations*, McGraw-Hill, New York (1968).
6. GITTUS, J. H., *Trans. ASME, J. Eng. Matls. and Technology*, **98**, 52 (1976).
7. COOK, T. S., RAU, C. A., and SMITH, E., *Trans. ASME, J. Eng. Matls. and Technology*, **98**, 180 (1976).
8. TOMKINS, B., *Trans. ASME, J. Eng. Matls. and Technology*, **97**, 289 (1975).
9. LARDNER, R. W., *Phil. Mag.*, **17**, 71 (1968).
10. DYSON, B. F., *Can. Met. Quarterly*, **13**, 237 (1974).
11. ROSEN, S. L., *Fundamental Principles of Polymeric Materials for Practicing Engineers*, Barnes and Noble, New York (1971).
12. GITTUS, J. H., *Creep, Viscoelasticity and Creep Fracture in Solids*, Halsted Press, Wiley, New York (1975).
13. HONEYCOMBE, R. W. K., *The Plastic Deformation of Metals*, Ed. Arnold, London (1968).

14. BARRETT, C. S., and MASSALSKI, T. B., *Structure of Metals*, 3rd Ed., McGraw-Hill, New York (1966).
15. CHURCHMAN, A. T., *Proc. Roy. Soc.*, **A226**, 216 (1954).
16. ROSI, F. D., DUBE, C. A., and ALEXANDER, B. H., *Journal of Metals*, **5**, 257 (1957).
17. TUER, G. L., and KAUFMANN, A. R., *The Metal Beryllium*, ASM, Metals Park, Ohio (1955).
18. LEE, H. T., and BRICK, R. M., *Trans. ASM*, **48**, 1003 (1956).
19. PRICE, P. B., in *Electron Microscopy and the Strength of Crystals*, p. 41, (G. Thomas and J. Washburn, eds.) Interscience, New York (1963).
20. BELL, R. L., and CAHN, R. W., *Proc. Roy. Soc.*, **A239**, 494 (1957).
21. SCHMID, E., and BOAS, W., *Kristallplastizität*, Springer Verlag, Berlin (1935).
22. MACKENZIE, J. K., Thesis, University of Bristol (1949).
23. TAYLOR, G. I., *Proc. Roy. Soc.*, **A145**, 362 (1934).
24. POLANYI, M., *Z. Physik*, **89**, 660 (1934).
25. OROWAN, E., *Z. Physik*, **89**, 605, 614, 634 (1934).
26. PEIERLS, R., *Proc. Phys. Soc.*, **52**, 34 (1940).
27. NABARRO, F. R. N., *Proc. Phys. Soc.*, **59**, 256 (1947).
28. McLEAN, D., *Grain Boundaries in Metals*, Clarendon Press, Oxford (1957).
29. HIRSCH, P. B., HORNE, R. W., and WHELAN, M. J., *Phil. Mag.*, **1**, 677 (1956).
30. KUHLMANN-WILSDORF, D., and WILSDORF, K. G. F., in *Electron Microscopy and the Strength of Crystals*, p. 575 (G. Thomas and J. Washburn, eds.) Interscience, New York (1963).
31. HIRSCH, P. G., HOWIE, A., NICHOLSON, R. B., PASHLEY, D. W., and WHELAN, M. J., *Electron Microscopy of Thin Crystals*, Butterworths, London (1965).
32. NABARRO, F. R. N., quoted by COTTRELL (Ref. [*2*]).
33. HEAD, A. K., *Phil. Mag.*, **44**, 223 (1953).
34. HARPER, S., and COTTRELL, A. H., *Proc. Phys. Soc.*, **63B**, 331 (1950).
35. PICKUS, M. R., and PARKER, E. R., *Trans. AIME*, **191**, 792 (1951).
36. DILLAMORE, I. L., and SMALLMAN, R. E., *Phil. Mag.*, **12**, 191 (1965).
37. McLEAN, D., *Mechanical Properties of Metals*, Wiley, New York (1962).
38. FRANK, F. C., and READ, W. T., *Phys. Rev.*, **79**, 722 (1950).
39. SMALLMAN, R. E., *Modern Physical Metallurgy*, Butterworths, London (1962).
40. HIRTH, J. P., *Met. Trans.*, **3**, 3045 (1972).
41. LOMER, W. M., and NYE, J. F., *Proc. Roy. Soc.*, **A212**, 576 (1952).
42. NICHOLSON, R. B., in *Grain Boundaries in Engineering Materials*, p. 669 (J. L. Walter, J. H. Westbrook, and D. A. Woodford, eds.) Claitor's, Baton Rouge, Louisiania (1975).
43. ESHELBY, J. D., FRANK, F. C., and NABARRO, F. R. N., *Phil. Mag.*, **42**, 351 (1951).
44. HALL, E. O., *Proc. Phys. Soc.*, **B64**, 742, 747 (1951).
45. PETCH, N. J., *JISI*, **173**, 25 (1953).
46. SEEGER, A., DIEHL, J., MADER, J., and REBSTOCK, K., *Phil. Mag.*, **42**, 351 (1957).
47. ANDRADE, E. N. DA C., and HENDERSON, D., *Phil. Trans. Roy. Soc.*, **A244**, 177 (1951).

48. McEvily, A. J., and Johnston, T. L., *Int. J. Fract. Mech.*, **3**, 45 (1967).
49. Orowan, E., *Nature*, **149**, 643 (1942).
50. von Mises, R., *Z. angew, Math. u. Mech.*, **8**, 161 (1929).
51. Sachs, G., *Z. VDI*, **72**, 734 (1928).
52. Kochendörfer, A., *Plastische Eigenschaften von Kristallen*, Springer, Berlin (1941).
53. Taylor, G. I., *Journal of the Institute of Metals*, **62**, 307 (1938).
54. Bishop, J. F. W., and Hill, R., *Phil. Mag.*, **42**, 414 (1951); **42**, 1298 (1951).
55. Thornburg, D. R., and Piehler, H. R., *Met. Trans.*, **6A**, 1511 (1975).
56. van Houtten, P., and Aernoudt, E., *Z. Metallkde.*, **66**, 202 (1975).
57. van den Beukel, A., *Phys. Stat. Solidi (a)*, **30**, 197 (1975).
58. Bauschinger, J., *Mitt. mech. tech. Lab. München*, 1 (1886).
59. Backofen, W. A., *Deformation Processing*, Addison-Welsey, Reading, Mass. (1972).
60. Underwood, F. A., *Textures in Metal Sheets*, Macdonald, London (1961).
61. Dillamore, I. L., and Roberts, W. T., *Met. Reviews*, **10**, 271 (1965).
62. *Recrystallization, Grain Growth and Textures*, ASM, Metals Park, Ohio (1966).

5

ANELASTICITY AND DAMPING IN METALS

5.1 INTRODUCTION

An important aspect of the deformation behavior of solids which has been ignored up to this point in the text is *anelastic* behavior. When a stress is applied to a solid, it will deform elastically if the stress is sufficiently low, while plastic or permanent deformation may be expected at higher stresses. However, solids deviate in two ways from ideal Hookean elastic behavior below the limiting value of elastic strain: first, recovery of elastic strain does not take place instantaneously on unloading; and second, the relationship between stress and strain is rate dependent. Thus, we do not have ideal elasticity in real solids, and they exhibit some degree of *anelasticity*, which may be defined as "the property of a solid in virtue of which stress and strain are not uniquely related in the preplastic range" [*1*].

Anelastic effects can have important consequences in situations where loading is repetitive or cyclic, as the phase lag which exists between stress and resultant strain causes energy absorption or *damping* in the material. Anelastic phenomena are important factors in the fatigue behavior of materials, as the internal damping of the materials themselves may serve to prevent the buildup of dangerously large resonance-induced stresses under the action of a small forcing stress [*2*]. Similarly, a material's ability to absorb, and hence not to transmit, vibration or noise within a system such as a gear train, is affected by its anelastic behavior or damping capacity. The good sound absorbing properties of gray cast iron because of its high damping capacity and its consequent use in components such as cylinder blocks in automotive applications, is a well-known example.

When a metal deforms by creep at elevated temperature, there is an anelastic or recoverable component of the creep strain, and this is of considerable importance in the practical situations in which load is decreased or reversed during creep. This matter is discussed further in Chapter 11, when dealing with dislocation models for creep deformation.

Internal friction measurements, which provide a quantitative evaluation of the anelastic behavior of a material, normally over a range of frequencies or temperature, or both, serve to supply much useful information concerning the internal structure of materials, including mobile dislocation density and the precipitation effects which take place during the aging of some alloys. In addition, such tests may be used to provide information relating to the activation energy for diffusion, or for the "viscosity" of grain boundaries, i.e., the ease and magnitude of grain boundary sliding [*1, 3*].

The stress-strain behavior of many non-metallic materials is markedly rate-dependent, much more so than for metals. Hence, *viscoelastic behavior*, which may be described as a generalization of elastic and viscous behavior, becomes of particular importance in considering how such materials behave under dynamic loading conditions. The formal mathematical models which are set out in the following section apply equally well to non-metals such as polymers and to metals, but the subsequent discussion of structural aspects and of application of the models is restricted to metallic materials only. A detailed discussion of the viscoelastic behavior of non-metals is provided by Gittus [*4*].

5.2 VISCOELASTIC MODELS

Basic models, composed of ideal linear elastic and linear viscous elements, have been used for many years to represent the behavior of real solids, and are a useful device in predicting the response of viscoelastic solids to dynamic stress or strain. The two basic mechanical components are a spring and a dashpot. The former represents a Hookean solid whose constitutive equation is $\tau_E = G\gamma_E$, where G is the shear modulus and is constant, while the latter represents a linear viscous or Newtonian fluid whose constitutive equation is $\tau_D = \eta\dot{\gamma}_D$, where η is a viscosity term and is again constant[1].

---

[1] In the discussion, shear stress and shear strain are used. These may be replaced by tensile stress and strain, with appropriate alteration to the modulus and viscosity terms.

In this section we shall consider only a few simple elements and models[2] under simple loading situations. For more detailed discussion of viscoelasticity the reader should refer to more specialized works, such as those of Bland [5] and Flügge [6].

5.2.1 The Maxwell Element

Figure 5.1 shows the Maxwell element, composed of a spring and dashpot in series. This represents a Newtonian liquid which exhibits an elastic component. The applied stress, τ, is carried by both spring and dashpot. Thus,

$$\tau = \tau_E = \tau_D \tag{5.1}$$

and the total strain in the element, γ, is the sum of the component strains, i.e.,

$$\gamma = \gamma_E + \gamma_D \tag{5.2}$$

Elastic strain is confined to the spring and is related to the applied stress by

$$\tau = G\gamma_E \tag{5.3}$$

while plastic deformation takes place entirely in the dashpot; thus

$$\tau = \eta\dot{\gamma}_D \tag{5.4}$$

Differentiating Eqs. (5.2) and (5.3) with respect to time, and substituting in Eq. (5.4), we obtain

$$\tau = \eta\dot{\gamma} - (\eta/G)\dot{\tau} \tag{5.5}$$

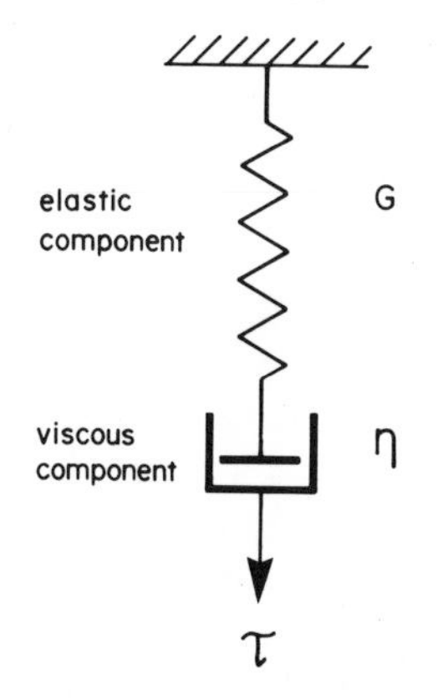

Fig. 5.1 The Maxwell element

[2] Following Bland [5], we use the term "model" only when considering a structure composed of more than two basic elements.

and the strain rate resulting from application of load is given by

$$\dot{\gamma} = \dot{\tau}/G + \tau/\eta \tag{5.6}$$

Integrating Eq. (5.6) from $t = 0$ to $t = t$,

$$\gamma = \tau/G + \int_0^t (\tau/\eta)\, dt \tag{5.7}$$

and, for τ = constant = τ_o, we have

$$\gamma = (\tau_o/G)(1 + Gt/\eta) \tag{5.8}$$

Thus, we see from Eqs. (5.6) and (5.8) that under constant stress the Maxwell element has a constant rate of deformation.

Consider the element subjected to an instantaneous strain γ_o, which is maintained constant. Only the spring can respond instantaneously, deforming so that at $t = 0$, $\gamma_o = \tau/G$. For $\gamma = \gamma_o$ = constant, we obtain from Eq. (5.6)

$$\dot{\tau} = -G\tau/\eta \tag{5.9}$$

which is rewritten as

$$\dot{\tau} = -\tau/\lambda \tag{5.10}$$

where $\lambda = \eta/G$ has the dimensions of time and is termed the *relaxation time*. The solution of Eq. (5.10), with the boundary condition $\tau = G\gamma_o$ at $t = 0$, is

$$\tau = G\gamma_o \exp(-t/\lambda) \tag{5.11}$$

Thus we see that the stress decays exponentially, and λ is the time required for the stress to decay by a factor of 1/e.

Figure 5.2 shows the response of the Maxwell element to constant stress and constant strain.

Now consider the element subjected to a sinusoidal stress. Substituting τ =

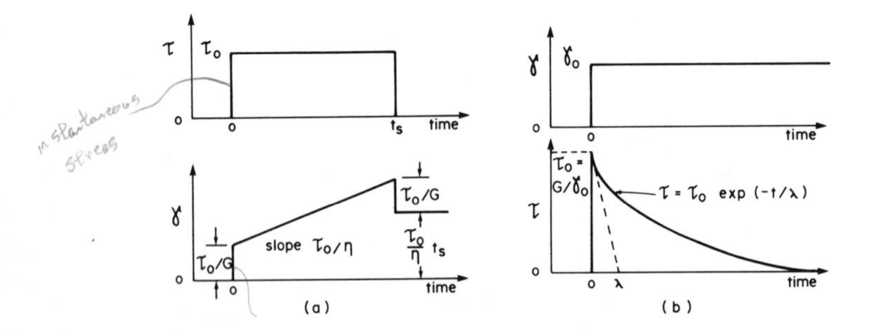

Fig. 5.2 Response of Maxwell element (a) to constant stress, (b) to constant strain

τ_o cos ωt in Eq. (5.7), we obtain

$$\gamma = (\tau_o/G) \cos \omega t + (\tau_o/\omega\eta) \sin \omega t + C \qquad (5.12)$$

where C is a constant. If $\gamma = 0$ at $t = 0$, the strain at time t is given by

$$\gamma = (\tau_o/G) [\cos \omega t + (1/\omega\lambda) \sin \omega t - 1] \qquad (5.13)$$

and this may be rearranged by adding the two vectors represented by the sine and cosine waves, these being perpendicular to each other, to give

$$\gamma = (\tau_o/G) [(1 + \omega^2\lambda^2)^{1/2} \cos(\omega t - \delta)/\omega\lambda - 1] \qquad (5.14)$$

where

$$\delta = \tan^{-1} (1/\omega\lambda)$$

Rearranging,

$$\gamma = (\tau_o/G) [\cos (\omega t - \delta)/\cos\delta - 1] \qquad (5.15)$$

Stress and strain are out of phase, and the phase difference is a function of frequency, decreasing as frequency increases. At zero frequency the element continues to creep owing to continued movement of the piston in the dashpot.

We may note here that because of the fact that applied stress and resultant strain are out of phase, energy is dissipated during each cycle, the magnitude of loss depending on the value of the phase lag. This matter is considered further in Section 5.3.

5.2.2 The Voigt or Kelvin Element

The Voigt, Voigt-Kelvin, or Kelvin element, to give it its alternative names, consists of a spring and dashpot in parallel, as shown in Fig. 5.3. The applied

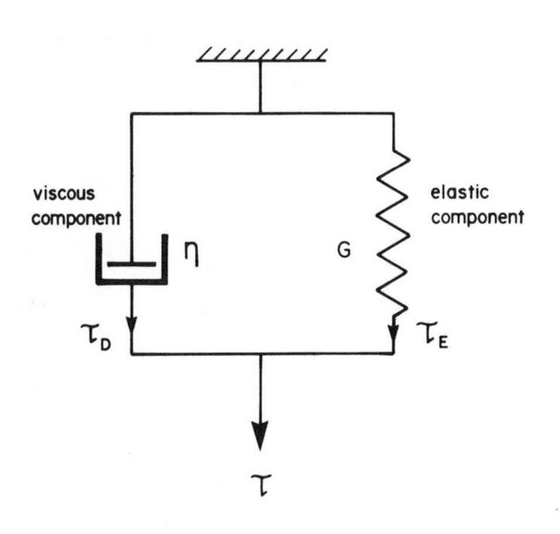

Fig. 5.3 The Voigt element

stress is given by the sum of the stresses in the two components, and their strains are equal. Hence,

$$\tau = \tau_E + \tau_D \tag{5.16}$$

and

$$\gamma = \gamma_E = \gamma_D \tag{5.17}$$

The differential equation for the Voigt element is

$$\tau = G\gamma + \eta\dot{\gamma} \tag{5.18}$$

and the general solution for Eq. (5.18), where τ is a function of time, is

$$\gamma = \exp(-t/\lambda)\,[\gamma_o + (1/\eta)\textstyle\int \tau \exp(t/\lambda)\,dt] \tag{5.19}$$

If the element is subjected to a constant stress of magnitude τ_o, we obtain from Eq. (5.19)

$$\gamma = \tau_o/G + (\gamma_o - \tau_o/G)\exp(-t/\lambda) \tag{5.20}$$

and for $\gamma = 0$ at $t = 0$ ($\gamma_o = 0$),

$$\gamma = (\tau_o/G)\,[1 - \exp(-t/\lambda)] \tag{5.21}$$

The behavior of the Voigt element to a steady stress is illustrated in Fig. 5.4.

If we attempt to apply an instantaneous strain, the stress required is infinite, because of the need to move the piston in the dashpot instantaneously.

Now consider the application of a sinusoidal stress $\tau = \tau_o \sin \omega t$. Substituting in Eq. (5.19) and integrating, we obtain

$$\gamma = \exp(-t/\lambda)\,[\gamma_o + (\tau_o/G(\omega^2\lambda^2 + 1)^{1/2})\exp(t/\lambda)\cos(\omega t - \delta) - \tau_o\,\omega\lambda/G\,(\omega^2\lambda^2 + 1)] \tag{5.22}$$

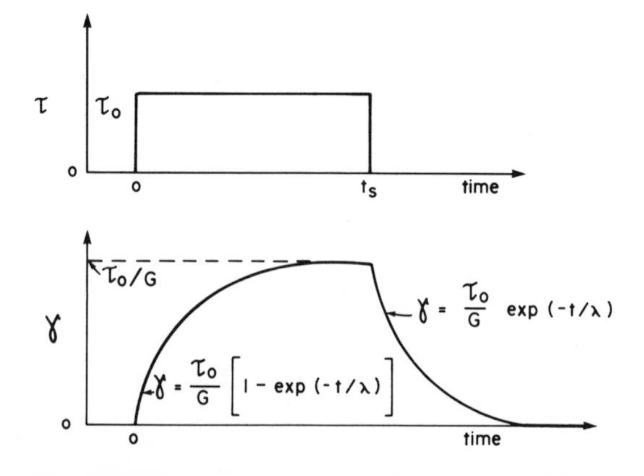

Fig. 5.4 Response of Voigt element to a steady stress

where

$$\tan \delta = -1/\omega\lambda$$

If the initial strain, γ_o, is zero at $t = 0$,

$$\gamma = \tau_o \left[\cos (\omega t - \delta) - \omega\lambda \exp (-t/\lambda)/(\omega^2\lambda^2 + 1)^{1/2}\right]/ G\, (\omega^2\lambda^2 + 1)^{1/2} \qquad (5.23)$$

In this case, the strain lags the stress by an amount which increases as frequency increases, while at zero frequency the element behaves elastically. In Eq. (5.23), the first term within the square brackets represents the steady state component of the cyclic strain, while the second represents the transient component of the response.

5.2.3 The Standard Linear Solid

Neither the Maxwell nor the Voigt elements can model the behavior of real materials in a fully realistic manner. A much closer representation of actual behavior is given by the *standard linear solid* which consists of a Maxwell element in parallel with a second spring, and this is illustrated in Fig. 5.5. It can model all the slow creep responses of a material with the exception of steady state creep, and on instantaneous loading or unloading there is an instantaneous change of strain, as observed in practice.

Defining the components of the Maxwell element by

$$\tau_1 = G_1\gamma_1 \text{ and } \tau_1 = \eta\dot{\gamma}_3$$

and the second spring by $\tau_2 = G_2\gamma_2$, the total stress on the solid is given by

$$\tau = \tau_1 + \tau_2 \qquad (5.24)$$

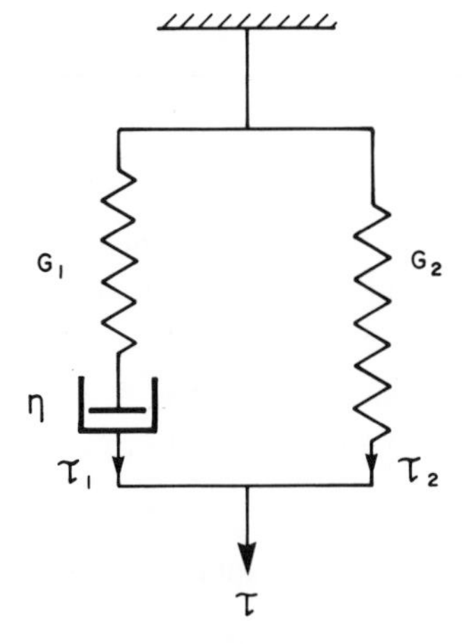

Fig. 5.5 The standard linear solid

and the total rate of strain is given by

$$\dot{\gamma} = \dot{\gamma}_1 + \dot{\gamma}_3 = \dot{\gamma}_2 \tag{5.25}$$

Thus,

$$\dot{\gamma} = \dot{\tau}_1 / G_1 + \tau_1 / \eta = \dot{\tau}_2 / G_2 \tag{5.26}$$

The general equation may be written in the form

$$\tau + \lambda_1 \dot{\tau} = G_2 \{\gamma + \eta[(G_1 + G_2)/G_1 G_2]\dot{\gamma}\} \tag{5.27}$$

where λ_1 is the relaxation time for the Maxwell element ($= \eta / G_1$).

From Eq. (5.27), we see that if an applied stress is maintained constant for an infinite time, the dashpot will relax completely, so transfering all load to the second spring in parallel with it. Thus, we define the *relaxed modulus*, G_R, as

$$G_R = \tau / \gamma = G_2 \tag{5.28}$$

If no time is allowed for relaxation, or if the frequency with which a stress is applied is very high, the *unrelaxed modulus*, G_U, may be defined as

$$G_U = G_1 + G_2 \tag{5.29}$$

We also define the relaxation time of the strain at constant stress, being the time it takes to attain its equilibrium value, as λ_2, and the solution of Eq. (5.27) corresponding to this situation where a stress τ_o was applied suddenly at $t = 0$ producing a strain γ_o, is

$$\gamma = \tau_o / G_R + (\gamma_o - \tau_o / G_R) \exp(-t / \lambda_2) \tag{5.30}$$

where $\lambda_2 = \lambda_1 \, G_U / G_R$.

Consider the model subjected to a sinusoidal stress. If the component of stress in the Maxwell element is $\tau_1 \cos \omega t$, the strain in the spring is $\gamma_1 = (\tau_1 / G) \cos \omega t$, and in the dashpot $\gamma_3 = (\tau_1 / \eta\omega) \sin \omega t$. Thus, for the model, the total strain is given by

$$\gamma = (\tau_1 / G) \cos \omega t + (\tau_1 / \eta\omega) \sin \omega t \tag{5.31}$$

The rotating vector diagram for the component stresses is shown in Fig. 5.6, and it is seen that the total strain in the model, γ, lags behind the applied stress, τ, which is the vector sum of τ_1 and τ_2, by the angle δ. This is given by

$$\tan \delta = \omega(\lambda_2 - \lambda_1)/(1 + \omega^2 \lambda_1 \lambda_2) \tag{5.32}$$

If we now introduce the geometric mean of the two relaxation times,

$$\bar{\lambda} = (\lambda_1 \lambda_2)^{1/2} \tag{5.33}$$

and the geometric mean of the two moduli,

$$\bar{G} = (G_U G_R)^{1/2} \tag{5.34}$$

we can rearrange Eq. (5.32) in the form

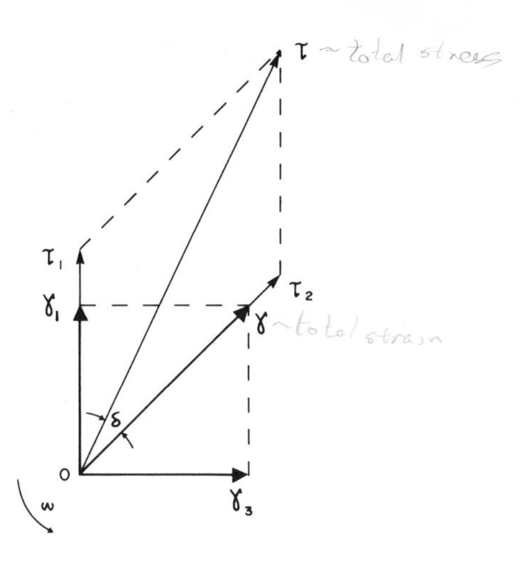

Fig. 5.6 Rotating vector diagram for component stresses and strains in standard linear solid subjected to a sinusoidal stress

$$\tan \delta = [(G_U - G_R)/\bar{G}]\, \omega\bar{\lambda}/[1 + (\omega\bar{\lambda})^2] \tag{5.35}$$

making use of the relation $\lambda_2/\lambda_1 = G_U/G_R$. Differentiating Eq. (5.35) and setting it equal to zero, we obtain the maximum value of $\tan \delta$ when $\omega\bar{\lambda}$ is unity.

The ratio of the amplitudes of stress and strain for the standard linear solid when under cyclic stressing is given by

$$\tau/\gamma = G_R\,[(1 + \omega^2\lambda_2^2)/(1 + \omega^2\lambda_1^2)]^{1/2} \tag{5.36}$$

and when $\omega = 1/\bar{\lambda}$,

$$\tau/\gamma = G_R(G_U/G_R)^{1/2} = \bar{G} \tag{5.37}$$

Thus, when the phase difference between stress and strain is a maximum, the ratio of the stress and strain amplitudes is given by $\bar{G}$.

The dependence on frequency of the phase angle, δ, and the modulus of the standard linear solid are shown in Fig. 5.7, and it should be noted that there is the significant difference between this model and the Maxwell and Voigt elements that in neither of the latter does the phase angle pass through a maximum value as the frequency of stressing is varied.

5.3 INTERNAL FRICTION

When the applied stress and resultant strain are out of phase, there is internal damping or *internal friction* absorbing energy during a stress cycle.

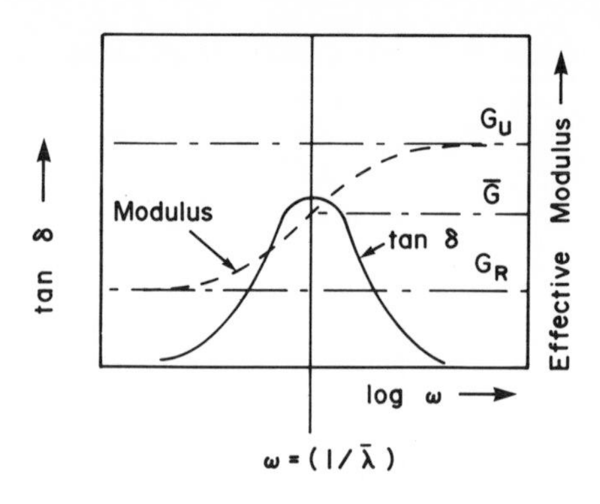

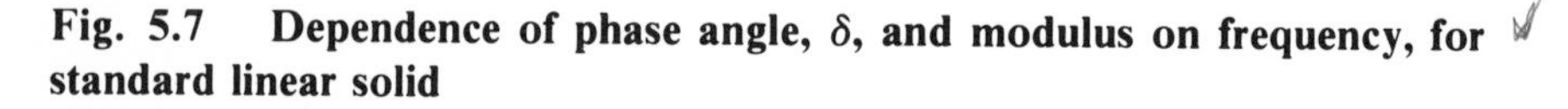

Fig. 5.7 Dependence of phase angle, δ, and modulus on frequency, for standard linear solid

Obviously, there may be external frictional effects such as air damping, but in the present discussion we shall assume these to be negligible and that all damping is internal.

The amount of damping or the magnitude of internal friction may be measured in a number of ways, one such measure being the phase angle δ.

Consider a periodic stress applied to a damped system. The stress, equal to $\tau_o \cos \omega t$, will produce a periodic strain, $\gamma_o \sin(\omega t - \delta)$, and the stress-strain curve will have the form of an elliptical closed loop. Hence the work done per cycle, or the energy absorbed per cycle, is

$$\Delta E = \int_0^{2\pi/\omega} \tau_o (\cos \omega t)\, \gamma_o\, \omega \sin(\omega t - \delta)\, dt$$

$$= \pi \tau_o \gamma_o \sin \delta \qquad (5.38)$$

For small angles, $\sin \delta \simeq \delta$: hence the energy loss per cycle is directly proportional to the phase angle δ, to a first approximation. The maximum energy of the system is the energy at the maximum value of strain, and is given by

$$E = (\gamma_o \tau_o \cos \delta)/2 \qquad (5.39)$$

Hence, the *relative damping*, is

$$\Delta E / E = 2\pi \tan\delta \simeq 2\pi\, \delta \qquad (5.40)$$

In a freely vibrating system, the decay in amplitude is another measure of internal friction. In such a system the amplitude of the successive oscillations decays according to

$$a = a_o \exp(-\beta t) \qquad (5.41)$$

where β is a constant depending on the characteristics of the system. Hence, the *logarithmic decrement* or the logarithm of the ratio of successive amplitudes is

$$\Lambda = \ln (a_1/a_2) \tag{5.42}$$

For a small degree of damping

$$a_2/a_1 = \exp(-\Lambda) \simeq 1 - \Lambda \tag{5.43}$$

Hence,

$$\Lambda \simeq \Delta a/a \tag{5.44}$$

We may note that the energy of vibration is proportional to the square of the amplitude. Putting $r = a_1/a_2 = \exp(\Lambda)$, we have

$$E_1/E_2 = r^2 = (a_1/a_2)^2$$

Thus

$$\Delta E/E_2 = (E_1 - E_2)/E_2 = r^2 - 1$$

or

$$\Delta E/E = \exp(2\Lambda) - 1 \simeq 2\Lambda$$

and, from Eq. (5.40),

$$\Delta E/E = 2\Lambda = 2\pi\delta \tag{5.45}$$

Another method of specifying damping in a system is in terms of the *bandwidth*, Q^{-1} which is defined as $\Delta f/f_o$, where f_o is the resonant frequency of the system ($= \omega_o/2\pi$) and $\Delta f = f_1 - f_2$, where f_1 and f_2 are the frequencies at which the value of γ falls to $\gamma_{max}/\sqrt{2}$ (see Fig. 5.8). The differential equation for the forced vibration of a damped mechanical system has the form

$$I\ddot{\gamma} + \eta\dot{\gamma} + G\gamma = \tau_o \cos \omega t \tag{5.46}$$

where I is the inertia. Solution of this for the condition where the strain amplitude reaches a maximum gives us the value of the bandwidth

$$Q^{-1} = \eta/\omega_o I \tag{5.47}$$

We see that the bandwidth is a measure of the damping coefficient, η.

From solution of the differential equation for the free-oscillation of the damped system discussed above, we can evaluate the logarithmic decrement in the form

$$\Lambda = \pi\eta/\omega_o I \tag{5.48}$$

Hence, from Eqs. (5.47) and (5.48)

$$\Lambda = \pi Q^{-1} \tag{5.49}$$

The various measures of damping or internal friction occurring in a standard linear solid may be related as follows:

$$\delta = (1/2\pi)(\Delta E/E) = \Lambda/\pi = Q^{-1} \tag{5.50}$$

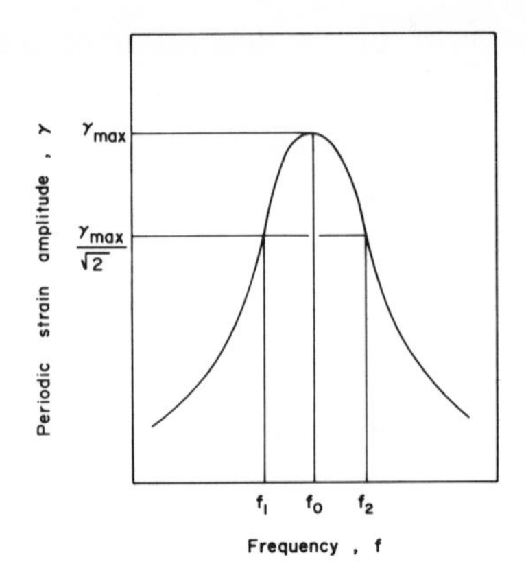

Fig. 5.8 Frequency dependence of amplitude, showing resonance peak at f_0

The quantities in Eq. (5.50) depend on frequency, but are independent of amplitude. The identities are approximately correct for a small degree of damping, as they were derived on this assumption: they may be seriously in error where damping is large.

In real materials, there are many different sources of damping which may occur, and each one has a characteristic frequency dependence. Our model of a standard linear solid is grossly simplified, and a real material might be better thought of as an interconnected network of spring and dashpot elements, with a spectrum of damping peaks. Figure 5.9 shows the form of such a series of peaks, and at any specific frequency the equalities given in Eq. (5.50) hold true for any linear viscoelastic material [7].

5.3.1 Mechanisms of Internal Friction in Metals

Many separate mechanisms give rise to internal friction in metals, and a number of these are shown on Fig. 5.9: each mechanism produces a characteristic internal friction or relaxation peak which, for a specified temperature, occurs at a particular frequency. As a corollary, for a constant frequency, there is a particular temperature at which internal friction is a maximum, for each mechanism. It is worth noting also that anelastic processes cause changes in the modulus of elasticity of a metal, as is apparent from consideration of the behavior of the standard linear solid subjected to a sinusoidal stress as illustrated in Fig. 5.7.

Point defects such as interstitials and substitutional atoms can produce internal friction effects. The best known is probably the Snoek effect [8] originally discovered for carbon lying interstitially in α-iron. Interstitials lie in

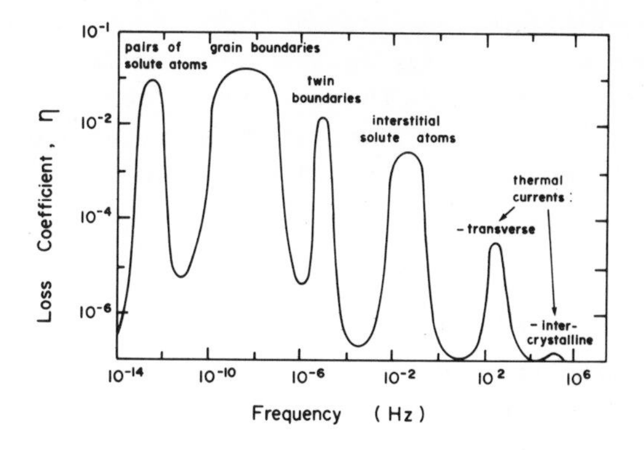

Fig. 5.9 Schematic diagram of a typical spectrum of damping peaks. (After Zener [1])

sites at the centers of the cell edges and the centers of the cell faces in the BCC iron unit cell. Their presence causes the unit cell to be distorted, extending it along the direction in which the interstitial lies and contracting it in the transverse directions. In the absence of an applied stress the interstitials are distributed uniformly in all three directions, but when a stress is applied in a specific direction, the interstitials rearrange themselves to lie preferentially along the newly extended direction, reducing internal energy and giving rise to additional deformation. If the stress is applied at high frequency, virtually no jumping of interstitials can occur: if it is applied at very low frequency, the rearrangement can take place almost completely. In both cases stress and strain are in phase and no damping occurs. However, at intermediate frequencies they will be out of phase, and the value of the frequency at the peak of the internal friction curve provides a measure of the jump frequency, while the magnitude of the peak is indicative of the number of interstitials present. The diffusion rate depends on temperature according to the Arrhenius equation

$$D = D_o \exp(-Q/RT)$$

where D is the diffusivity, Q the activation energy, and R the universal gas constant, and we can consider the diffusion rate equation in terms of relaxation times in the form

$$\lambda = \lambda_o \exp(Q/RT) \qquad (5.51)$$

where λ_o is a constant.

Up to now we have considered λ as being fixed, and damping, which is a function of $\omega\lambda$, was regarded as a function of ω only. But we see now that damping is a function of $[\omega \exp(Q/RT)]$. Hence we can obtain the same type of damping response to temperature for a fixed frequency, as is obtained when frequency is varied at fixed temperature.

Consider two damping curves obtained at two separate frequencies f_1 and f_2, over a range of temperatures. The peak damping occurs at temperatures T_1 and T_2, respectively. The *function* of temperature and frequency for peak damping is the same in both cases, and we may write

$$\ln f_1 + Q/RT_1 = \ln f_2 + Q/RT_2$$

or

$$(Q/R)(1/T_1 - 1/T_2) = \ln (f_2/f_1) \tag{5.52}$$

Thus, if we let $\Delta(1/T)$ represent the displacement of the two peaks plotted on the basis of inverse temperature,

$$(Q/R)\,\Delta\,(1/T) = \ln(f_2/f_1) \tag{5.53}$$

The activation energy, Q, can be determined by measuring the displacement $\Delta(1/T)$ for the two peaks obtained at specified frequencies. The method is particularly useful for the measurement of activation energies at low temperatures where the diffusion rate is low.

Other point defect relaxation peaks may be caused by the presence of vacancy pairs, impurity pairs, or an impurity associated with a vacancy, all of which can constitute elastic dipoles, lower the local symmetry of the lattice, and give rise to anelastic deformation [9].

Grain boundaries, being relatively disordered, have viscous properties to some degree, and from comparison of tests made on polycrystalline and single crystal specimens, it has been demonstrated that they give rise to a relaxation peak. The work of Kê [10] has been particularly significant in this area.

Grains are considered to slide over each other to some extent, with a viscosity, η, defined as

$$\eta = \tau t/v \tag{5.54}$$

where v is the velocity of the two crystals sliding on each side of a boundary of thickness t under the action of the shear stress τ. The anelastic nature of the strain arises because a stress concentration builds up at the junctions where several grain boundaries meet, and this restores the sliding areas of the boundaries in time after removal of the external stress. Generally, Kê-type tests using torsional oscillation are used to evaluate η, and the value obtained at a specified temperature can be extrapolated to the melting point using the relation

$$\eta_T = \eta_o \exp(-Q/RT) \tag{5.55}$$

where Q is the activation energy for grain boundary sliding. The values of viscosity obtained at the melting point by extrapolation are close to the measured values of viscosity of the liquid near to the melting point, supporting the idea of an essentially "liquid" nature to the high-angle grain boundaries [10].

Stress-induced ordering effects can also cause a relaxation peak, and internal friction is also modified by aging effects taking place in a precipitation hardening alloy. Thus, such tests provide a sensitive and non-destructive means of following aging phenomena [*11, 12*].

Anelastic effects and consequent relaxation peaks are also associated with electronic effects occurring at very low temperatures at frequencies in the MHz range [*7*], and with thermoelastic effects. The latter may occur on a macro or a micro scale and are due to temperature gradients produced when stressing is non-uniform or when the orientation differences between grains give rise to differing mechanical and thermal properties in the direction of the principal stresses in the material. If deformation is non-uniform and adiabatic (rapid loading), then temperature differences will be produced which produce heat flow unless the load is released rapidly. For high frequency loading, little heat flows and the damping effect is small: at very low frequency, the conditions are essentially isothermal and, again, damping is small. However, at some intermediate frequency where the period of stressing is of the same order as the time for heat flow across the specimen, significant energy loss may occur. On a micro scale, heat may flow across grain boundaries because of the local temperature variations produced in a nominally uniform stress field, and again a relaxation peak may be found at a particular frequency [*1*].

Anelastic damping may also be produced by eddy-currents generated in ferromagnetic materials subjected to cyclic stress, but the magnitude of the effect is small [*7*]. However, magnetoelastic static hysteresis, caused by the motion of domain boundaries in a stressed crystal, is an important effect in dissipating energy in high damping capacity materials. As it does not give rise to a relaxation peak as a function of frequency, it is not anelastic in nature [*7*].

An important group of internal friction effects relate to dislocations in metals, and we may group these effects under the general heading *dislocation damping*. When an ultrasonic plane wave passes through a solid containing no defects, it is attenuated very little: however, when dislocations are present, attenuation takes place and its magnitude depends on the mobility of the dislocations, this in turn depending on their interaction with other dislocations, point defects and phonons. If the dislocations are immobile, no attenuation occurs; thus, internal friction measurements — generally using ultrasonic techniques — are an extremely useful means of studying *mobile* dislocations. As these are the dislocations which contribute to creep deformation, the technique has important implications for the study of dislocation creep mechanisms which are discussed in Section 11.2.1.

The most generally accepted model for dislocation damping is that due to Granato and Lücke [*13*] and illustrated in Fig. 5.10. This is able to account successfully for both amplitude-independent dislocation damping as observed at low stress amplitudes and amplitude-dependent damping as observed at high stresses. Dislocations are considered as strings in a viscous medium, and the equation of motion can be written as [*14*]

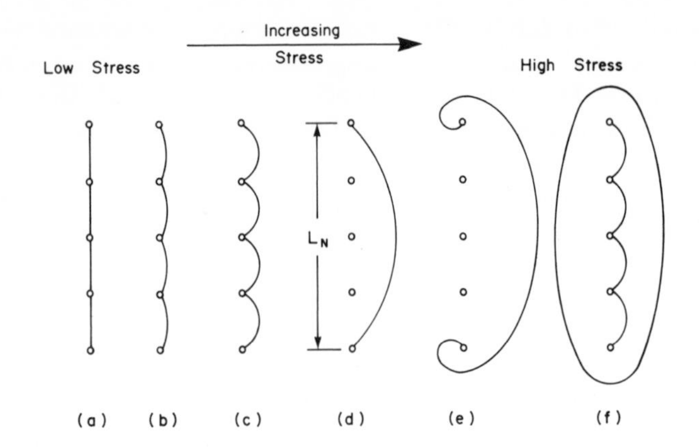

Fig. 5.10 Granato and Lücke model for dislocation damping

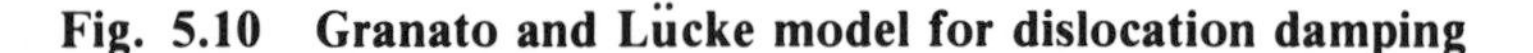

$$m\partial^2 y/\partial t^2 + \eta\partial y/\partial t - T_L\partial^2 y/\partial x^2 = b\sigma \quad (5.56)$$

where m is the effective mass per unit length, y is the lateral displacement measured from the equilibrium position at a distance x along the dislocation, η is a viscous damping constant, T_L is tension, b is Burgers vector and σ is the applied stress. Solution of Eq. (5.56) with the boundary conditions $y = 0$ at $x = 0$ and $x = l$, these being pinning points, gives y as a function of the frequency of the applied stress. Hence, the decrement can be evaluated and is found to be directly proportional to frequency, and the damping versus frequency curve contains a typical resonance peak. The dislocation is moving back and forth under the applied stress, dissipating energy because of the lattice friction.

The amplitude-dependent damping may be interpreted on the basis of the breakaway of the dislocation from the soft pinning points as shown in Figs. 5.10(d) and 5.10(e), while it is still held at the nodes or hard pinning points, L_N apart. The stress-strain curve for the dislocation is shown in Fig. 5.11, and it is seen that when the applied stress becomes large there is a sudden large increase in strain at breakaway. On removal of the stress, the strain returns to zero and the resulting hysteresis loss is shaded in Fig. 5.11. Thus, there is a loss which is dependent on strain amplitude and independent of frequency.

At higher stresses still, the hard pinning points act to facilitate the generation of new dislocations by the Frank-Read process and the magnitude of damping and degree of non-linearity increase sharply, and we may reach a situation characteristic of fatigue damage with plastic deformation occurring during each stress cycle.

Prior plastic deformation causes the appearance of additional damping peaks. The Bordoni peak occurs in FCC metals at a temperature within the

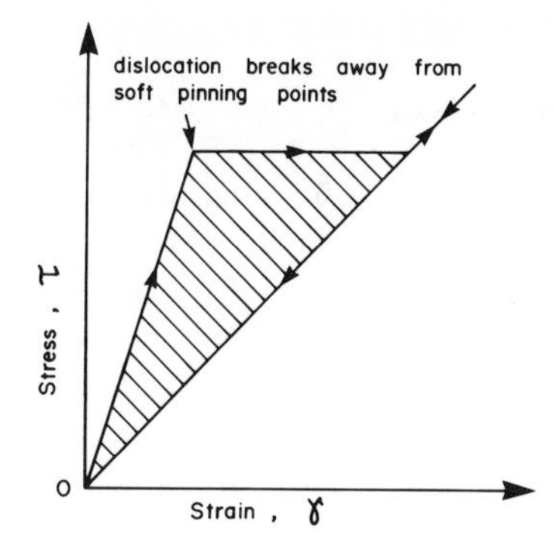

Fig. 5.11 Stress-strain curve for dislocation shown in Fig. 5.10

range 50-100 K after small amounts of plastic deformation and, although several theories have been proposed, the most commonly accepted is that a dislocation segment in a crystal moves under stress from its position in a "Peierls well" to lie in the next well with the formation of a pair of kinks [*15*]: the suggested mechanism is shown in Fig. 5.12. After complete recrystallization, Bordoni peaks disappear.

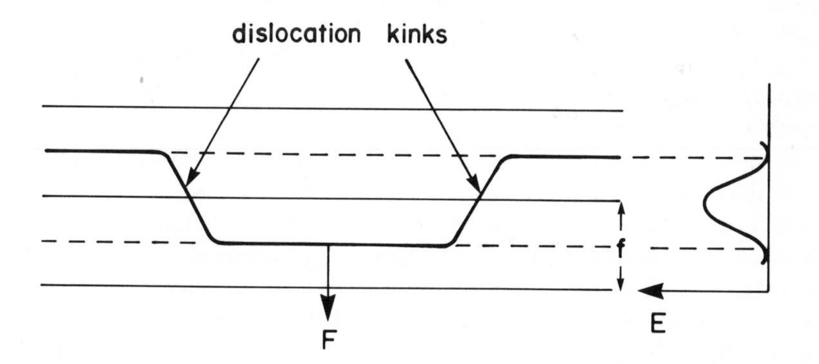

Fig. 5.12 Dislocation segment having moved to an adjacent Peierls well with the formation of two kinks

Hasiguti peaks [*16*] are also found in FCC metals just below room temperature after plastic deformation. They anneal out very easily, unlike the Bordoni peaks, and it has been suggested that dislocations pinned at nodes and point defects may be unpinned by a combination of stress and thermal effects [*17*].

5.4 DAMPING IN ENGINEERING ALLOYS

Damping in engineering alloys is of importance in the suppression of noise and in the extension of fatigue life by reducing the amplitude of vibration under resonant conditions of forcing. Generally, in an engineering application the strains are larger than those which correspond to the anelastic effects discussed previously, at least in those alloys specifically intended to have a large damping effect, although the anelastic micromechanisms do make an important contribution to damping. In addition to "internal" damping mechanisms, many engineering systems utilize "external" damping, depending on slip at a joint in a structure or at the interfaces between particles or agglomerates. However, the present brief discussion is concerned only with internal damping within a continuous member of essentially uniform material.

In many ways the subject is a neglected one both from the point of view of the practicing engineer and in terms of research activity, and much of the collected data on specific materials, together with a detailed exposition of damping phenomena, are to be found in the work of Lazan [*7*].

The most significant mechanisms of internal damping in high-damping alloys are reported to be plastic deformation, strain-induced reorientation of the crystal lattice, and magnetoelastic hysteresis referred to briefly in the preceding section [*18*]. The first is achieved through irreversible damage to the component which may lead to failure by fatigue, despite the reduction in vibration; the second is of importance in certain alloys such as manganese-copper ones which transform from an FCC solid to a face-centered tetragonal structure by reversible diffusionless shear; while the third is of importance in a number of ferromagnetic alloys, including manganese-copper, cobalt, nickel and iron base alloys.

In specifying the damping capacity of materials, it is important to define several quantities, and the terminology of Lazan [*7*] is followed. First, it should be noted that a particular test may not give the damping properties of the *material* itself, but may give the damping properties of the *system* used in making the test, or the damping properties of the *member* being tested. In many cases the published information regarding damping properties is not sufficiently detailed with regard to test methods and specimens for the damping properties of the material to be specified with any degree of accuracy, and this must be taken account of in making use of published data. In his extensive data compilation, Lazan [*7*] specified whether the values listed are for a member or the material, and also provides available information concerning test conditions.

The *damping energy* of a material (or a member) is defined as the energy absorbed per cycle by unit volume of the material (or by the entire member;

system, model, etc.). Thus the damping energy of a material, D, is

$$D = \oint \tau d\gamma$$

or

$$D = \oint \sigma d\epsilon \qquad (5.57)$$

the stress and strain measurements depending on the loading mode. The strain energy, U, is defined as the area under the $\tau_{mid} - \gamma$ or $\sigma_{mid} - \epsilon$ curve from 0 to maximum strain, so that the strain energy per cycle is $2\pi U$. Hence, the damping properties of a material may be defined in terms of the *loss coefficient*, η, as

$$\eta = D/2\pi U \qquad (5.58)$$

The bandwidth, Q^{-1}, is identical to the loss coefficient, η, and this may be related to the other measures of damping by Eq. (5.50), although for the larger values of δ found in engineering materials and systems under large amplitudes of strain, we should write $\eta = \tan \delta$.

Another term used frequently in earlier papers was *specific damping capacity*, ψ. This is defined as D/U, so is equal to $2\pi\eta$.

For linear materials, the loss coefficient for the member, η_S, is equal to η, and the bandwidth Q_S^{-1} (or the *quality factor*, Q_S) for the system equals Q^{-1} (or Q) for the member. For nonlinear materials the differences between the values of material and member may be substantial.

Damping in engineering materials is not very much affected by frequency, but the damping energy is affected by stress amplitude. Generally this can be written as

$$D = J\sigma_a^n \qquad (5.59)$$

where J and n are material constants and σ_a is the stress amplitude: J is termed the *damping constant* and n the *damping exponent*. In metals at low stress amplitudes, n is approximately 2 in many cases, and the relative energy units are independent of stress. However, at high stress levels, nonlinear behavior is sometimes observed, and n may exceed 15. The relative damping units may become very dependent on stress or strain amplitude, and it is found preferable under these conditions to express damping properties in terms of unit damping energy, D, rather than relative damping units [7].

Rather than provide a small selection of data for the damping properties of specific alloys, the reader in search of design data is recommended to consult the detailed list compiled by Lazan [7]. Again, the inadequacy of sufficient well-controlled studies in this area must be emphasized, and the matter is of particular importance when the ever more stringent requirements on noise control are considered.

REFERENCES

1. ZENER, C., *Elasticity and Anelasticity of Metals*, University of Chicago Press, Chicago (1948).
2. KENNEDY, A. J., *Processes of Creep and Fatigue in Metals*, Oliver and Boyd, Edinburgh (1962).
3. McLEAN, D., *Grain Boundaries in Metals*, Clarendon Press, Oxford (1957).
4. GITTUS, J. H., *Creep, Viscoelasticity and Creep Fracture in Metals*, Halsted Press, Wiley, New York (1975).
5. BLAND, D. R., *The Theory of Linear Viscoelasticity*, Pergamon, Oxford (1960).
6. FLÜGGE, W., *Viscoelasticity*, 2nd Ed., Springer-Verlag, New York (1975).
7. LAZAN, B. J., *Damping of Materials and Members in Structural Mechanics*, Pergamon, Oxford (1968).
8. SNOEK, J. L., *Physica,* **8**, 711 (1941).
9. NOWICK, A. S., *Internal Friction, Damping, and Cyclic Plasticity*, STP 378, p. 21, ASTM, Philadelphia (1965).
10. KÊ, T. S., *Phys. Rev.,* **71**, 533 (1947).
11. BERRY, B. S., and NOWICK, A. S., NACA Tech. Note 4225 (1958).
12. KRISHNADEV, M. R., DELAGE, L., GALIBOIS, A., and LE MAY, I., *Metallography,* **6**, 425 (1973).
13. GRANATO, A., and LÜCKE, K., *J. Appl. Phys.,* **27**, 583, 789 (1956).
14. GRANATO, A., *Internal Friction, Damping, and Cyclic Plasticity,* STP 378, p. 88, ASTM, Philadelphia (1965).
15. SEEGER, A., *Phil. Mag.,* **1**, 651 (1956).
16. KOIWA, M., and HASIGUTI, R. R., *Acta Met.,* **11**, 1215 (1963).
17. KOIWA, M., and HASIGUTI, R. R., *Acta Met.,* **13**, 1219 (1965).
18. BIRCHON, D., *The Engineer*, **222**, 207 (1966).

6

STRENGTHENING MECHANISMS AND STRENGTH-STRUCTURE RELATIONS

6.1 INTRODUCTION

In this chapter the various strengthening mechanisms of importance for metals and alloys are reviewed, first in a qualitative manner, and subsequently in terms of the quantitative relationships which may be derived between strength and microstructural features. While an understanding of the qualitative relationships between strength and microstructure is sufficient in many cases, there is currently considerable interest in the development of detailed quantitative relationships in order to design alloys, and the thermomechanical processing of them, to have optimum and specified properties. For complete specification of optimum processing conditions and microstructure, it is of course necessary to consider properties other than strength alone, including ductility, toughness, fatigue resistance, etc., but to do this would require a much more extensive coverage than is possible here. However, it is hoped that the principles of alloy design procedures, at least, will become clear from Sections 6.2 and 6.3. It should be emphasized that a full understanding of such relationships is essential if micromechanical modeling of the processes of deformation and fracture is to be conducted satisfactorily.

In addition to allowing one to make some quantitative prediction of properties on the basis of microstructural features, consideration of the quantitative relationships between structure and properties enables the physical nature of the various strengthening mechanisms to be understood more fully.

Following on from this, some of the processing variables that can be employed to produce high strength materials on the basis of the mechanisms available are discussed in Section 6.4. While most of the commercial processes are qualitatively based, consideration may be given to optimizing processing on the basis of quantitative evaluation of the various contributions to strength.

In the final section of the chapter, the mechanisms of radiation damage are discussed, together with the accompanying changes taking place in the mechanical properties of metals and alloys. Radiation damage and resultant hardening are topics of considerable importance to many engineers and metallurgists in connection with the general expansion in nuclear power productive capacity which is taking place today. While it might reasonably be argued that a complete and separate chapter should have been devoted to this topic, it is considered more appropriate to discuss the basic effects here, as they are closely related to the strengthening mechanisms which depend on the introduction of defects into the crystal lattice.

6.2 QUALITATIVE REVIEW OF STRENGTHENING MECHANISMS

In this discussion the term "strength" is taken generally as a measure of a material's resistance to plastic flow, i.e., of the stress required to produce some measureable and specified permanent deformation. It may also be used at times to mean the ultimate tensile strength (UTS), compressive failure strength, or fracture strength of a material, and these are of interest to the engineer and metallurgist in connection with solids of high strength but limited ductility, as well as in the specification of many engineering alloys in which the UTS may be given as a specified minimum or range, or where the minimum ratio of yield to ultimate is specified, as is frequently done, for example, in linepipe steels. In any event, this chapter is concerned primarily with methods of raising yield strength, and these are important in strong (brittle) solids as well as in the more conventional (relatively) ductile engineering materials, as the former must not be able to deform plastically before reaching their required high strength.

All strengthening mechanisms have this in common: they are required to reduce dislocation mobility and to increase the stress necessary to move a dislocation through an appreciable distance within the material. At high temperatures additional factors may become important, such as the inhibiting of grain boundary sliding, and these are discussed in Chapter 11 when dealing with creep deformation.

The principal methods of strengthening crystals are as follows:

1) Solid solution strengthening;
2) Precipitation or dispersion strengthening;
3) Grain size refinement;
4) Substructure formation;
5) Strain (or work) hardening;
6) Dispersion of a second phase (including reinforcement with strong fibers).

In practice, several strengthening mechanisms are frequently combined in a commercial alloy, and complex thermal and mechanical treatments may be conducted in order to increase strength. However, the resulting strength increment over that of the base metal can be partitioned between the mechanisms listed.

These methods of strengthening, together with processing procedures designed to produce high strength, are covered in the excellent volume complied by Kelly and Nicholson [*1*], which must be regarded as an essential and basic reference for this subject. Here, we shall consider the qualitative aspects of each strengthening process in turn, except that fibrous composite materials are dealt with separately in Chapter 7.

Solid solution or matrix strengthening by alloying is an effective way to increase the resistance of a metal to the motion of a dislocation. Pure metals have low yield stresses, and the addition of either substitutional or interstitial atoms to the lattice causes a stress field to be set up around each impurity atom: these stress fields interact with those surrounding dislocations as discussed in Chapter 4, causing the latter's motion to be impeded. If the solute atoms were randomly dispersed, the strengthening effect would be small, as, for a straight dislocation and random solute distribution, there would be a balance between the positive and negative forces exerted by the solute atoms. In practice this does not take place; (a) because there is never a completely random distribution, and (b) because a dislocation line will bend under the influence of attractive and repulsive forces. The magnitude of the strengthening effect is a function of the binding energy between the solute atom and the dislocation, and of the solute concentration. Thus it is limited by the amount of solute which can be maintained in solution in the matrix.

When a crystal structure is ordered, additional work must be done in passing a dislocation through it to supply the energy required by the resultant production of disorder. *Short range ordering* provides an increase in flow stress, but *long range ordering* can lead to a reduction in this strengthening, as the *superlattice dislocations* move in pairs or groups, and passage of a second dislocation will restore the ordering destroyed by the first, thus moving under a much reduced stress. In such alloys, which are generally brittle, strength is highest at an intermediate degree of ordering [*2*].

The presence of vacancies produces local distortions in the crystal lattice similar to those arising from substitutional atoms. Hence, an alloy with an excess of quenched-in vacancies or vacancies produced by irradiation, will have an increased strength over the same alloy containing its equilibrium concentration of vacancies. This is of great technological importance in the nuclear engineering field, particularly as with such potentially useful increases in strength occurring in service, there is a concomitant (and undesirable) decrease in ductility. This matter is discussed further in Section 6.5.

One of the technologically most important strengthening mechanisms is *precipitation* or *dispersion hardening*. We distinguish between the two terms by considering the way in which a distribution of fine second phase particles is introduced into the matrix. In a precipitation hardened alloy, the particles were originally in solid solution and have precipitated owing to a thermal treatment and from consideration of phase equilibrium: in a dispersion hardened alloy, the particles are introduced mechanically by dispersing them in the melt; by causing chemical reaction to take place in the melt, so producing insoluble particles; or by sintering a metallic powder containing an oxide dispersion. Common examples of dispersion hardened alloys are thoria-dispersed (TD) nickel and cobalt alloys, and sintered aluminum powder (SAP). Here, we shall first examine the precipitation (or age) hardening process, before discussing the strengthening mechanisms in terms of dislocation-particle interaction, this being essentially the same for either precipitation or dispersion hardened materials.

Age hardening was discovered accidently by Wilm [*3*], who found that the hardness of a quenched aluminum alloy increased with time. Martin [*4*] has reviewed the historical developments as well as the mechanisms of aging and the property changes arising from it, and his useful text also provides substantial extracts from the classical papers on this subject. Here, we may note that the first essential for age hardening is that there be a decreasing solubility of a (generally hard) constituent with decrease in temperature, although other factors do enter into the picture, it being important, for example, that the precipitate formed be coherent with the matrix in the first instance. We can show from classical nucleation theory that the fineness of the precipitate formed on aging a supersaturated solid solution obtained by rapid cooling is dependent on aging temperature, the precipitation being finer the lower the aging temperature, as indicated in Fig. 6.1.

The decomposition of a supersaturated solid solution frequently involves several stages, a typical sequence on aging being:

supersaturated solid solution → zones →
intermediate precipitate → equilibrium precipitate.

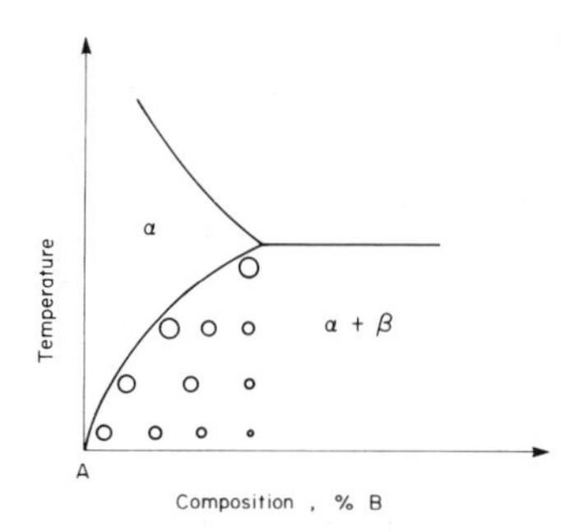

Fig. 6.1 Schematic showing precipitate size variation with aging temperature

The zones are essentially solute-rich clusters within the matrix lattice, and are fully coherent with this, while the intermediate precipitate may be partially coherent. Normally, the equilibrium precipitate is noncoherent with the matrix.

Typical aging curves for a precipitation hardened alloy are shown in Fig. 6.2. In general, loss of coherency and formation of the equilibrium precipitate occurs beyond peak hardness or strength, and the strength continues to fall as these precipitates coarsen during overaging.

Strengthening occurs when a moving dislocation impinges on the particles dispersed either by aging or mechanically. Depending on the strength of the particles, the dislocation may be able to cut through them [Fig. 6.3(a)], or may be forced to loop around them (Orowan looping [*6*]) [Fig. 6.3(b)]. Coherent particles possess a surrounding strain field, and this decreases their effective spacing, as seen by the moving dislocation, providing increased resistance to the latter's motion.

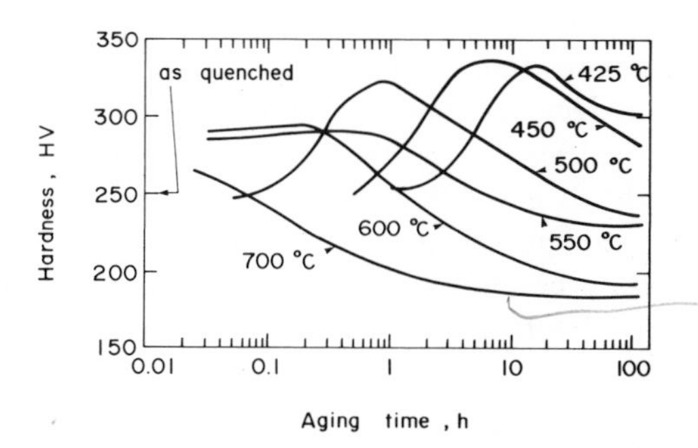

Fig. 6.2 Age hardening curves for a Cu-containing steel. After Krishnadev and Le May [*5*]

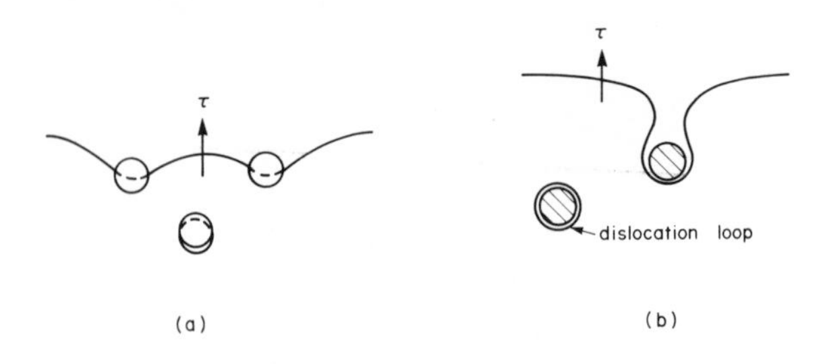

Fig. 6.3 Dislocation-precipitate interactions: (a) dislocation cutting weak particles; (b) Orowan looping around strong particles

There is an obvious limitation to the service temperature at which precipitation hardened alloys can be used. This must be less than the aging temperature which would give rise to precipitate coarsening and loss of strength, and it is important to choose systems having very stable precipitates for elevated temperature service. However, by judicious design of the alloy, it may be possible to have precipitation taking place during long-time elevated temperature service, giving rise to increased strength and creep resistance *in situ*. In the case of a dispersion hardened alloy, no similar temperature limitation exists, as the dispersed particles are chosen as being insoluble in the matrix.

As indicated in Chapter 4, *grain boundaries* act as a barrier to dislocation movement, a sufficient build up of dislocations within a grain being required to provide the stress necessary to initiate slip in an adjacent grain before measurable plastic deformation can take place. Hence, a reduction in grain size produces an increase in the number of barriers to dislocation movement, a reduction in the pile-up length within the grains, and an increase in the stress at which observable deformation occurs (the yield stress). Alloys having high strengths can be produced by making the grain size extremely small, for example by electrodeposition, vapor deposition, splat cooling or similar rapid cooling followed by recrystallization, but it is difficult, as a rule, to produce the grain sizes of significantly less than 5 μm that are required to give a large increase in strength[1]. It is worth noting here that small grain size and good high temperature strength properties (i.e., good creep resistance) are *not* synonymous, as grain boundary sliding can be a major component of high temperature creep deformation, and its proportion will increase with

[1] Steels are an exception to this, by virtue of the ferrite grain refinement occurring during the $\gamma \rightarrow \alpha$ transformation. This, and cycling through the transformation temperature, are discussed further in Section 6.4.

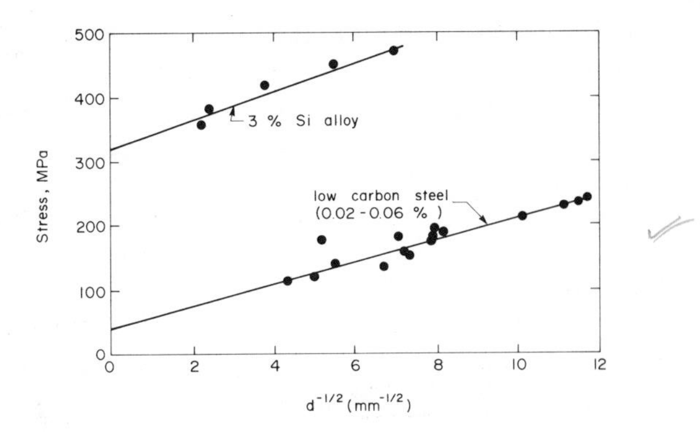

Fig. 6.4 Effect of grain size on lower yield point of low carbon steels. After Tegart [7]

reduction in grain size. The effect of grain size on the room temperature lower yield strength of low carbon steels is illustrated in Fig. 6.4.

In part because of the considerable *grain boundary strengthening* which can occur[2], it is common in specifying engineering alloys to designate the allowable range of grain size. Rather than specify this in terms of mean diameter (in mm or μm), grain size is frequently and conveniently specified using the procedure and charts developed by the American Society for Testing and Materials (ASTM)[3]. In this procedure, grain size is specified by a number n, defined in the expression

$$N = 2^{n-1} \tag{6.1}$$

where N is the number of grains per square inch at a magnification of 100X.

Subgrain or *substructure strengthening* operates in a similar manner to grain boundary strengthening. The term "substructure" may be used to cover all dislocation substructures, including grain boundaries, subgrains, and dislocations distributed from plastic deformation or phase transformation. Here, and in Section 6.3.4., we use it primarily to designate subgrains or a dislocation cell structure. Grain boundary effects and dislocations arising from cold work are discussed separately, while the effect of forest dislocations produced by phase transformation can be considered in terms of an increase in

[2] Other effects of grain size on ductility, toughness, ductile-brittle transition temperature, etc., may be of equal or greater importance, depending on the alloy and the application.

[3] ASTM E112-63, "Standard Methods for Estimating the Average Grain Size of Metals".

dislocation density, which is related to flow stress in the discussion of Section 6.3.5 dealing with strain hardening.

Dislocation cells or subgrains are formed during many processes, for example when a cold worked metal is annealed, and a considerable increase in yield stress over that of the material without substructure may result. Of particular and practical interest is that ductility does not fall significantly, as would be the case for an equivalent strength increase by cold working. Although less effective than grain refinement in increasing strength, the fact that very fine subgrains can be produced more than compensates for this.

Strain hardening of metals and alloys is historically the oldest method of hardening them, dating from man's earliest use of metals. Nevertheless, it is still a most important method of strengthening. It is of particular relevance where the service temperature is sufficiently low that recovery processes are insignificant, and most high strength aluminum alloys are hardened in this way. The higher the dislocation density, the more forest dislocations there are for mobile dislocations to cut through, and the higher the stress for plastic deformation. Alloys which have been strengthened by solid solution hardening, precipitation or dispersion hardening, grain size refinement, or substructure formation, can all be cold worked subsequently to provide increased strength, and it is appropriate to discuss briefly their strain hardening behavior.

In the case of polycrystalline alloys strengthened by solid solution hardening, grain refinement or subgrains, the stress-strain curve has a parabolic nature, and the effect of strain hardening is essentially additive to the other strengthening components. The situation is somewhat different in the case of precipitation hardened alloys, however.

To illustrate this, consider the shear stress-shear strain curves for four similarly oriented single crystals of Al-4.5 wt.% Cu alloy shown in Fig. 6.5 [*8*]. The four specimens were heat treated as follows[4]:

1) Air quenched to give a supersaturated solid solution;
2) Aged 2 days at 130°C to produce fine coherent G.P.1 zones of about 10 pm diameter;
3) Aged 27½ h at 190°C to produce G.P.2 zones and θ' precipitates: optimum strengthening;
4) Overaged and slowly cooled from 350°C: coarse θ ($CuAl_2$) precipitates.

It is seen that the supersaturated solid solution (curve 1) behaves as do other solid solution single crystals, with a large degree of Stage I hardening after a

---

[4] In the Al-Cu system, Guinier-Preston (G.P.) 1 zones form first, being plate-like clusters of Cu atoms on the {100} aluminum matrix. These are followed by G.P.2 (or θ'') coherent precipitates, which in turn transform to θ' semi-coherent precipitates. Finally the θ phase ($CuAl_2$) noncoherent precipitates are formed on overaging [*9*].

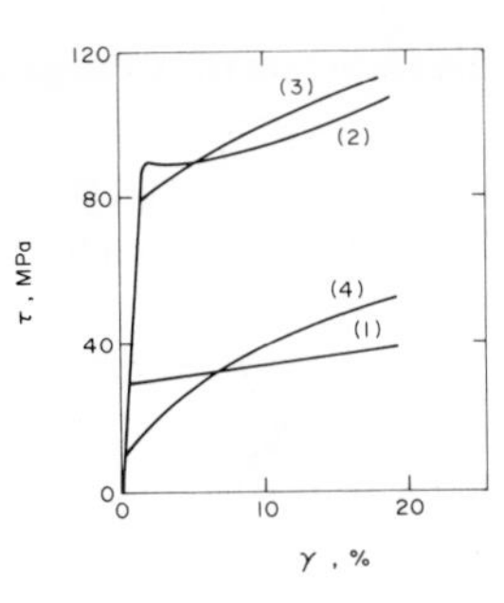

Fig. 6.5 Stress-strain curves for Al-4.5 wt. % Cu single crystals aged to different degrees. After Greetham and Honeycombe [8]

sharp yield. The specimen aged at 130° C (curve 2) has a greatly increased yield strength, but exhibits a sharp yield with a drop in flow stress before work hardening at a low rate comparable to the supersaturated material. The optimum heat treatment (giving peak hardness) produces a lower initial yield strength (curve 3), but greatly enhanced work hardening of a parabolic nature, while in overaged material (curve 4), although the strength is reduced greatly, we see similar parabolic work hardening.

These observations may be rationalized on the basis of simple models for dislocation cutting of coherent particles and Orowan looping around noncoherent overaged ones. In the former case, once precipitates have been sheared, their resistance to the passage of subsequent dislocations will fall, giving rise to the drop in flow stress shown in curve 2, before Stage I hardening commences. With the presence of θ' and θ precipitates (curves 3 and 4) dislocation looping commences, and each subsequent dislocation will find its passage more difficult, because of the residual dislocation loops left behind around each particle (see Fig. 6.6). With coherent precipitates, the slip lines

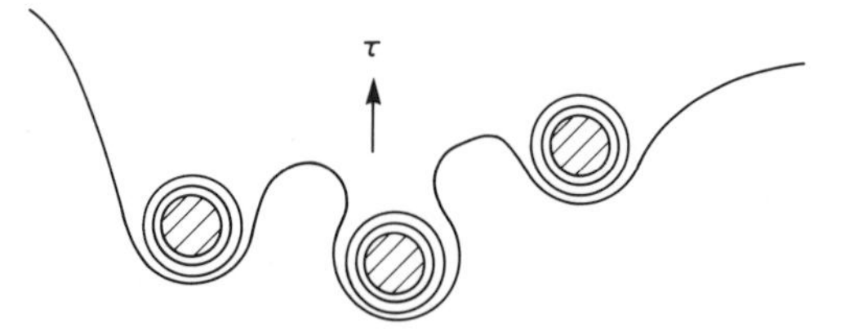

Fig. 6.6 Dislocation movement impeded by the presence of loops from prior dislocations

are seen to be long and straight, while in overaged material widespread cross slip takes place, confirming the mechanisms postulated [*10*].

The principles illustrated are important in that they emphasize the relative instability of underaged alloys. Should such a material be deformed locally by reason of a stress raiser, a local reduction in strength may take place: slightly overaged alloys have much more stable mechanical properties. It should be noted also that dispersion hardened alloys behave in a manner similar to overaged alloys, as strengthening is due primarily to Orowan looping.

As indicated earlier, pure metals are inherently weak, but may be strengthened by providing barriers to dislocation movement, such as grain boundaries. More effective still are *phase boundaries*, as the material on the two sides of these will have different orientations and Burger's vectors for slip, and possibly different crystal structures also. Grains in a single phase alloy have only an orientation difference between them. Many second phases which can be introduced to produce strengthening are also hard and have structures in which dislocation movement is difficult, as, for example, cementite in steel. We distinguish here between hardening due to a fine dispersion (as already discussed), and due to a distribution of larger particles or grains of the second phase.

Normally, strengthening due to a second phase is additive to that from solid solution strengthening in the matrix; however the effect of a second phase is complex, and depends, among other factors, on its shape, size and distribution, as well as on the strength, ductility and strain hardening characteristics of both matrix and the second phase, together with their interfacial energy. Empirical relations can be used to predict strength, most notably the simple *Rule of Mixtures* (see Section 6.3.6), but to illustrate the difficulty of quantifying such strengthening, Fig. 6.7 illustrates the effect of

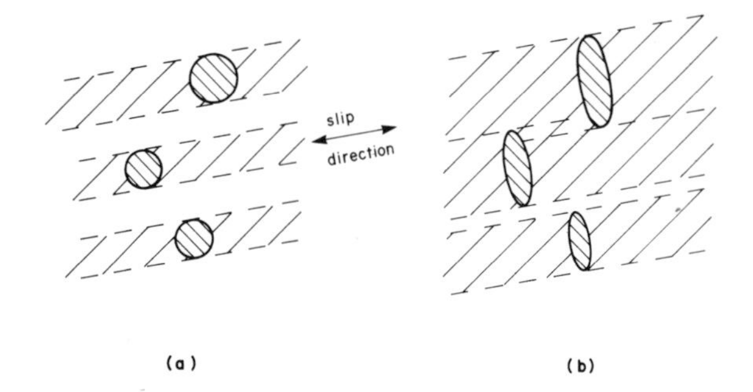

Fig. 6.7 Effect of particle shape on dislocation movement: (a) spherical particles; (b) elongated particles. The shaded areas are regions where dislocation movement is impeded. After Felbeck [*11*]

shape of second phase particles on the impeding of matrix dislocations. With the same volume fraction of second phase, flattening of the particles impedes the dislocation movement considerably, unless slip were to take place in a direction parallel to the elongated particles, when the original strength would be reduced. Carrying this to extremes, we can visualize the effect of cementite plates in pearlite, and of the fibers in fiber composite materials, the latter being considered separately and in some detail in Chapter 7.

6.3 QUANTITATIVE MICROSTRUCTURE-STRENGTH RELATIONS

6.3.1 Solid Solution Strengthening

(a) Random or disordered solid solutions

A dislocation in a solid solution in which the solute atoms are dispersed on an individual or atomic basis will interact with these atoms, on the basis of the interaction of their mutual stress fields, and we shall consider the case where this interaction is positive and dislocations are attracted to solute atoms. The magnitude of the interaction with the dislocation line will depend on the concentration of solute, and on the size difference between solute and matrix atoms. Such interactions will cause the dislocation to be curved in a periodic manner with wavelength λ, where λ is the average separation distance between solute atoms. The dislocation's curvature will be given by

$$1/\rho \simeq 2\tau_o/Gb \tag{6.2}$$

from consideration of the line tension $T \simeq Gb^2/2$ (Eq. 4.12) and the geometry of Fig. 6.8, where an internal stress τ_o acts on the dislocation from the solute atoms *near* to it. For a straight line of finite length, L, the resultant of the internal stresses has a value given by (Ref. [*12*], p. 380)

$$\tau_i(L) \simeq [G\eta b/2L]c^{2/3} \tag{6.3}$$

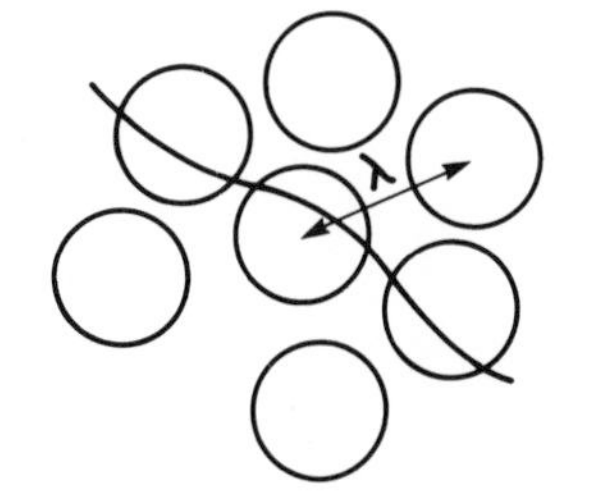

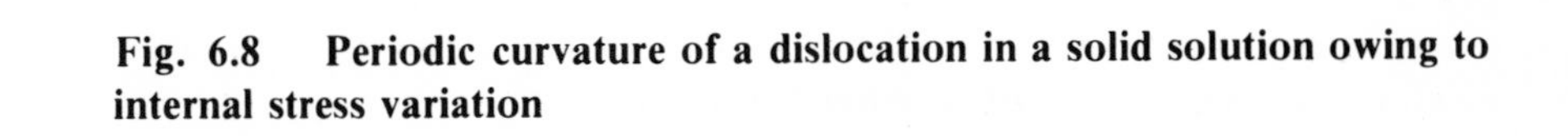
Fig. 6.8 Periodic curvature of a dislocation in a solid solution owing to internal stress variation

where η is the size factor ($= |\Delta| / R$, where R is the radius of the solute atom and R-Δ the radius of the cavity in the matrix which it occupies), and c is the concentration of the solute. It is assumed that size effects predominate over differences in elastic constants in developing Eq. (6.3).

Solute atoms *on* the dislocation will interact more strongly with it, to produce much larger curvatures than given by Eq. (6.2) on the basis of internal stress. Hence the dislocation may be expected to take up a configuration as shown in Fig. 6.9, where it is pinned by solute atoms at A, B, C, etc. It can lie in the zig-zag pattern shown because the increase in strain energy from the increased length is more than offset by the decrease in energy at the pinning points. For atomic obstacles, there are c/b^2 per unit area of the slip plane, and the decrease in energy per unit length in moving from the straight position (XX') to intersect with these atoms is given by

$$E_B = U_o / y \tag{6.4}$$

where U_o is the binding energy of the dislocation to a solute atom. The increase in strain energy from the increased length of the dislocation is given by

$$E_L = [Gb^2/2y]\,[(x^2 + y^2)^{1/2} - y] \simeq Gb^2x^2/4y^2 \tag{6.5}$$

Minimizing $E_L - E_B$, we have

$$x^3 = U_o / Gc \tag{6.6}$$

where $c = b^2/xy$. Now, $U_o \simeq Gb^3\eta/2$ where it depends on a size effect (Ref. [*12*], p. 380), thus

$$x \simeq (\eta/2c)^{1/3} b \tag{6.7}$$

It may be seen that x is a very small quantity and is normally equal to a small multiple of b at most: hence the straight portion of the dislocation will always be unstable, and will move to a pinned and zig-zagged position.

If a stress τ is applied, the portion ABC may be moved independently of the rest of the dislocation to $AB'C$, and the stress may be evaluated for x small by

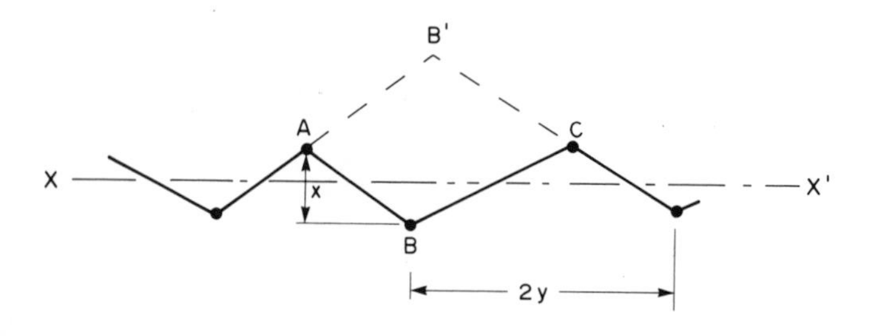

Fig. 6.9 Dislocation pinned at solute atoms

considering the stress required to move the segment to the intermediate position of highest energy, AC. We evaluate this by equating the work done by the applied stress to the increase in energy of the system, $(E_B - 2E_L)\, 2y$. Hence

$$2\,[\tau - \tau_i\,(y)]\; bxy = (E_B - 2E_L)\; 2y \qquad (6.8)$$

The small internal stress $\tau_i(y)$ from Eq. (6.3), with $L = y$, may be neglected, and hence from substitution of Eqs. (6.4), (6.5),(6.6) and (6.7), we obtain

$$\tau = U_0 / 2bxy$$

or

$$\tau = (U_0 / 2b^3)c \qquad (6.9)$$

It may be seen that hardening (or strengthening) is directly proportional to the binding energy and the concentration of the solute atoms in dilute solid solutions. Such behavior is observed in solid solutions of FCC metals at low temperature, however strengthening decreases rapidly with increase in temperature above some critical value, as the dislocations are then able to break away from the solute atoms easily from thermal agitation, and the treatment of Diehl [*13*] will be followed to illustrate this.

Consider a dislocation held up at obstacles as shown in Fig. 6.10(a). Due to the applied stress τ, the force on an obstacle is

$$F = \tau\, bL$$

Without thermal agitation the dislocation will be in the equilibrium position x_1 in Fig. 6.10(b), with its interaction force $f = F$. The hatched area in the figure represents the activation energy ΔG required to overcome the obstacle, and this quantity determines the strain rate $\dot{\gamma}$ of the material,

$$\dot{\gamma} = \dot{\gamma}_0 \exp(-\Delta G / kT) \qquad (6.10)$$

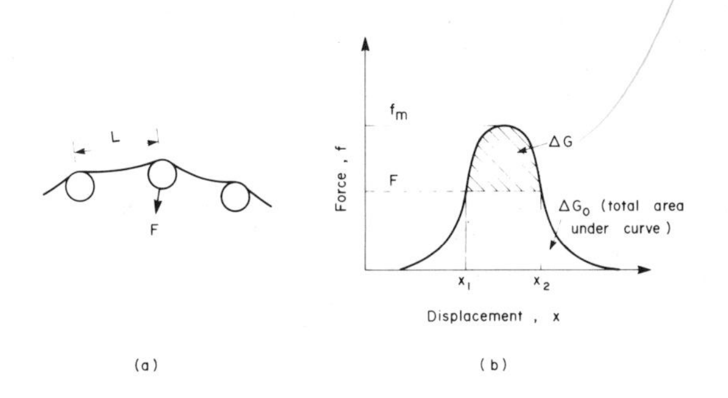

Fig. 6.10 (a) Dislocation held up at obstacles. (b) Force-distance relationship

Hence

$$\Delta G(\tau) = kT \log (\dot{\gamma}_0/\dot{\gamma}) \tag{6.11}$$

where k is Boltzmann's constant and T is the absolute temperature. For a random distribution of obstacles all giving rise to the same relations for force, $\dot{\gamma}_0$ should be a constant including the attempt frequency of the dislocations and geometrical factors.

ΔG will depend on F and hence on τ [Fig. 6.10(b)], and the zero stress activation energy ΔG_0 is the total area under the curve and can be considered to be a characteristic value for the obstacle. Hence, if the relation between f and x, or the dependence of ΔG on τ is known, the dependence of τ on the deformation temperature T, the strain rate $\dot{\gamma}$, and the obstacle density, can be derived by solving Eq. (6.11) for τ.

If the temperature and strain rate are constant, and τ only enters into Eq. (6.11) through the term τbL, then τ should be proportional to $1/L$. For a random distribution of barriers $1/L$ will be proportional to the square root of the area density of obstacles in the slip plane, N_A, which is proportional to the volume density (or concentration) of the obstacles, c. Hence the effective shear stress is expected to vary linearly with the square root of the concentration of the obstacle atoms:

$$\tau \propto c^{1/2} \tag{6.12}$$

Eq. (6.12) has been found to hold for neutron irradiated copper and other irradiated metals where dislocations overcome barriers of condensed vacancies by thermally activated processes, this being discussed further in Section 6.5.2.

In two extreme cases of dispersed barrier hardening, thermal activation becomes unimportant. These are where the obstacles become large or where the maximum interaction force f_m is high. In the former case, the temperature and strain rate dependence of τ become small as the F-level (and consequently τ) must vary little with temperature in accounting for variation of ΔG with T. Hence thermal agitation contributes little, and the dislocation will move only if $F = f_m$ or if $\tau = \tau_m$ where

$$\tau_m = f_m/bL$$

In the second case, τ may become sufficiently large that the dislocations will bow out to pass between particles by an Orowan-type mechanism, and Eq. (4.14) will hold[5]:

$$\tau = Gb/L$$

[5] See Section 6.3.2 for more rigorous discussion of strengthening by means of dispersed barriers.

It is observed that the low temperature hardening of BCC metals containing interstitially dissolved elements is much greater and less temperature sensitive than for FCC metals. This arises because the binding energy U_o is larger (e.g., $\sim 1.2 \times 10^{-19}$ J for C in Fe [*14*]) as compared to that for substitutional atoms in FCC metals ($\sim 0.16 \times 10^{-19}$ J [*15*]), and is responsible in part for the low-temperature brittleness in such materials, as the dislocations may be so strongly pinned that brittle fracture can result when stress is applied [*16*]. (See Chapter 8).

(b) Ordered solid solutions

When an alloy has a *locally ordered* structure or *short range order*, the passage of a dislocation will destroy this, and the consequent increase in the energy on the slip plane will cause an increase in the flow stress of the alloy over that for disordered material. Concentrated solid solutions are likely to exhibit considerable short range order, while superlattice alloys which have been cooled rapidly may have a generally disordered state, but they retain some short range order.

The stress increment required to move a dislocation through the alloy is [*17*]

$$\tau = E/b \tag{6.13}$$

where E is the energy required per unit surface area, and b is the magnitude of the Burgers vector of the latice. Flinn [*18*] has determined the energy associated with short range order in a binary solid solution (of A and B) as

$$E = Nzm_Am_B\nu\ \alpha \tag{6.14}$$

where N is the number of atoms in the lattice, z is the coordination number, m_A and m_B are the mole fractions of A and B respectively, ν is the net interaction energy ($= \nu_{AB} - (\nu_{AA} + \nu_{BB})/2$, where ν is the interaction energy for the atom pairs designated), and α is the short range order coefficient. The total energy E is shared among $Nz/2$ nearest neighbor bonds, thus the average energy per nearest neighbor due to local order is

$$\epsilon = 2\ E/Nz = 2\ m_A\ m_B\ \nu\ \alpha \tag{6.15}$$

We may obtain ν from the relation [*19*]

$$\alpha/(1 - \alpha)^2 = m_A\ m_B\ [\exp\ (2\nu/kT) - 1] \tag{6.16}$$

where α is the experimentally determined local order coefficient, k is Boltzmann's constant and T the absolute temperature. The energy to disrupt short range order, per unit area of slip, is obtained by dividing the energy increase per atom in the slip plane by the area per atom in this plane, and for the FCC lattice, where 2 of 3 neighbors are unchanged in the (111) slip plane, we have [*18*]

$$E = (8/\sqrt{3})\ (m_A\ m_B\ \nu\ \alpha/a^2) \tag{6.17}$$

where a is the lattice parameter. This is equated to the work done in moving the dislocation, τb, hence

$$\tau = 16(2/3)^{1/2} (m_A m_B \nu \alpha/a^3)$$

or

$$\tau = 16(2/3)^{1/2} [c(1 - c)\nu \alpha/a^3] \tag{6.18}$$

where c is the concentration (molar fraction) of the solute.

All the terms of Eq. (6.18) are temperature independent: hence short range order strengthening is athermal. However, because α depends on annealing temperature (ordering decreasing with increase in temperature), the strength increment due to short range ordering is dependent on the thermal history of the material.

The strengthening effect of *long range order* is more complex, such a structure containing domain walls between ordered regions, as shown in Fig. 6.11(a). Passage of a dislocation will require that it cuts the *antiphase boundaries*, increasing the area of these as shown in Fig. 6.11(b). Stoloff [2] has reviewed the state of knowledge of strengthening in such ordered systems, and analysis has shown that the yield stress peaks at an intermediate degree of order at both room temperature and elevated temperature, this peak corresponding to a transition from the glide of superdislocations (which consist of unit dislocations separated by a strip of antiphase boundary) in a more ordered structure, to glide of unit dislocations in a less ordered configuration [*20*].

6.3.2 Precipitation Strengthening and Dispersion Strengthening

When a moving dislocation encounters a series of obstacles it will be bent to some angle $0 \leqslant \phi \leqslant \pi$ under the action of the applied stress, before it can

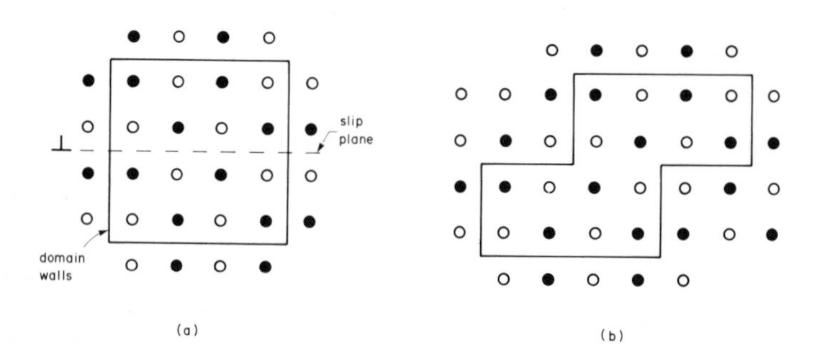

Fig. 6.11 Increase in antiphase boundary area in a long range ordered alloy when cut by a dislocation. (a) Before passage of dislocation: (b) after dislocation movement

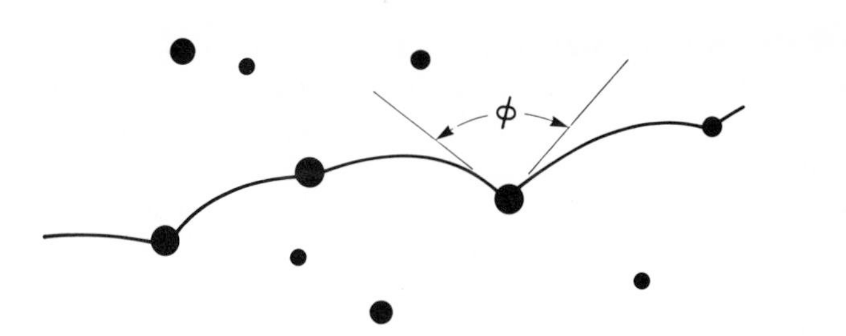

Fig. 6.12 Dislocation held up at obstacles showing definition of bowing angle

continue through the crystal (Fig. 6.12). If the obstacles are sufficiently strong, the dislocation may bow out until it bypasses them by the Orowan mechanism [*21*]: alternatively, if the obstacles have a lower strength, the dislocation may be able to cut through the particles at some value of $\phi > 0$, this value depending on the obstacle strength.

At the critical condition where the dislocation breaks away from the particle, the strength of the obstacle, F, may be stated as

$$F = 2\ T \cos\ (\phi_c/2) \tag{6.19}$$

where T is the line tension of the dislocation = $Gb^2/2$ [Eq. (4.12)], and ϕ_c is the critical angle. The stress required to cause the dislocation to break away from the obstacle is then obtained from

$$\tau_c\ b\ \lambda = F$$

as

$$\tau_c = F/b\lambda = 2\ T \cos\ (\phi_c/2)/b\lambda \tag{6.20}$$

where λ is the effective interparticle spacing.

Considerable controversy has taken place in recent years concerning the value of the effective interparticle spacing, and many refinements of the calculations for T, the dislocation line tension, have been made. For order of magnitude calculations, a constant line tension = $Gb^2/2$ is satisfactory [*21*]: for more refined calculations the variation of energy as the dislocation bends can be accounted for by the de Wit-Koehler model [*22*], and a detailed review of this and other more rigorous evaluations of T is given by Brown and Ham [*21*], while a brief discussion of the de Wit-Koehler model is presented in the present section when discussing *order hardening*.

The effective spacing λ met by a dislocation moving on a slip plane is related to the number of particles per unit area which intersect this plane, n_A, by the relation [*23*]

$$\lambda = B/(n_A)^{1/2} = BL \tag{6.21}$$

where B is a constant close to unity and the particle spacing $L = (1/n_A)^{1/2}$. For a regular square array of impenetrable particles $B = 1$, while $B = 1.07$ for a hexagonal array. Foreman and Makin [*24*] have examined the passage of a dislocation through a two-dimensional array of point obstacles generated by computer. The dislocation is represented by circular arcs between the points, the radius of curvature being given by

$$R = T/\tau b \tag{6.22}$$

For an obstacle at which $\phi < \phi_c$, the dislocation is assumed to have broken through, moving on to the next obstacle. It will take up a stable position in contact with obstacles at all of which $\phi > \phi_c$, and the stress is then increased and the process repeated. Foreman and Makin showed that there is a well-defined stress at which the dislocation can move long distances, and this is taken to be the flow stress for the alloy. At stresses greater than this, the dislocation will not be held up at any obstacles.

The value of B was determined as 1.23 for the random array, but considering that in practice a completely random array is not normally to be expected, a somewhat lower value may be appropriate, and a value of $B = 1.2$ has been suggested [*23*].

Another important result of Foreman and Makin's studies is an understanding of the mechanism by which obstacles are bypassed. For $\phi_c \simeq 0$, corresponding to strong particles, the dislocation follows a path of easy movement by the Orowan process as shown in Fig. 6.13. "Difficult" groups of particles may be surrounded by dislocation loops which have been pinched off as the dislocation moves through on either side of them. For weaker particles, where $\phi_c \simeq \pi$, the dislocation moves by a "sidewise unzipping process" as shown in Fig. 6.14, this being termed the *Friedel process* [*21*].

Brown and Ham [*21*] have presented an analysis of strengthening due to weak obstacles based on the work of Friedel [*12*]. A dislocation restrained at points O, P, and Q is shown in Fig. 6.15. If the dislocation overcomes the obstacle at P and unzips to a new equilibrium position P', it passes through an area $Z = OPQP'$, while still being retained at O and Q. Hence, in passing through this area, the dislocation meets one obstacle only on average, and there is a steady state during the unzipping process, such that

$$Z\, n_A = 1 \tag{6.23}$$

If we assume $OP = PQ = L'$, which is the effective obstacle spacing, and consider a nearly straight dislocation where β_1, $\beta_2 = 2\beta_1$ and δ are all small, then

$$Z = 4\beta_1^3 R^2 = (\phi_c')^3 R^2/2 \tag{6.24}$$

where $2\beta_1 = \phi_c' = \pi - \phi_c$, and R is given by Eq. (6.22). Substituting for Z and for $n_A = 1/L^2$ in Eq. (6.23), and substituting for R from Eq. (6.22), with $T = Gb^2/2$,

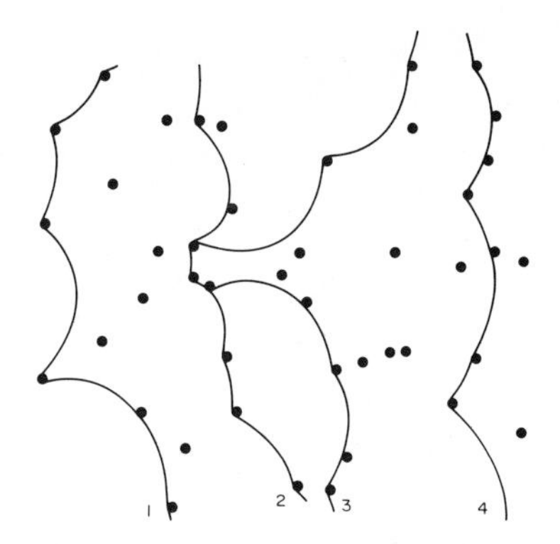

Fig. 6.13 Schematic showing dislocation progressing through an array of obstacles by the Orowan mechanism. Numbers indicate successive positions

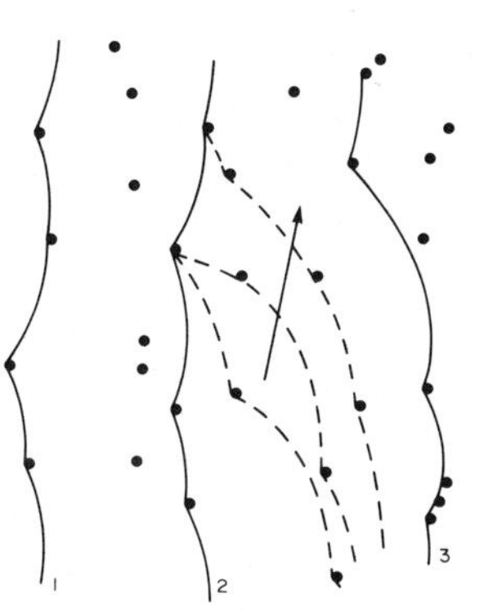

Fig. 6.14 Dislocation moving from position 2 by the Friedel process, overcoming weaker obstacles

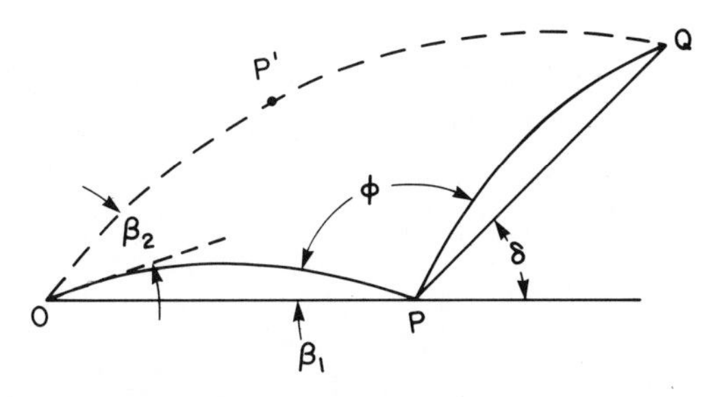

Fig. 6.15 The geometry for the Friedel process. After Brown and Ham [*21*]

we obtain the Friedel relation

$$\tau_c \simeq (Gb/L)\,(\phi_c'/2)^{3/2} \tag{6.25a}$$

or

$$\tau_c \simeq (Gb/L)\,[\cos(\phi_c/2)]^{3/2} \tag{6.25b}$$

and the effective obstacle spacing L' is given by

$$L' = L\,(\phi_c'/2)^{-1/2} \tag{6.26}$$

Foreman and Makin [24] proposed an empirical equation on the basis on their computer generated data, in the form

$$\tau_c = (Gb/L)\,[\cos(\phi/2)]^{3/2}\,(0.80 + \phi/5\pi) \tag{6.27}$$

Brown and Ham [21] have shown that this cannot be correct for $\phi \rightarrow 0$, as it shows $d\tau_c/d\phi \neq 0$ for $\phi = 0$, and that in this condition, with the Orowan process dominating, a small change in ϕ will produce a second order change only in the area swept out by the dislocation. They suggest that a better representation is given by

$$\tau_c = (0.8Gb/L)\cos(\phi/2), \qquad \phi \leqslant 100° \tag{6.28a}$$

$$\tau_c = (Gb/L)\,[\cos(\phi/2)]^{3/2}, \qquad \phi \geqslant 100° \tag{6.28b}$$

(a) Overaged and underaged alloys

In overaged alloys the precipitate particles are generally large, relatively strong and noncoherent with the matrix. When the particles act as impenetrable barriers, the dislocations bow out and the alloy's strength may be evaluated quantitatively on the basis of the Orowan equation [Eq. (4.14)] refined to take account of the statistical factor discussed above, more precise estimates of the line tension, and to include the effect of the mutual interaction of the bowed-out dislocation arms on each side of the particle, as suggested by Ashby [25]. The currently accepted version of the Orowan equation, taking into account these modifications, is

$$\tau = [0.83Gb/2\pi\,(1-\nu)^{1/2}]\ln(l/r_o)/(L-l) \tag{6.29}$$

where l is the particle diameter, r_o is the dislocation core radius and the other symbols are as defined in the foregoing section. When the particles are not impenetrable but are broken through, Eqs. (6.28) may be applied.

It is generally assumed that in overaged alloys the bowing angle, ϕ, is constant, and this may be evaluated on the basis of the straight line relationship between stress increment and inverse particle size ($1/l$) or inverse interparticle spacing ($1/L$), typical plots being shown in Fig. 6.16. Rewriting Eqs. (6.28) in terms of strengthening increment $\Delta\tau$, and substituting for L in terms of the square lattice relationship [30], we obtain

$$\Delta\tau = (1.6Gb/l)\,(f/\pi)^{1/2}\cos(\phi/2), \qquad \phi \leqslant 100° \tag{6.30a}$$

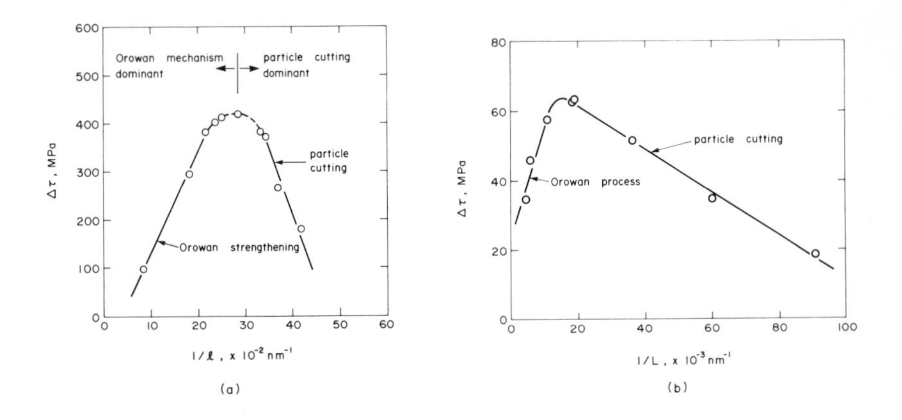

Fig. 6.16 (a) Precipitation strengthening increment as a function of inverse particle size. After Ref. [*26*] using the data of Scarlin and Edington [*27*]. (b) Strengthening increment versus inverse interparticle spacing. After Ref. [*28*] using the data of Phillips [*29*]

and

$$\Delta\tau = (2Gb/l)\,(f/\pi)^{1/2}\,[\cos(\phi/2)]^{3/2}, \qquad \phi \geqslant 100° \quad (6.30b)$$

Hence, by substituting the known values of the various parameters and equating the slope in Eqs. (6.28) or (6.30) with the experimentally determined slope, ϕ may be evaluated. Data for a number of alloys are presented in Fig. 6.17 [*31*], from which it may be seen that ϕ depends on the ratio of the shear moduli of the precipitates and matrix (G'/G), bowing to the extent considered in the original Orowan theory occurring only when $G'/G = 0$ or $G'/G > \sim 0.95$, providing the particles are noncoherent in the latter case[6].

In the case of underaged alloys, the observed variation in yield stress increment with $1/l$ or $1/L$ is linear over a considerable range of particle sizes or spacings (see Fig. 6.16), the slope being of opposite sign to that for overaged material. In order that Eqs. (6.25) or (6.28b) for the Friedel relation apply, there must be a systematic variation in the value of ϕ_c, this decreasing continuously on aging at constant volume fraction, except perhaps where strengthening is based on mechanisms arising from disorder at the precipitate surface, when it may be independent of particle size [*46*]. In order to apply Eqs. (6.20) or (6.28), either the force F required to cut the particles or the angle

---

[6] Point 14 on Fig. 6.17 was taken from data for a Cu-0.35 w/o Cr alloy [*45*] in which the particles are coherent or semicoherent with the matrix. Hence cutting, rather than looping, takes place.

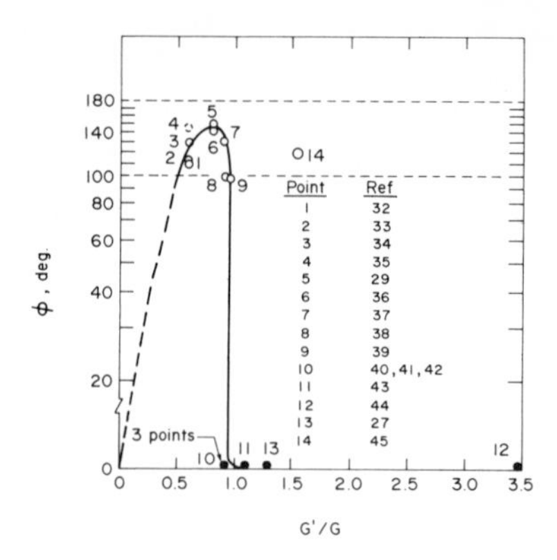

Fig. 6.17 Variation in bowing angle with *G′/G* [*31*]. Key to points gives references for data sources

ϕ at cutting must be computed and substituted appropriately.

The shearing force, *F*, may arise in a number of ways, as follows [*23*]:

- Elastic coherency strains between particle and matrix: *coherency hardening*.
- Energy requirement to produce additional interface between particle and matrix: *surface hardening*.
- Differences in the SFE of particle and matrix: *stacking fault hardening*.
- Differences in the elastic moduli of particle and matrix: *modulus hardening*.
- The work required to create an internal or antiphase boundary in the case of internally ordered particles: *order hardening*.

Many theories have been derived for the various strengthening mechanisms, and the most accepted versions are given in Table 6.1. Unfortunately, from an analytical viewpoint, it is difficult to find a *single* mechanism which applies to a particular alloy system, as more than one mechanism may contribute to hardening. The matter is further confused because most models are based on the assumption of a single particle size, whereas a distribution of sizes obtains in practice. In checking for the validity of a particular model, both the linearity of strength increment against a particular structural parameter (such as $l^{1/2} f^{1/2}$) and the slope of the plot must be examined, but this is not necessarily sufficient to discriminate completely between different mechanisms [*23*]. More detailed comparison of models with experimental data for various alloy systems is made by Brown and Ham [*21*].

More detailed examination of the different models for strengthening by

Table 6.1 Theories of Hardening by Means of Deformable Particles

Mechanism	Authors [Ref.]	Strength Increment, $\Delta\tau$
Coherency Hardening	Gerold and Haberkorn [47]	$\beta G\epsilon^{3/2} (l/2b)^{1/2} f^{1/2}$
	Gleiter [48]	$11.8\ G\ \epsilon^{3/2} (l/2b)^{1/2} f^{5/6}$
Surface Hardening	Kelly and Nicholson [49]	$(\sqrt{6}/\pi)\ \gamma_s (2f/l)$
	Harkness and Hren [50]	$(60.4/b^3)(T/b)^{1/2} (\gamma_s)^{3/2} (l/2)^2 f^{1/2}$
Stacking Fault Hardening	Hirsch and Kelly [51]	$(4/\pi)(2/3\pi)^{1/2} [(\gamma_m - \gamma_p)/b] [3K(\alpha) \ln (\gamma_m/\gamma_p)/T]^{1/3} (2\bar{\omega}/l)(1 - 3\pi\bar{\omega}/16l) f^{2/3},\ l >> \bar{\omega}$
	Gerold and Hartman [52]	$(8/\pi)^{1/2}\ G\ [(\gamma_m - \gamma_p)/Gb]^{3/2} (l/2b)^{1/2} f^{1/2} I_m$
Modulus Hardening	Kelly [53]	$(\Delta G/4\pi^2)(3\Delta G/Gb)^{1/2} [0.8 - 0.143 \ln (l/2b)]^{3/2} (l/2)^{1/2} f^{1/2}$
	Weeks *et al.* [54]	$(\Delta Gb/2\pi\lambda) [(\pi^2/12) + \ln (l/r_o)]$
Order Hardening	Gleiter and Hornbogen [55]	$(0.28\ \gamma_o^{3/2} r_o^{1/2} f^{1/3})/b^2 G^{1/2}$
	Brown and Ham [21]	$(\gamma_o/2b) [(4\ \gamma_o f r_s/\pi T)^{1/2} - f],\quad \pi Tf/4\gamma_o < r_s < T/\gamma_o$ $(\gamma_o/2b) [(4f/\pi)^{1/2} - f],\quad r_s > T/\gamma_o$

β = constant (3 for edge dislocation, 1 for screw); ϵ = matrix particle misfit; γ_s = matrix/particle interfacial energy; γ_m = matrix SFE; γ_p = precipitate SFE; $K(\alpha)$ = partial dislocation separation force times the separation distance; $\bar{\omega}$ = mean SF ribbon width; I_m is a complex function of l and γ_p; ΔG = difference in shear moduli of matrix and precipitate; γ_o = specific antiphase boundary energy; r_s = average dimension of a particle intersected by a slip band; other symbols as defined previously in the text.

particle cutting will not be given here with the important exception of order hardening, which is of particular relevance to nickel-base superalloys strengthened by γ' precipitation.

(b) Order hardening

In examining this situation, and in order to follow the analysis of Ham [*56*] and Brown and Ham [*21*] for the strengthening increment, we shall first consider the free line tension or de Wit-Koehler model for dislocation line tension [*22*] referred to briefly at the beginning of Section 6.3.2. In this case we consider the changes in orientation and resulting energy changes as a dislocation bends. Consider an imaginary segment of the dislocation which is free to change in orientation. In order to prevent this rotating to a lower energy orientation, a couple must be applied in addition to the collinear tension forces. This results in a total force which must be applied to the curved segment of [*22, 57*][7]

$$T(\theta) = U(\theta) + d^2U(\theta)/d\theta^2 \tag{6.31}$$

where $T(\theta)$ is the de Wit-Koehler approximation for the line tension and $U(\theta)$ is the energy of the straight dislocation of unit length if it makes an angle θ with its Burgers vector.

For a long straight dislocation on the slip plane of an isotropic crystal, the elastic energy per unit length is given by [*58, 59*]

$$U(\theta) = (Gb^2/4\pi)\,[(1 - \nu\cos^2\theta)/(1 - \nu)]\ln(R/r_o) \tag{6.32}$$

where θ is the angle between the dislocation line and the Burgers vector, this equation being a more exact form than is given by Eq. (4.11a). Combining Eqs. (6.31) and (6.32), we obtain

$$T = (Gb^2/4\pi)\,[(1 + \nu - 3\nu\sin^2\theta)/(1 - \nu)]\ln(R/r_o) \tag{6.33}$$

Under the influence of an applied stress, the dislocation will no longer be a circular arc, but its radius of curvature at every point will be given by

$$\rho = T/\tau b$$

Thus, for an edge dislocation the model predicts a much smaller radius than for a screw dislocation, the factor being $(1 - 2\nu)/(1 + \nu)$.

Consider the dislocation meeting a set of *weak* obstacles, which distort it only slightly. Instead of Eq. (6.25b), based on constancy of line tension, we obtain

$$\tau_c = (2T/bL)\,[\cos(\phi_c/2)]^{3/2} \tag{6.34}$$

---

[7] The derivations of the various equations quoted here are given in the Appendix to the paper by Brown and Ham [*21*].

The force on the obstacle is given by [21]

$$F = 2T \cos (\phi/2) \tag{6.35}$$

and combining Eqs. (6.34) and (6.35), we obtain the flow stress as

$$\tau_c = (2/bLT^{1/2})\,(F/2)^{3/2} \tag{6.36}$$

Thus the effective obstacle spacing for the de Wit-Koehler model becomes

$$L' = (2T/\tau_c b)\,(\phi_c'/2) = (2TL^2/\tau_c b)^{1/3} \tag{6.37}$$

Consider the dislocation cutting an ordered particle. Neglecting the particle-matrix interfacial energy, we can balance the force $\tau_I b$ on the first dislocation shown in Fig. 6.18 against the antiphase boundary (APB) energy created, this being $2r_s\,\gamma_o/L_I$, where γ_o is the specific APB energy, r_s is the average dimension of a particle intersected by a slip band [$= (2/3)^{1/2} r_v$]. Hence

$$\tau_I = 2r_s\,\gamma_o/L_I b \tag{6.38}$$

Now $n_A = f/\pi r_s^2$, so from Eq. (6.21), $L = (\pi/f)^{1/2} r_s$. Substituting for L in Eq. (6.37), we obtain

$$L_I = (2T\pi r_s^2/f\tau_I b)^{1/3} \tag{6.39}$$

Substituting for L_I in Eq. (6.38), we have the stress required to push the dislocation through the precipitate as

$$\tau_I = (\gamma_o^{3/2}/b)\,(4fr_s/\pi T)^{1/2} \tag{6.40}$$

From Eq. (6.40) it is seen that, for constant volume fraction f, τ_I increases with increase in r_s, the dislocation increasing in flexibility as it interacts with coarser particles, and the increase in τ_I is proportional to the square root of r_s, as opposed to the linear dependence on $-1/r$ (or $-1/l$) observed for particle cutting and discussed previously.

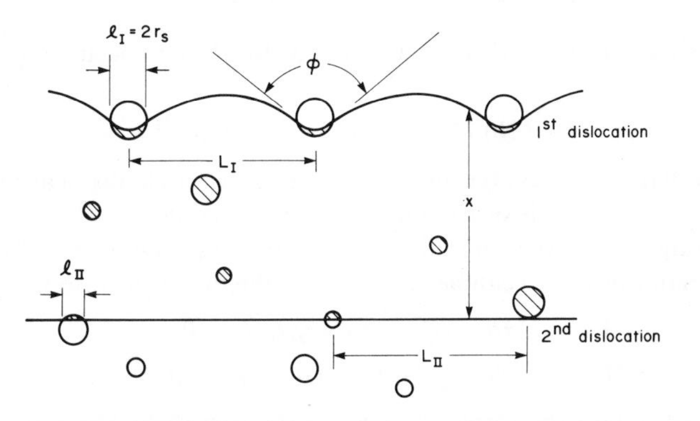

Fig. 6.18 A dislocation pair intersecting ordered particles. The shaded areas represent APB in the particles

Now consider superlattice dislocation pairs interacting with the particles (see Fig. 6.18), following the principles of Gleiter and Hornbogen [55]. The second dislocation will be pulled forward by the APB present in all particles cut by the first one. For dislocations having the same shape and whose separation x is relatively small (but greater than r_s), the second dislocation may lie outside the particles. Under equilibrium, the repulsive force between the dislocations will balance the forward stress on the second one. Thus (see Section 4.3.5)

$$\tau_{II}\, b = Gb^2/2\pi x \tag{6.41a}$$

for screw dislocations, or

$$\tau_{II}\, b = Gb^2/2\pi(1-\nu)x \tag{6.41b}$$

for edge ones. τ_{II} will equal the applied stress τ in this situation, and the stress on the first dislocation $\tau_I = 2\tau$, because of the effect of the second one.

Now, the fraction of the first dislocation line cutting the precipitates and touching APB is given by

$$2r_s/L_I = (4\gamma_o f r_s/\pi T)^{1/2}\;,\quad \pi Tf/4\gamma_o \leqslant r_s \leqslant T/\gamma_o \tag{6.42}$$

and at the upper limit, corresponding to Orowan bowing, $r_s = T/\gamma_o$ and $L_I = L$, or

$$2r_s/L_I \rightarrow 2r_s/L = (4f/\pi)^{1/2} \tag{6.43}$$

For this condition, the stress required to shear a particle by dislocation I acting alone is

$$\tau_I = (\gamma_o/b)\,(4f/\pi)^{1/2} \tag{6.44}$$

The lower limit of the fraction of dislocation line touching APB corresponds to a straight dislocation and very fine precipitates, or

$$2r_s/L_I = f \qquad r_s \leqslant \pi Tf/4\gamma_o \tag{6.45}$$

Thus for shear of the particles by the first dislocation to occur [Eq. (6.43)], we may write

$$\tau_{II} = (\gamma_o/2b)\,(2r_s/L) = (\gamma_o/2b)\,(4f/\pi)^{1/2} \tag{6.46}$$

this being half the stress required for shear by a single dislocation only.

Where the second dislocation does come into contact with the APB and is nearly straight, as is the general case, the obstacles become less effective still. The forces in Fig. (6.18) can be balanced for the two dislocations, as follows:-

$$\text{I} \qquad \tau b + Gb^2/2\pi x - \gamma_o l_I/L_I = 0 \tag{6.47a}$$

$$\text{II} \qquad \tau b + \gamma_o l_{II}/L_{II} - Gb^2/2\pi x = 0 \tag{6.47b}$$

Solving, we obtain the forward stress on the first dislocation as

$$2\tau b + \gamma_o l_{II}/L_{II} = \gamma_o l_I/L_I \tag{6.48}$$

And since dislocation II is straight during shear of particles by dislocation I, we can substitute from Eqs. (6.45) and (6.42) for l_{II}/L_{II} and l_I/L_I respectively, so that

$$2\tau b + \gamma_0 f = (4\gamma_0 f r_s/\pi T)^{1/2}\gamma_0 \qquad (6.49)$$

or

$$\tau = (\gamma_0/2b)\,[(4\gamma_0 f r_s/\pi T)^{1/2} - f] \qquad (6.50)$$

This equation gives us the applied stress required to cut the particles by means of paired dislocations.

For $r_s > T\gamma_0$ and stresses less than the stress for bowing, we modify Eq. (6.50) by substituting Eq. (6.43) into Eq. (6.48) to give

$$\tau = (\gamma_0/2b)\,[(4f/\pi)^{1/2} - f] \qquad (6.51)$$

When $r_s \leqslant \pi/Tf/4\gamma_0$, $\tau = 0$, and in the less extreme situation the applied stress required to cut the particles falls off as the amount of APB touched by the second dislocation is increased.

If we substitute the simplified line tension expression, $T = Gb^2/2$, into Eq. (6.50), we obtain

$$\tau_c = (\gamma_0/2b)\,[(8\gamma_0 f r_s/\pi G b^2)^{1/2} - f] \qquad (6.52)$$

The model has been used successfully to predict the behavior of a number of superalloys, both iron and nickel based [*60*], and the main points are illustrated schematically in Fig. 6.19.

(c) Dispersion hardening

In a dispersion hardened material, such as a thoriated nickel alloy, the particles act as barriers to dislocation movement, and they can be bypassed in four ways: by Orowan bowing between them; cross slip around them; particle shearing; or climb over them.

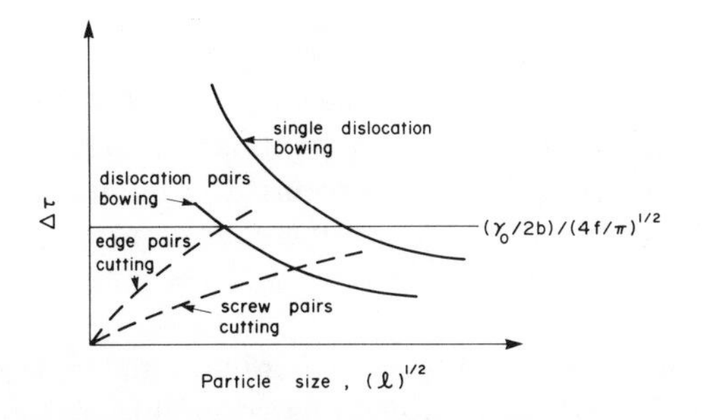

Fig. 6.19 Schematic illustration of relations between order hardening and Orowan strengthening as a function of particle size

Observations which have been made of such materials in the deformed condition using transmission electron microscopy techniques, indicate that particles are not sheared, and that the Orowan mechanism is the most important one at temperatures below $0.5 T_m$. Consequently, the yield strength, τ_Y, of dispersion hardened materials can generally be predicted using the Orowan relation [*6*]. These remarks are in agreement with the data regarding dislocation bowing shown in Fig. 6.17 for Cu-SiO_2 and Cu-Al_2O_3, but the observations of Kupcis *et al.* [*44*] for single crystals of Cu containing fine coherent α-Al_2O_3 particles smaller than 14 nm should also be noted. They found that the particles appeared to be broken, and Lui and Le May [*31*] have computed the critical value of ϕ, the dislocation bowing angle, to be ~160°, despite a modulus ratio (G'/G) of 3.5. This observation suggests that very fine coherent particles in a dispersion hardened alloy may be cut by dislocations. In general, however, τ_Y is given by

$$\tau_Y \simeq \tau_m + Gb/L \tag{6.53}$$

from Eq. (4.14), where τ_m is the matrix yield stress, and L the mean planar interparticle spacing. The more exact formulations of Orowan strengthening given by Ashby [*61*] should, however, be noted:

$$\tau_Y = \tau_m + (1/1.18)\,(Gb/2\pi)\,[1/(L - 2r_s)]\,\ln\,(r_s/b) \tag{6.54a}$$

for edge dislocations, and

$$\tau_Y = \tau_m + [1/(1 - \nu)]\,(1/1.18)\,(Gb/2\pi)\,[1/(L - 2r_s)]\,\ln\,(r_s/b) \tag{6.54b}$$

for screw dislocations, where r_s is the planar particle radius.

In some cases dislocations may cross slip past particles after bowing round them to some extent, but Ashby [*61*] concluded that Orowan bypassing was the controlling step in such materials, any cross slip occurring at stresses of the order of the Orowan stress.

As noted above, particle shear does not normally occur with hard particles [*62*], except in special cases [*44*], but Ansell [*63*] did propose a model in which hard particles are sheared. In this, the yield stress is given by

$$\tau_Y \simeq \tau_m + G'(2r_s)/4C(L - 2r_s) \tag{6.55}$$

where G' is the shear modulus of the particles, and C is a constant ($\simeq$30 for defect-free particles). This expression has not been found to hold in practice so well as the Orowan relation, and there seems little doubt that the latter should be preferred except in the special cases where the particles are fine and cutting does occur, in which the strength increment can be given by Eqs. (6.28a) and (6.28b).

Dislocation climb occurs in dispersion hardened materials at temperatures greater than 0.5 T_m, and creep theories based on this mechanism have been developed for such metals [*64-66*]. However, none of these is sufficiently well developed to allow correct prediction of the stress dependency (n) of the deformation rate in the relation for steady state creep

$$\dot{\epsilon} \propto \sigma^n D_{sd} \tag{6.56}$$

where D_{sd} is the self diffusivity of the matrix.

(d) UTS—particle size relationships

The discussion of strength-structure relationships throughout Section 6.3.2 has been confined to dependence of *yield strength* on particle size and spacing, and there is little quantitative information available relating the UTS and microstructural features of alloys. Recently, however, Lui and Le May [*67, 68*] have reported on a linear relationship between UTS and the square root of inverse carbide size in several steels, for an approximately constant volume fraction of carbide and a given particle shape. This matter is referred to further in Section 8.2.2 as it is concerned more with fracture phenomena.

6.3.3 Grain Size Strengthening

Yield stress and grain size have been found to be related by the Hall-Petch equation [*69, 70*] for a wide range of metals and alloys. This states that

$$\sigma_Y = \sigma_o + k_Y d^{-1/2} \tag{6.57}$$

where σ_Y is the yield stress, σ_o and k_Y are constants, and d is the average grain diameter. This relationship has been observed to hold for the lower yield point in iron alloys (the stress required to propagate a Lüder's band along a test specimen) as well as the yield stress in other BCC, FCC, and HCP metals.

The explanation of the Hall-Petch equation is readily seen in terms of the barrier effect of grain boundaries to slip propagating from grain to grain in polycrystalline metal. At the end of a slip band restrained at a grain boundary, a stress concentration builds up owing to dislocation pile-up until the stress is sufficiently high to initiate a Frank-Read source in the adjacent grain. This stress is τ_c in Eq. (4.29).

The flow stress σ_ϵ can be expressed similarly as

$$\sigma_\epsilon = \sigma_{o\epsilon} + k_\epsilon d^{-1/2} \tag{6.58}$$

where $\sigma_{o\epsilon}$ and k_ϵ are constants which depend on strain. From Eq. (4.31), the shear stress on a slip band which is about to induce slip in a neighboring grain is

$$\tau = \tau_o + k_s d^{-1/2}$$

where τ_o is approximately the resolved easy glide shear stress for a single crystal[8], and k_s is a constant corresponding to the stress concentration required to propagate slip across the grain boundary.

---

[8] It may be somewhat greater than this because of multiple slip being required to deform the aggregate, from geometrical considerations.

For a polycrystal we may write the flow stress as [*71*]

$$\sigma_{\epsilon} = m\tau = m\,(\tau_o + k_s\, d^{-1/2}) \tag{6.59}$$

where m is the orientation or Taylor factor relating single crystal shear strength to polycrystal strength. Comparing Eqs. (6.58) and (6.59),

$$\left.\begin{aligned} \sigma_{o\epsilon} &= m\,\tau_o \\ k_{\epsilon} &= m\,k_s \end{aligned}\right\} \tag{6.60}$$

Because of the similarity between a slip band and a microcrack in terms of relaxation of shear stress, the maximum shear stress ahead of a slip band under a shear stress τ is given by $(\tau - \tau_o)\,(d/r)^{1/2}$ [*72*], and is in the plane of the slip band. This must reach a value τ_{max} in order to initiate slip at distance r, hence

$$k_s = \tau_{max}\, r^{1/2} \tag{6.61}$$

In the adjacent crystal, it is possible that the source will not lie in the maximum shear stress plane ahead of the slip band, and the magnitude of τ_{max} must be sufficient to operate the source, with a stress τ_d being reached in the required slip plane at the source. The distribution of stress is described in terms of the orientation factor m, so that, on average,

$$\tau_{max} = m\,\tau_d \tag{6.62}$$

From Eqs. (6.61) and (6.62)

$$k_s = m\,\tau_d r^{1/2} \tag{6.63a}$$

and from Eq. (6.77)

$$k_{\epsilon} = m^2\,\tau_d\, r^{1/2} \tag{6.63b}$$

From the foregoing analysis, the influence of various factors on the grain size dependence of flow (or yield) stress can be determined. For a limited number of slip systems m will be larger, and there will be a greater dependence of flow stress on grain size. The value of τ_d depends on the shear modulus G and the locking of dislocations, and such locking will raise τ_d and the grain size dependence of flow stress. Also, the distance r at which slip is initiated in an adjacent grain will be determined by dislocation density where internal Frank-Read sources operate.

The values of these constants for a number of metals are listed in Table 6.2, and it may be seen that for FCC metals, where m is not large because of the large number of slip systems[9], and τ_d is small because locking does not occur, then k_{ϵ} is small, as is also the dependence of yield stress on grain size and, indeed, this effectively disappears at larger strains than are shown. In the case

[9] A value of 3.1 is normally considered appropriate for FCC metals [*73*].

Table 6.2 Hall-Petch Constants for Room Temperature Deformation [71]

Metal or Alloy	σ_o, $\sigma_{o\epsilon}$ (MPa)	k_Y, k_ϵ (MPa . $m^{1/2}$)
BCC		
Mild steel, yield point	71	0.74
Mild steel, ϵ = 0.01	294	0.39
Swedish iron, yield point	47	0.71
Swedish iron, part decarburized, yield point	45	0.31
Swedish iron, no yield point, proportional limit	30	0.20
Molybdenum, yield point	108	1.77
Molybdenum, ϵ = 0.01	392	0.53
Niobium, yield point	69	0.04
FCC		
Copper, ϵ = 0.005	26	0.11
Aluminum, ϵ = 0.005	16	0.07
Silver, ϵ = 0.005	37	0.07
70/30 Brass, yield point	45	0.31
70/30 Brass, ϵ = 0.2	337	0.34
Al-3.5% Mg, yield point	49	0.26
HCP		
Zinc, ϵ = 0.005	32	0.22
Zinc, ϵ = 0.175	72	0.36
Magnesium, ϵ = 0.002	6.9	0.03
Titanium, yield point	78	0.40
Zirconium, ϵ = 0.002	29	0.25

of 70/30 brass, there is increased locking, and the grain size dependence persists to larger values of strain, and the value of k_ϵ is higher than for Cu or Al.

In BCC metals, where interstitial atoms are present, large values of τ_d and, hence, of k_ϵ are found, despite the fact that m is lower than for FCC metals, owing to the larger number of slip systems.

There is no locking in HCP metals, hence τ_d will be small. However, because of the very limited number of slip systems, m is large[10] and, hence so is k_ϵ in most instances.

[10] A value of m = 6.5 has been found to be appropriate for HCP metals [74].

Johnson [75] re-examined the validity of the Hall-Petch analysis as it pertains to BCC metals and, from consideration of the effects of grain size, temperature, and strain rate on their lower yield point, concluded that any one of these variables can change the rate-controlling deformation mechanism. He concluded that models developed on the basis of one mechanism alone, as have been used to interpret the Hall-Petch equation, are inadequate, and that there is probably an "inherent grain size effect" caused by variation in the width of the Lüder's front with grain size, which can account for part, at least, of the grain size dependence of yield stress. However, the overall utility of the Hall-Petch equation is well founded, and it forms a useful and practical basis on which to assess effects of grain size on the strength of commercial alloys.

6.3.4 Substructure Strengthening

Subgrain boundaries act in a similar manner to grain boundaries, restricting the movement of other dislocations, and they have been found to produce a strength increment having the form of the Hall-Petch equation [76, 77]. Thus, the yield stress is given by the relation

$$\sigma_Y = \sigma_o + k'_Y + d_s^{-1/2} \tag{6.64}$$

similar to Eq. (6.57), but where k'_Y is the subgrain equivalent of the Hall-Petch slope, and d_s is the subgrain (or cell) diameter.

Equation (6.64) implies that the stress required to initiate yielding at the subgrain boundary is not very sensitive to the misorientation between subgrains [78]. The value of k'_Y is less than that of k_Y (the Hall-Petch slope) for a similar material in polycrystalline form, but substructure-free, by a factor of from 2 to 5 [79]. However, as the size of subgrains can be controlled by hot or cold working, and as cells in the size range 0.5-5 μm can be produced, such processing does allow the production of material of very high strength, while still having a relatively large grain size.

The sub-boundaries can be strengthened by pinning with solute atoms, and strain-aging effects may be observed in alloys strengthened in this way. Segregation of solute to the boundaries also serves to increase the value of k'_Y, by requiring a higher stress to be applied at the boundary to force dislocations out of it.

In the high temperature deformation of metals, a cell-structure is frequently formed, this being associated with *dynamic recovery* which allows a simultaneous softening or recovery process to occur together with the plastic deformation. During both creep and hot working a steady state condition is achieved, whereby the flow stress remains constant during constant strain rate deformation over large strains. This implies constancy of structure, and Wong *et al.* [80] noted that the shape and size of the cells formed remains constant even after large deformations. Thus it appears that the cell walls under such conditions must undergo continuous rearrangement [81]. Figure 6.20 illustrates a typical substructure produced during hot working.

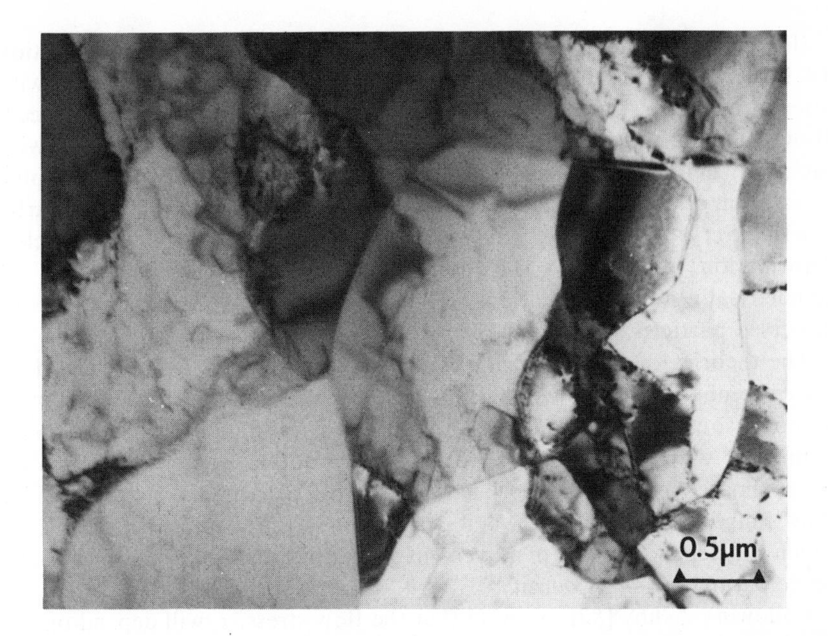

Fig. 6.20 Substructure produced by hot working of Type 304 stainless steel at 1000°C. $\dot{\epsilon}$ = 0.07 s^{-1}. Courtesy of Professor H. J. McQueen, Concordia University

As noted in Section 6.2, the dislocations produced by phase transformation can be treated quantitatively in terms of the equations for substructure strengthening (by means of subgrain boundaries) or for strain hardening. If the phase transformation gives rise to a cell structure, Eq. (6.64) may be applied: on the other hand, if a forest of random dislocations is produced, we can use the relations developed in Section 6.3.5.

6.3.5 Strain Hardening

The dislocations produced during plastic deformation may be stored subsequently in the crystal lattice during straining for two reasons. First, they may be required for geometrical reasons to allow compatibility between various parts of the specimen (e.g., in bending or in the deformation of a non-homogeneous alloy); second, they may accumulate by trapping each other in a random way. These latter dislocations are not required for shape change, for example in the uniform deformation of a homogeneous alloy. The two classes are termed *geometrically necessary dislocations* and *statistically stored dislocations*, respectively.

If we consider the deformation of a single crystal, as discussed in Section 4.4.1, there are three distinct stages, and the work hardening rate varies with orientation and strain, statistically stored dislocations being produced. However, in a polycrystalline metal, strain will be non-homogeneous, with each grain deforming by a different amount, depending on its crystallographic orientation and on the restraints imposed on it by adjacent grains. Similarly, in a single crystal or a polycrystalline solid reinforced with dispersed particles, deformation will always be non-homogeneous, and the number of geometrically stored dislocations will increase with increase in the number of dispersed particles.

The theories of work hardening in single crystals have been reviewed by Honeycombe [*10*]. Here we shall discuss a simple theory applicable to non-homogeneous deformation, thus to pure polycrystalline metals or ones reinforced with a dispersion of particles of another phase. To obtain an approximate picture of the work hardening of such materials, we can consider short range interactions only, and describe the dislocation substructure, on which will depend the ease of movement of a mobile dislocation, by means of a single parameter — its density.

Following Ashby [*82*], we note that the flow stress, τ, will depend on the elastic modulus, G, the Burgers vector, b, and the average dislocation density, $\bar{\rho}$. The only dimensionless relation of these quantities is

$$\tau / G = \alpha \, (b^2 \, \bar{\rho})^m \qquad (6.65)$$

where α and m are constants.

An applied stress, τ, exerts a force τb (see Section 4.3.3.) on a dislocation. Its motion is opposed by other dislocations, and the interaction force is proportional to b^2 [see Eq. (4.16)] for parallel dislocations, and this is true for intersecting ones also [*83*]. Hence $\tau \propto b$, and $m = 1/2$. Thus

$$\tau = \tau_0 + \alpha \, Gb(\bar{\rho})^{1/2} \qquad (6.66)$$

where τ_0 is the matrix yield strength or friction stress. Detailed calculation gives a value of $\alpha \simeq 0.2$ for FCC metals and $\alpha \simeq 0.4$ for BCC metals [*84*].

In a plastically non-homogeneous material, the geometrically necessary dislocation density, ρ^G, is given by the following equation [*82*],

$$\rho^G = (1/\lambda^G) \, (4\gamma/b) \qquad (6.67)$$

where λ^G is the *geometric slip distance* and is a characteristic of the microstructure, and γ is the shear strain. Such a material will have a parabolic stress-strain curve, as is observed in practice for dispersion hardened alloys [*30*], and for all polycrystals.

λ^G for well separated, equiaxed inclusions is given by r/f, where $2r$ is the diameter of an inclusion and f represents their volume fraction, while for plate-like particles of size comparable to their spacing L, $\lambda^G = L$ [*82*]. For a material without such particles or other obstacles, the grain size, d, is proportional to the geometric slip distance. Hence, we may write the yield stress as

$$\tau = \tau_0 + \alpha_1 G (b\gamma/d)^{1/2} \tag{6.68}$$

which is the Hall-Petch equation [cf. Eq. (6.57)].

The total mean dislocation density, $\bar{\rho}$, can be used to make approximate estimates of yield stress from Eq. (6.66). Ashby [*82*] demonstrated that the two components of dislocation density, ρ^G and ρ^S (the statistically stored dislocation density), appear to superimpose, with the larger contributor essentially controlling the flow stress. It is possible to estimate ρ^G in a quantitative manner, and this is the major component in the deformation of two-phase and polycrystalline materials. However, it is difficult to evaluate ρ^S in a quantitative manner, but this is, fortunately, of less importance in developing constitutive equations based on dislocation theory.

6.3.6 Two Phase Alloys

A special case of two phase alloys was dealt with at length in Section 6.3.2, in the discussion of precipitation or dispersion hardened alloys. Here we shall consider the more general case where there is a distribution of two phases in a material. Such a dispersion can have many forms, and we may consider a distribution of discrete particles, a continuous network, or lamellae of the second phase, and once more this may be either soft or hard with respect to the matrix.

An *estimate* of the overall mechanical properties of an alloy containing two ductile phases can be obtained by taking a weighted average of the properties of the two components, using the *Rule of Mixtures*.

If we assume that the strain in each phase is equal, we can write the average stress, $\bar{\sigma}$, in terms of the volume fractions of the two phases, f_1 and f_2 respectively, as

$$\bar{\sigma} = f_1\sigma_1 + f_2\sigma_2 \tag{6.69}$$

where σ_1 and σ_2 are the stresses in the two phases.

Alternatively, if we assume that the two phases are stressed equally, they will each be subject to different strains, ϵ_1 and ϵ_2, and we can write the average strain, $\bar{\epsilon}$, as

$$\bar{\epsilon} = f_1\epsilon_1 + f_2\epsilon_2 \tag{6.70}$$

Figure 6.21 illustrates the flow stress curves estimated using the two methods. Normally, the actual flow stress curve will lie between these two.

It will be realized that neither of the two expressions given above is completely sound physically, as strain is neither homogeneous as required by Eq. (6.69), nor is there a sharp discontinuity in it at the boundaries between the two phases as required by Eq. (6.70). In practice it is found that, with a

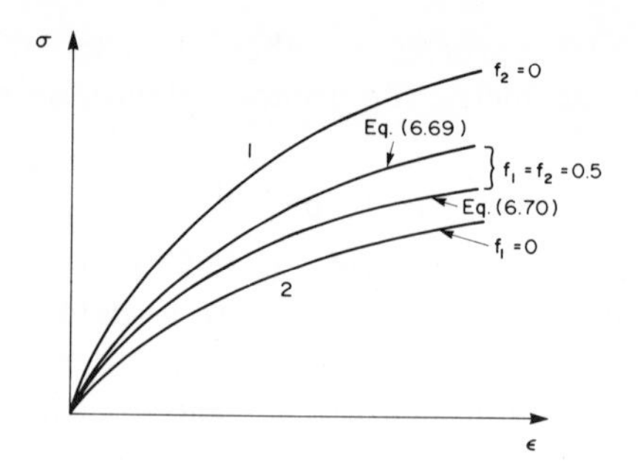

Fig. 6.21 Flow stress curves based on equal stress and equal strain in a two phase alloy. After Unkel [*85*]

dispersion of a harder phase in a softer matrix, the dispersed phase has larger strains at the interfaces than in its interior [*86*].

The strengthening effect of more brittle phases was studied by Gensamer [*87*], in terms of Fe_3C in steels. He found that, for both ferritic-pearlitic steels and tempered steels containing spheroidized carbides, the yield stress was proportional to the logarithm of the ferrite mean free path, the relationship breaking down only at very large mean free paths. Some useful approximate relationships for steels are given by Felbeck [*11*].

6.3.7 Combined Strengthening Mechanisms and Relations

Where an alloy is strengthened by more than a single mechanism, e. g., by a combination of solid solution and precipitation hardening, it may be possible to combine the mathematical expressions already derived for the various components of strengthening, to give an equation relating the strength of the complex alloy to its various microstructural components. We can write the yield strength of the alloy as

$$\sigma_Y = \sigma_o + \Delta\sigma_A + \Delta\sigma_P + \Delta\sigma_G + \Delta\sigma_S + \Delta\sigma_\epsilon \tag{6.71}$$

where σ_o is the lattice friction stress (Peierls-Nabarro force), and $\Delta\sigma_A$, $\Delta\sigma_P$, $\Delta\sigma_G$, $\Delta\sigma_S$ and $\Delta\sigma_\epsilon$ are the respective strength increments due to matrix solution hardening, precipitates or dispersed particles, grain boundaries, subgrain boundaries, and forest dislocations. If there are two phases present (excepting the dispersed or precipitated particles), it is necessary to apply these increments to the appropriate volume fraction only.

Unfortunately, there is not a large body of quantitative data available so that we can optimize the design and heat treatment of alloys in this way. There is also the problem that some component mechanisms do not act in a directly additive way, subsequent deformation of a precipitation hardened alloy containing fine coherent precipitates being one example which has been discussed already in a general way in Section 6.2. Nonetheless, development of such combined relations is receiving considerable attention today [*23*], as most commercial alloys do depend on several strengthening mechanisms (see Table 6.3).

As an example, consider the case of an overaged, precipitation hardened alloy, where the matrix yield strength, σ_o', may be taken to be independent of any solid solution effects. The yield strength, σ_Y, of the alloy can be expressed in terms of Eqs.(6.28a) and (6.28b) as

$$\sigma_Y = \sigma_o' + (0.8\ m\ Gb/L) \cos(\phi/2), \qquad \phi \leqslant 100° \quad (6.72a)$$

and

$$\sigma_Y = \sigma_o' + (mGb/L)\,[\cos(\phi/2)]^{3/2}, \qquad \phi \geqslant 100° \quad (6.72b)$$

where m is the Taylor factor [*73*] (see Section 4.4.2), and the other terms are as defined previously.

If the square lattice relationship [*30*], $L = (l/2)(\pi/f)^{1/2}$, is used, l being the particle diameter and f the volume fraction, we can rewrite these equations as

$$\sigma_Y = \sigma_o' + (1.6m\ Gb/l)\,(f/\pi)^{1/2} \cos(\phi/2), \qquad \phi \leqslant 100° \quad (6.73a)$$

and

$$\sigma_Y = \sigma_o' + (2m\ Gb/l)\,(f/\pi)^{1/2}\,[\cos(\phi/2)]^{3/2}, \qquad \phi \geqslant 100° \quad (6.73b)$$

For overaged materials f is virtually constant, and assuming ϕ to be independent of particle size as discussed in Section 6.3.2, it may be seen that the increment in yield stress due to precipitation, $\sigma_Y - \sigma_o'$, is a linear function of $1/l$, as has been observed in practice [*45*].

The relationship between grain size and matrix yield strength can be represented by

$$\sigma_o' = \sigma_o + k_Y\, d^{-1/2}$$

from Eq. (6.57), where σ_o signifies the resistance to flow *within* a grain. Substituting this into Eqs. (6.73a) and (6.73b), we obtain

$$\sigma_Y = \sigma_o + k_Y\, d^{-1/2} + (1.6m\ Gb/l)\,(f/\pi)^{1/2} \cos(\phi/2), \qquad \phi \leqslant 100° \quad (6.74a)$$

and

$$\sigma_Y = \sigma_o + k_Y\, d^{-1/2} + (2m\ Gb/l)\,(f/\pi)^{1/2}\,[\cos(\phi/2]^{3/2}, \qquad \phi \geqslant 100° \quad (6.74b)$$

Equations (6.74a) and (6.74b) show the combined effects of grain size and precipitation hardening on the basis that these are directly additive, and the data of Rezek [*45*] for a Cu-0.35 wt.% Cr alloy can be utilized to demonstrate that this is so for the particular alloy studied, at least. Lui and Le May [*88*] have shown that these data can be replotted to show a linear relationship between the critical stress for yield and interparticle spacing or inverse precipitate size in the absence of grain boundaries (i. e., with $d^{-1/2} = 0$). Figure 6.22 shows Rezek's data replotted to show that yield stress depends linearly on

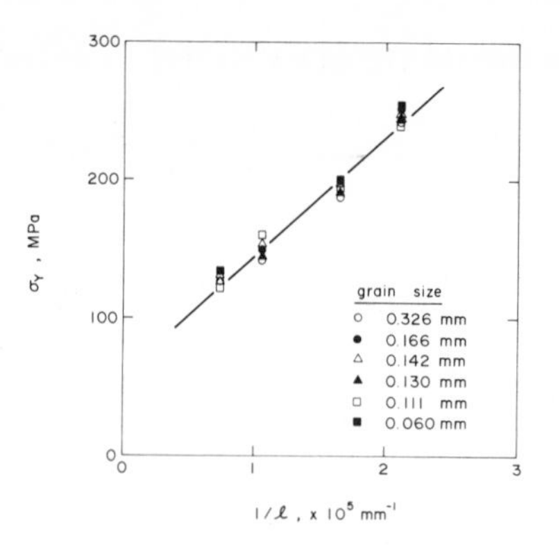

Fig. 6.22 Data of Rezek [*45*] replotted to show the effect of inverse particle size for Cu-0.35 wt. % Cr

inverse precipitate size, and indicates also that the effect of grain size is a very secondary one.

Equation (6.71) obviously can be expanded in terms of each of its component parts, but in practice it may be difficult to separate the various strengthening components explicitly, even when interaction does not take place. In this connection, the discussion by Christian [*89*] of the contribution of the various strengthening mechanisms to martensite is apposite.

A matter of some importance relates to the study of strengthening due to two different dispersions of deformable particles introduced by precipitation reactions. Such dispersions are present in an increasing number of high strength commercial alloys, and neither models nor theory have yet been developed to cover this situation, which would allow better definition of processing and alloying parameters. One other point of relevance may be mentioned. It has been suggested that there may be useful possibilities in providing a range of particle sizes, or even a bimodal particle size distribution [*79*]. In the case of an alloy strengthened by means of ordered precipitates, and in which the dislocations move in pairs, the strength is reduced from that which would obtain with single dislocations (see the discussion of order hardening in Section 6.3.2). Provision of a distribution of particle sizes might be expected to disrupt the orderly movement of dislocation pairs, so increasing strength. Such bimodal size distributions are easily produced in, for example, the Ni-based superalloys strengthed by γ' or γ'' precipitates, and the matter appears worthy of further and more detailed study.

Some further discussion of combined strengthening mechanisms is given in the following section, when dealing with the processing of some commercial alloys for high strength.

6.4 PROCESSING FOR HIGH STRENGTH

In order to produce high strength levels in commercial alloys, we must design the composition, heat treatment, and mechanical processing to take best advantage of the various strengthening mechanisms which have been discussed, and which can be applied to the specific system of concern. Most commercial alloys involve more than a single strengthening mechanism, and Table 6.3 illustrates this by listing the relevant strengthening mechanisms for a number of materials: also, Table 6.4 indicates the principal strengthening mechanisms operative for a number of materials used at elevated temperature. In the design of an alloy and the heat treatment to produce high strength level, matters are complicated to some degree because the factors which favor high strength at low temperatures do not necessarily favor good high temperature creep strength or resistance to fatigue. For example, while a high volume fraction of γ' precipitates is desirable for high yield and tensile strength in Ni-base superalloys, the creep strength of the γ' itself is very poor, and the relative fatigue resistance falls with increasing volume fraction of precipitates [*60*]. Thus, each individual alloy system must be assessed carefully, but it is instructive to examine some processing schedules used to produce high strength alloys, as they demonstrate the principles already discussed.

6.4.1 Processing Variables for Commercial Alloys

A number of specific alloys have been discussed in passing, but here we shall look at the overall processing variables and provide examples of a few specific cases.

First, having fixed upon the composition of the alloy, we must differentiate between production of a strong microstructure directly from the melt or by transformation from the solid state. The former case has been reviewed by Davies [*91*], and involves careful control of the solidification process, and we shall consider it first.

A major advantage of the production of components by controlled solidification is that quite complex shapes can be produced directly, so avoiding the costs and difficulties involved in intermediate shaping and machining operations. This is used to advantage in the aerospace industry, for example, in the casting of gas turbine blades. Another advantage is that it is possible, in many cases, to ensure that preferred orientation of microstructural features is accomplished, to provide optimum in-service properties.

Grain boundaries constitute a major source of *weakness* at high temperature, and considerable attention has been paid to the production of cast gas turbine blades having directional solidification to produce long grains running along the length of the blade, or more importantly, single crystal blades. Such components are subsequently hardened by heat treatment to

Table 6.3 Strengthening Mechanisms in Commercial Alloys [*90*]

Alloy and Condition	Approx. Yield Stress (MPa)	Strengthening Mechanisms
Al Alloys		
Al-Cu-Mg; heat treated, cold worked (2024-T3)	345	Strain hardening, precipitation
Al-Zn-Mg-Cu; annealed (7075-0)	100	Solid solution
Al-Zn-Mg-Cu; heat treated (7075-T6)	500	Precipitation
Cu Alloys		
Cu (electrolytic); annealed (0.04% O)	70	Strain hardening
Cu (OFHC); cold worked	275	Strain hardening
Cu-Zn (α-brass); annealed	240	Solid solution
Cu-Be; heat treated	965	Precipitation
Ni Alloys		
Ni-Cr-Fe (Inconel); cold worked	1035	Strain, solution hardening
Ni-Mo-Fe (Hastelloy B); heat treated	275	Solution hardening
Ni-Co-Cr-Mo-Ti-Al (Udimet 700); heat treated	825 (at 650° C)	Precipitation
Fe Alloys		
Fe (0.01% C), annealed	170	Solution hardening, grain boundary
Fe-C (0.2% C); annealed	310	Solution hardening, grain boundary
Fe-Ni-Mo-Mn-Cr-C (4340); heat treated	1380	Solution hardening, grain boundary, substructure
Maraging Steel 300; heat treated	2000	Precipitation, grain boundary, substructure
Tungsten Wire (cold drawn)	3725	Strain hardening, grain boundary

Table 6.4 Strengthening Mechanisms at Elevated Temperature [*90*]

Material	Principal Elevated Temperature Strengthening Mechanisms
Refractory Metals (Ta, W, Mo)	Intrinsic strength (Peierls-Nabarro force), grain boundary
Ni-based superalloys	Precipitation (γ' precipitates)
Co-based superalloys	Solid solution (W and Mo in solid solution)
TD-Nickel and TD-Nichrome	Dispersion (ThO_2 inclusions)
Sintered Aluminum Powder (SAP)	Dispersion (Al_2O_3 inclusions)
Oxides (e.g., Al_2O_3)	Intrinsic strength, precipitation

produce a fine γ' precipitate, and have much better creep resistance than do conventional polycrystalline blades, owing to the elimination of grain boundary sliding.

Controlled solidification can also be used to produce lamellar or rod-like eutectic structures in the case of alloys having appropriate phase diagrams. In general, we require an alloy having a eutectic reaction producing a hard, strong intermetallic phase and a solid solution, and the relative proportions of the two phases will depend on the composition of the alloy as well as on the phase diagram. Different volume fractions of the hard phase can be produced by directional solidification away from the eutectic point, without the production of separate grains of a primary phase, the eutectic-like structure being retained. Thus greater volume fractions of the rod-like phase can be obtained than are possible at the eutectic composition [*92*]. A number of such alloy systems have been studied, including Al-Al_3Ni [*91*], Nb-Nb_2C and Ta-Ta_2C [*93*], and Fe-Fe_2B [*94*]. Such alloys can achieve an effective combination of high strength and toughness, good fatigue resistance and high temperature strength, and hold considerable promise. Their strength can best be considered by treating them as composite materials, as discussed in Chapter 7.

Possibilities of improving strength by rapid solidification (for example by "splat cooling" [*95*]) have been examined for some time. Such processes produce metallic glasses, with no long range order in their structure. The properties of these are not particularly good for engineering purposes, as ductility is low: however it is possible to quench-in larger quantities of an element dissolved in the liquid phase than can be done using conventional solidification and cooling rates, and thus to produce much higher than normal volume fractions of precipitates, so providing higher strength in the aged condition.

Turning to strengthening in the solid state, the best known examples of processsing are the quenching of carbon steels to produce martensite, and the aging of aluminum alloys to produce a fine dispersed intermetallic precipitate, although both processes can be applied to a wide range of alloys having the appropriate phase diagrams (an allotropic transformation or eutectoid reaction in the first case, and solid solubility strongly dependent on temperature in the latter). However, we shall now discuss some of the methods of achieving higher strength in commercial materials in terms of processing variables and the operative mechanisms.

Dispersion or precipitation strengthening has already been discussed at some length, and a number of examples given of particular alloy systems in which this may be exploited. In practice, the greatest concern is to ensure that the cooling rate is sufficiently high that uncontrolled precipitation in an age-hardening alloy does not occur. Increasing attention is being paid today to the precipitation hardening of steels, either by means of carbides or intermetallics.

Grain refinement can be accomplished when an alloy can be cycled through an $\alpha \rightarrow \beta$ type transformation, and the best known example is the fine grain size which can be achieved when steel is cycled through the $\alpha \rightarrow \gamma$ transformation. Each austenite grain transforms to several ferrite grains on cooling, whereas one austenite grain is formed from each ferrite grain on rapid heating. Great refinement in grain size of steels can be achieved in this way [*96*], with appropriate increases in yield strength in accordance with the Hall-Petch relation [Eq. (6.57)]. Control of grain growth requires rapid heating or, alternatively, it may be inhibited by the presence of second phase particles which act to pin grain boundaries, the Al in a killed steel, for example, forming small aluminum nitride particles which serve to control austenite grain size [*97*]. Fine grain sizes may also be produced in materials having a duplex structure, for example by warm working two-phase (α and β) 60/40 brass. Similar duplex structures can be produced in many alloys by hot working a eutectic structure. Again, such alloys obey the Hall-Petch equation.

We shall now review the various strengthening mechanisms that are operative in a quenched and tempered steel. Christian has discussed the strength of martensite in some detail [*89*], and the general conclusions only will be summarized here. Martensite depends for its strength on the interstitial carbon in solution, a fine uniform distribution of carbides and intermetallics, and the intrinsic lattice resistance to dislocation motion (this last point refers to low temperature strength). Grain size effects are generally relatively small. Strength is increased by low temperature aging to precipitate carbides, further tempering to coarsen these reducing hardness and strength.

A high strength steel with good ductility and toughness may be produced by using a very low carbon iron-based alloy containing 17-19% Ni, 7-9% Co, 3-

5% Mo, together with small quantities of Ti and Al. Such *maraging steels* can have strength levels of the order of 2700 MPa after heat treatment, which consists of a solution anneal above 800° C, air cooling to room temperature to produce massive martensite, and aging at ~480° C to give a fine dispersion of intermetallics. The hardenability of such alloys is high, and they may readily be machined or formed before aging, because the martensite contains little carbon. Additional processes involving mechanical working as well as thermal treatments are considered in Section 6.4.2.

Recently there has been greatly increased interest in, and use made of high strength, microalloyed steels, and Gladman *et al.* [*98*] have made a careful review of the structure-property relationships in such alloys, attempting to provide quantitative relationships between the microstructural parameters and the mechanical properties. Korchynsky and Stuart [*99*] have also developed graphical relationships connecting grain size, the strengthening increment due to dispersion, the yield strength and toughness in ferritic steels, while Bramfitt and Marder [*100*] have used an additive approach between the various strengthening components in examining the effects of low temperature rolling on strength in a low carbon steel. They rolled plates, starting in the austenitic condition, to finish at progressively lower temperatures down to 150° C, this giving structures ranging from fully recrystallized to cellular dislocation substructures.

Bramfitt and Marder proposed an equation for yield strength in the form

$$\sigma_Y = \sigma_o + k_Y [d^{-1/2} f_r + d_s^{-1/2} (1 - f_r) k'_Y/k_Y \quad (6.75)$$

where

k_Y = experimentally determined value from a plot of σ_Y versus (grain size)$^{-1/2}$ for fully recrystallized structures of grain size d,

k'_Y = experimentally determined value from a plot of σ_Y versus (cell size)$^{-1/2}$ for fully cellular structure of cell size d_s,

and f = the volume fraction of recrystallized ferrite.

This relationship fitted reasonably well with the experimental data, although Gladman *et al.* [*98*] suggest that an effective grain size should be used, taking into account all boundaries which act as slip barriers. This does not improve the fit of predicted with experimental data, however, and there is considerable work still required to quantify fully the properties of such mixed structures.

6.4.2 Thermo-Mechanical Treatment (TMT)

Increased attention has been paid in recent years to improving the strength and toughness of alloys by controlled hot working, such processes being termed thermomechanical treatment or processing (TMT or TMP). While most such developments have been concerned with ferrous alloys, TMP may be used to improve the properties of non-ferrous alloys also, the essential thing about such a process being that the structure (and, hence, the combination of properties) cannot be achieved by either mechanical working or heat treatment separately, but only by their simultaneous operation. Some specific examples will be cited here to illustrate the principles involved and the useful results achieved for ferrous alloys.

Irani and Latham [*101*] used a TMT to produce a structure similar to that of a quenched and well-tempered steel. Termed *isoforming*, it involves the deformation of steel during the austenite→pearlite transformation, as illustrated in Fig. 6.23 (a). After a critical deformation (~70% reduction) is reached, the pearlite structure is broken up from its normal lamellar form to produce a fine dispersion of spheroidized carbides. Thus a structure (and corresponding properties) approximating to a well-tempered steel is produced without quenching being required, although careful scheduling of the mechanical working is necessary.

Ausforming is an important process in producing high strength steels of good toughness[11], and is illustrated schematically in Fig. 6.23 (b). Deformation of the metastable austenite causes the formation of a uniform high dislocation density and the simultaneous precipitation (through strain aging) of a fine dispersion of alloy carbides. On cooling to room temperature, martensite with a very small plate size is produced, and the resulting steel, comprising very fine martensite of high dislocation density, has an outstanding combination of properties, these varying with alloy composition and processing variables.

TRIP steels constitute a particular group of ausformed steels. In these, after ausforming, the steel is cooled below a temperature at which martensite may be formed by means of mechanical deformation (M_D) and above its martensite start temperature (M_S). Deformation of the austenite at a temperature between M_D and M_S causes the formation of very fine strain induced martensite, and the resultant product is strong and very ductile. The term "TRIP steels" arises from the method used to increase the ductility, namely **TR**ansformation **I**nduced **P**lasticity [*102*].

Grange [*103*] has discussed the use of thermomechanical processing for the production of oriented or fibrous structures having greatly improved strength properties in the longitudinal direction, at some sacrifice of transverse properties. Figure 6.24 shows schematically three of these processes to produce (a) oriented coarse pearlite, and (b and c) fibrous ferrite-martensite mixtures.

---

[11] Such steels also have excellent fatigue resistance.

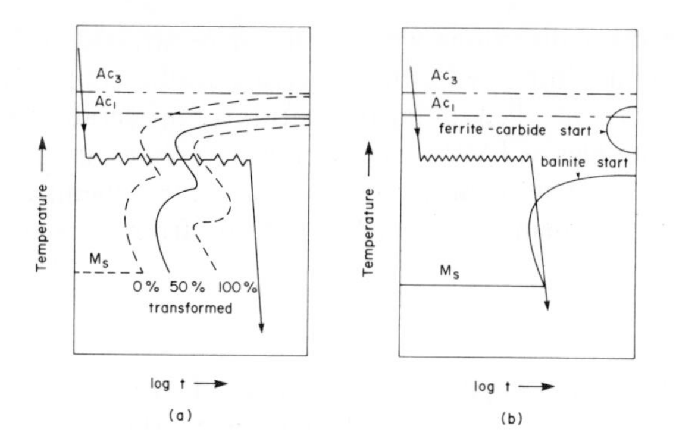

Fig. 6.23 Thermo-mechanical treatments. (a) Isoforming, involving periodic deformation and reheating during transformation from austenite to pearlite. (b) Ausforming, involving deformation of metastable austenite before transformation to martensite

In the first case, a 0.8 wt. % C steel is austenitized at high temperature to produce a coarse γ structure, and transformed just below A_1 to form very coarse pearlite. Mechanical working will cause the pearlite to become oriented, with the ferrite lamellae becoming cold worked and lacking in ductility. Rapid reheating to just above A_1 will cause transformation to austenite without allowing time for diffusion of carbon to take place to any extent, so that it remains partitioned between the prior ferrite and cementite lamellae. Air cooling will cause these to reform in their previous positions, giving an oriented pearlite structure, without any cold working to restrict ductility.

Figure 6.24 (b) shows the processing required to produce a fibrous ferrite-martensite mixture in a hypoeutectoid steel. Cold rolling of the hot rolled plate will produce an oriented, fibrous structure. Heating to above A_1 will allow the pearlite regions to transform to austenite, without giving time for carbon redistribution over long distances, or for loss of the oriented structure. Quenching will then produce martensite from the ~0.8% C austenite, in the form of long fibres in a ferrite matrix.

Finally, the process for AISI 4315 steel (applicable to similar alloys also), shown in Fig. 6.24 (c) involves holding at ~732° C to produce a mixed ferrite-austenite structure. On quenching this would produce an equiaxed ferrite-martensite mixture, however hot rolling will orient both ferrite and austenite, and subsequent quenching will produce a fibrous structure of ferrite and martensite.

In all of these processes the aim is to improve properties in the direction which will be subject to the greatest stress in the final product, with sacrifice of properties in other directions which will not be stressed so highly. Hence, effective strength is improved without increase in alloy content, but the cost of conducting such controlled rolling treatments is considerable. However, these and similar examples of TMP are becoming more widely used in practice. The longitudinal properties arising from the foregoing treatments are shown in the data of Table 6.5, being compared with those from conventional thermal treatments.

Table 6.5 Longitudinal Properties of Thermally and Thermomechanically Treated Steel [*103*]

(a) Oriented Coarse Pearlite in Strip Form (AISI 1086 Steel)

Structure	UTS (MPa)	0.2% Yield Strength (MPa)	Fracture Elongation* (%)
Random lamellae	670	336	10.5
Oriented lamellae	759	558	18.5

(b) AISI 1016 Rod

Treatment	UTS (MPa)	0.2% Yield Strength (MPa)	RA (%)	Fracture Elongation* (%)
Normalized: air cooled from 927° C	423	296	71	35
Quenched in brine from 927° C	1172	956	26	5
TMP + 25% cold work, tempered 204° C	1136	1109	30	5

(c) AISI 4315 Plate

Treatment	UTS (MPa)	0.2% Yield Strength (MPa)	Fracture Elongation* (%)	Hardness (HRC)
Quenched, tempered 204° C	992	670	9.5	29
TMP, tempered 204° C	1141	843	11.5	33.5

*All elongations based on a 25 mm gage length

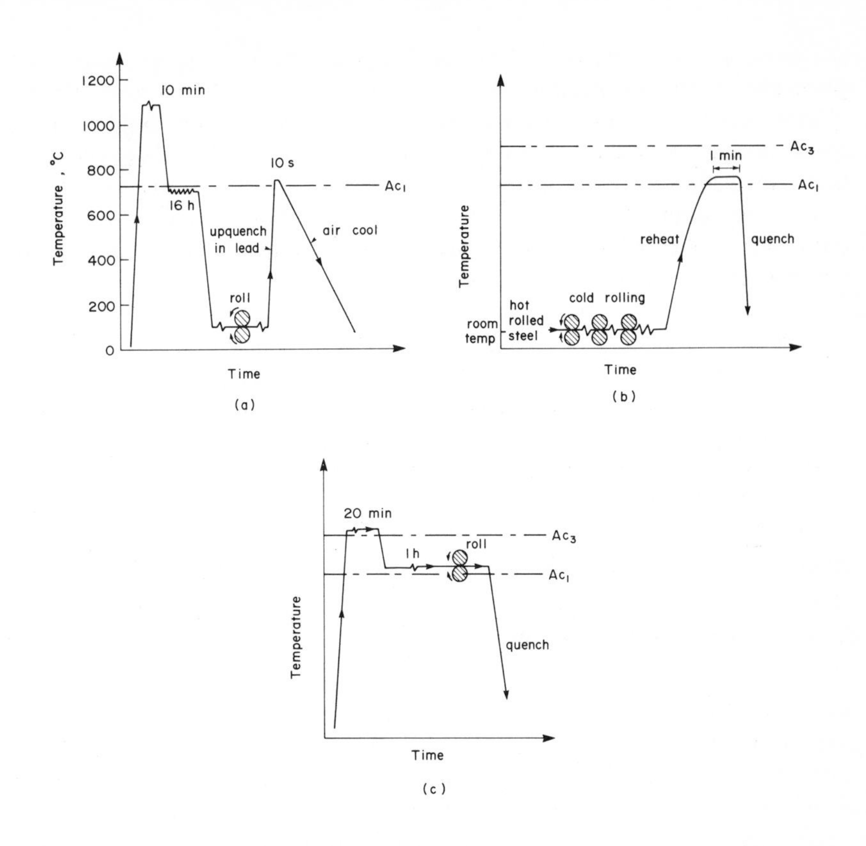

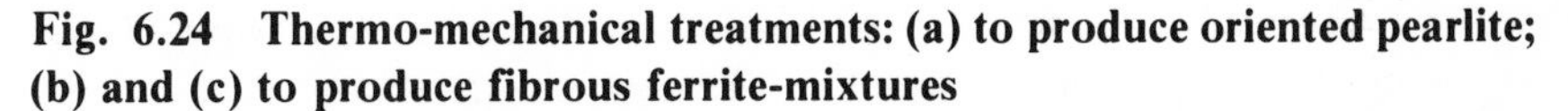

Fig. 6.24 Thermo-mechanical treatments: (a) to produce oriented pearlite; (b) and (c) to produce fibrous ferrite-mixtures

6.5 RADIATION DAMAGE AND RADIATION HARDENING

When crystalline materials are exposed to radiation in a nuclear reactor or a particle accelerator, several types of defect may be formed and, as a consequence of the localized loss of periodicity in the lattice, the material's properties may be altered. In complex alloys for use in reactor core structures or fuel element pins or assemblies, irradiation-induced precipitation (for example, the formation of $M_{23}C_6$ precipitates in austenitic stainless steels [*104*]) or dissolution of normally stable precipitates (for example, γ' dissolution in Ni-Al alloy or in Nimonic[12] PE16 [*105*]) may take place. The magnitude of these effects is dependent on many factors, principally temper-

[12] Trade Mark of Henry Wiggin and Co. Ltd., England.

ature and radiation flux, but including also prior cold working, grain size, and prior distribution of precipitates.

In general, radiation damage results in an increase in yield strength and a concomitant loss of ductility in metals, while enhanced creep and radiation-induced swelling may be expected. Consequently, it is imperative that such changes in properties during service be considered in the design of nuclear components, particularly for fast reactors where conditions of temperature and neutron flux density are both severe.

In non-metals, the consequence of exposure to radiation may be drastic, as in long-chain polymers which are embrittled to such an extent as to be mostly useless in reactor applications [*106*], while in an amorphous solid little damage can be done in terms of disordering a structure which is essentially completely disordered already. Graphite is widely used in gas-cooled reactors (including high temperature reactors) as a moderator and in structural components. In this case, the radiation damage which takes place and which would be expected to cause swelling and high stresses, also allows enhanced irradiation-creep so relieving these stresses and preventing fracture [*107*].

6.5.1 Mechanisms of Radiation Damage

Four main types of radiation damage may be identified: first, *electronic damage*, in which electrons are knocked out of their shells to higher energy states; second, *knock-on damage*, in which an atom is knocked out of its normal position, and may in turn knock out other atoms to form a *displacement spike* of localized damage, or may cause a high degree of agitation of other atoms which do not leave their stable positions, thus giving rise to a *thermal spike* where local heating has occurred to a temperature which may exceed the melting point; third, *transmutation damage*, caused by a neutron being captured by a nucleus and transforming it to a different chemical species, with accompanying release of energy, and possible release of an alpha particle or proton; and fourth, *void formation* under fast flux conditions, which will be discussed separately.

Electronic damage is virtually non-existent in metals because of the high mobility of their valence electrons, but is an important matter in materials such as semiconductors. Knock-on damage is caused primarily by *fast neutrons*, produced from fission reactions and having kinetic energies of the order of 3.2×10^{-13}J, and is an important mechanism in metals. Vacancy-interstitial pairs (Frenkel pairs) may be produced when an atom is knocked out of its lattice site [Fig. 6.25 (a)], and production of a spike of displaced atoms is illustrated in Fig. 6.25 (b). A thermal spike may cause recrystallization and phase change, or localized melting and solidification: in steel, hard brittle martensite may be formed in such localized regions. Transmutation damage is due largely to the absorption of *thermal neutrons*,

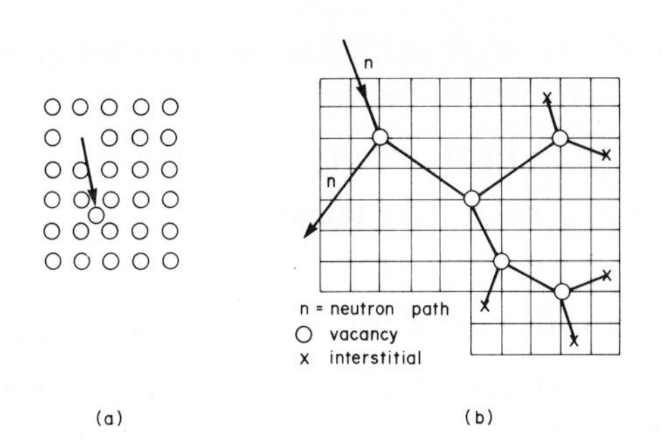

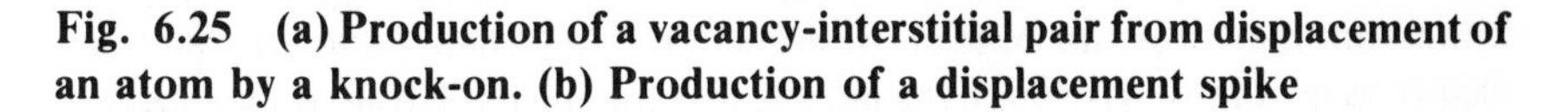

Fig. 6.25 (a) Production of a vacancy-interstitial pair from displacement of an atom by a knock-on. (b) Production of a displacement spike

and apart from the production of impurity atoms, the subsequent emission of helium nuclei (α-particles) or hydrogen nuclei (protons) can cause severe damage in terms of a displacement spike before the particle is brought to rest. Hydrogen or helium gas is also produced, the latter being much the more important factor in affecting the mechanical properties of the material.

An important consideration in examining damage accumulation in metals is the number of particles striking the material. Rather than considering the general picture which includes particles energized in an accelerator, the discussion will be confined to neutrons in a reactor. The treatment which follows is due to Stiegler and Weir [*108*].

The total flux of neutrons of all energies is given by

$$\phi = \int_0^\infty \phi(E)\, dE \tag{6.76}$$

expressed in neutrons.m^{-2}s^{-1}, and the time-integrated flux or *fluence* is

$$\int_0^t \phi(t)\, dt$$

The number of interactions between the neutrons in the flux and the metallic atoms is given by

$$N_r / N_o = \int \phi(E)\, \sigma(E)\, dE \tag{6.77}$$

where

N_r / N_o = fractional number of interactions per second
ϕ = neutron flux
σ = microscopic cross-section, a measure of the probability per neutron of an interaction (m^2)

The maximum energy transferred when a particle of mass m_1 and energy E_1

strikes a particle of mass m_2 at rest is given by the following expression, which is based on elastic interactions between hard spheres:

$$E_{max} = [4\, m_1\, m_2/(m_1 + m_2)^2]E_1 \tag{6.78}$$

and, since the neutron has a mass number of 1,

$$E_{max} \simeq 4\, E_1/A_2 \tag{6.79}$$

where A_2 is the mass number of the particle which is struck.

If the energy which is transferred to the struck atom exceeds some threshold value, the atom will be displaced, and may in turn interact with other atoms in the lattice before gradually being brought to rest. At high energies the primary knock-on may lose its outermost electrons, and then gradually pick up electrons as it moves through the lattice, so reducing its charge. Hence, at high energies we may expect the energy loss to be principally by ionization, while elastic collisions producing secondary displacements become more frequent as the energy of the knock-on decreases.

Kinchin and Pease [*109*] assume that there is a cutoff energy approximately equal to 1000 m_2, above which primary knock-on causes ionization only and below which it produces displacements only.

The number of additional atoms displaced for each primary knock-on is given by[13]

$$N_d = E/2E_d \qquad 2\, E_d < E < E_i \tag{6.80a}$$

and

$$N_d = E_i/2\, E_d \qquad E > E_i \tag{6.80b}$$

where

E is the energy of the primary knock-on
E_d is the displacement threshold energy
E_i is the ionization threshold energy.

Substituting values appropriate to typical thermal reactor conditions (ϕ = 10^9 neutrons . $m^{-2}\ s^{-1}$, $E_1 = 1.6 \times 10^{-13}$ J, $\sigma = 3 \times 10^{-28}\ m^2$), and for one year's exposure,

$$N_r/N_o \simeq 1 \times 10^{-3}$$

For iron [$m = 56$, $E_d \simeq 40 \times 10^{-19}$ J, $E_i \simeq 90 \times 10^{-16}$ J], $E_{max} \simeq 0.11 \times 10^{-13}$ J $> E_i$. Hence

$$N_d \simeq 10^3$$

Thus, we see that, on the basis of these simplified calculations, one in a

[13] Note the use of 2 E_d in the denominator. Since E_d is required to displace the atom originally in the site, 2 E_d is required if the impinging atom is to leave also.

thousand atoms may be expected to receive a primary displacement, while each atom in the structure can be expected to be displaced as a result of knock-on during one year. Hence it may be seen that radiation can introduce many defects (interstitials and vacancies) into thermal reactor core components. One of the factors mitigating its severity, however, is the annealing out of those defects which are thermodynamically unstable[14]. This is aided by the greatly increased diffusion rates which obtain in material subject to radiation and containing excess vacancies [*105*]. Nonetheless, from the analysis, we may expect there to be many single point defects, small clusters of point defects, and larger areas containing excess vacancies: because of the localized nature of the spike phenomenon, an inhomogeneous distribution is to be expected.

Our knowledge of radiation damage depends to a large extent on thin foil transmission electron microscopy, and the principal features which have been observed will be summarized here. In addition to indirect studies which involve the examination of thin foils produced from bulk samples maintained in a reactor or other neutron source for some time, much information has been developed by direct observation of the effects of bombardment of thin foils within the electron microscope using either high-energy electrons or ions.

When metals are irradiated at low temperatures ($< 0.3\ T_m$), black spots several nm in diameter are formed. These have been found to consist of defect clusters and loops [*110*], and more detailed studies have shown that two types of defect cluster exist [*111, 112*], large interstitial loops and small vacancy clusters. Point defect clusters can form even during low temperature irradiation down to 8 K, and dislocation loops can form at temperatures as low as 145 K in copper [*113*]: the diffusion which takes place to allow growth of these clusters and loops may be explained in terms of radiation-induced diffusion [*114*]. At higher temperatures the black spots become darker and more widely distributed, and as temperature is increased to the extent that thermally activated diffusion is rapid, they may not be observed at all. At such temperatures (e. g., 650° C and above for stainless steels) and under radiation conditions obtaining in thermal reactors, helium gas bubbles are to be found precipitated through the material, with preferential precipitation at grain boundaries. As indicated earlier, irradiation induced precipitation may also occur, with new or non-equilibrium phases being formed, or else existing precipitates may be redissolved: this matter of structural stability has been reviewed by Hudson [*105*].

Void formation

In fast flux conditions, as in fast breeder reactors in which the fluences to

---

[14] Vacancies are thermodynamically stable to an extent dependent on temperature — see Section 4.3.8. The number produced by radiation damage is normally in excess of this.

which components are exposed may be up to ~5 x 10^{26} neutrons.m^{-2}, it has been found that prolific void formation may take place, with swelling to the extent of a 15 percent volume increase in some cases [*115*]. The phenomenon was first reported by Cawthorne and Fulton [*116*] in 1967, and has since been observed in a wide range of metals and alloys over specific temperature ranges. The temperature range is different for different materials, and for annealed stainless steels in a fast reactor the peak void formation rate corresponds to about 525° C.

The subject of void formation was the theme of a 1971 conference [*117*], and an excellent review of observations of various materials, including the influence of fluence, temperature and structural variables on void development and stability, has been given by Bagley *et al.* [*118*]. It has been established that a high dosage of fast neutrons is not by itself a sufficient factor to cause void formation, but that a high rate of thermal migration is also required. The mechanism of formation involves the production of vacancies and interstitials by radiation, and their migration to sinks. Because of elastic interactions, interstitials migrate more rapidly to dislocations than do vacancies, leading to an excess of the latter which agglomerate to form voids. From the (n, α) reaction, helium atoms are present in the lattice, and these are attracted to the small voids, so stabilizing them and allowing them to act as sinks for additional vacancies. Void formation does not occur so readily at higher temperatures because the number of vacancies present from thermal equilibrium considerations becomes comparable to that from irradiation damage: at lower temperature the rate decreases because of decreased vacancy mobility and a higher rate of annihilation of interstitials by vacancy-interstitial recombination. Also of importance is dislocation density, since if this is high, dislocations may also act as vacancy sinks, so reducing void formation and swelling: if no dislocations were present there would be no preferential sinks for the interstitials and hence no void formation, since there would be no vacancy supersaturation.

The major effect of void formation, apart from swelling, is the high temperature embrittlement which it leads to.

6.5.2 Irradiation Hardening

Polycrystalline metals are hardened by exposure to radiation, as noted above, their hardness increasing with radiation dose. This is to be expected as the interaction of the point defects produced by radiation with dislocations will impede movement of the latter, in accordance with the discussions in Sections 6.2 and 6.3. It has been observed by many investigators that hardness and yield strength increase rapidly with neutron irradiation at first, but that the rate of increase slows down as the dose increases, and Fig. 6.26 shows some typical stress-strain curves obtained for copper irradiated to different degrees

[*119*]. It may be seen that the work hardening rate after yielding decreases after irradiation, with UTS increasing to a lesser extent than the YS. Some typical values of strength and ductility for various engineering alloys are given in Table 6.6.

While irradiation hardening might be considered a useful and beneficial phenomenon, the decrease in ductility may more than compensate for this. This decrease arises because the initial yield stress may be raised to such an extent that strong work hardening no longer occurs, and the material may reach the condition of tensile instability with little or no uniform elongation (see Section 3.1.3). Hence a fuel element or other structural component exposed to radiation will be subject to localized thinning at points where strain is concentrated, with consequent danger of rupture. In BCC metals, the increase in yield stress with fracture stress remaining relatively constant causes a large increase in ductile-brittle transition temperature, with consequent danger of brittle fracture in low temperature conditions.

Several equations have been proposed to describe the observed increase in yield stress, the three most important ones being described below.

First, if saturation effects are ignored, as at low doses, the number of defects per unit volume N_v may be taken as proportional to the neutron dose received, (ϕt). If the distribution of defects is assumed to be random, the average separation distance between defects on the slip plane will be proportional to $(N_v)^{-0.5}$ and, on the basis of hardening by a dispersion of barriers where the increase in yield stress depends inversely on the obstacle spacing (see Section 6.3.2), the strengthening increment will be proportional to the square root of

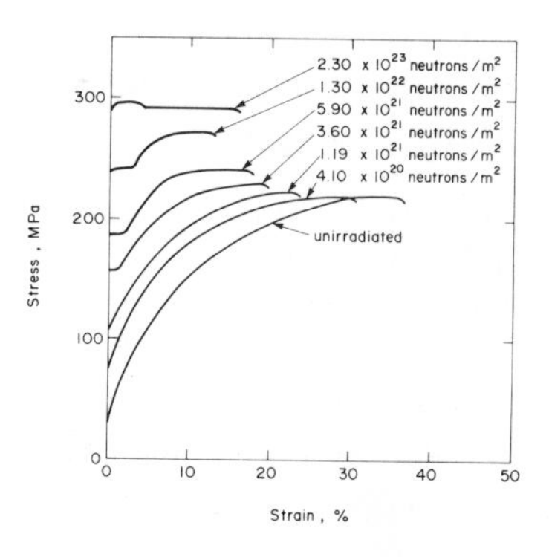

Fig. 6.26 Stress-strain curves of polycrystalline copper tested at room temperature after different irradiation doses. Radiation doses are in neutrons.m^{-2}. After Makin [*119*]

Table 6.6 Effect of Neutron Irradiation on the Strength and Ductility of Various Alloys [*120*]

Material	Integrated Fast Flux, ϕt (neutrons.m^{-2})	Radiation Temperature (K)	Tensile Strength (MPa)	Yield Strength (MPa)	Elongation (%)
Austenitic SS Type 304	0	-	576	235	65
	1.2×10^{25}	373	720	632	42
Low Carbon Steel A-212 (0.2% C)	0	-	517	276	25
	2×10^{23}	352	676	634	6
	1×10^{24}	352	800	752	4
	2×10^{23}	566	703	524	9
	2×10^{23}	677	579	386	14
Aluminum 6061-0	0	-	124	65	28.8
	1×10^{24}	339	257	177	22.4
Aluminum 6061-T6	0	-	310	265	17.5
	1×10^{24}	339	349	306	16.2
Zircaloy-2	0	-	276	155	13
	1×10^{24}	411	310	279	4

the neutron dose. Thus

$$\Delta\sigma \simeq A\ (\phi t)^{0.5} \tag{6.81}$$

Second, Blewitt *et al.* [*121*] have proposed, on the basis of irradiation hardening of copper, that strengthening should be given by

$$\Delta\sigma \simeq A\ (\phi t)^{1/3} \tag{6.82}$$

and the relative validity of Eqs. (6.81) and (6.82) has been discussed by many authors (e. g., [*122, 123*]). Certainly, there should be deviations from the $(\phi t)^{0.5}$ relationship at high doses as the proportional relationship assumed between the barrier spacing and yield stress increase becomes less accurate due to interaction between opposing arms of the bowing dislocation segment [*124*].

The third relation, proposed by Makin and Minter [*125*], is

$$\Delta\sigma = \alpha\,[1 - \exp(-\beta\phi t)]^{0.5} \tag{6.83}$$

where α and β are constants. This equation has been found to describe the behavior of a wide range of BCC, FCC and HCP metals and alloys [*126*]. In its derivation it is assumed that small obstacles can be overcome by dislocations, and that the increase in yield stress will be proportional to the square root of the obstacle concentration, c, so that[15]

$$\Delta\sigma = A\, c^{0.5} \qquad (6.84)$$

It is further assumed that the rate of obstacle production, $dc/d(\phi t)$, decreases linearly with increase in concentration. This assumption takes into account saturation effects, and assumes each defect to be surrounded by a volume V, within which no additional defects can be produced. Hence the constants of Eq. (6.83) are

$$\alpha = A\, V^{-0.5}$$

$$\beta = \Sigma\, V$$

where Σ is the macroscopic cross-section for the production of the defects. Under conditions of low doses, Eq. (6.83) reduces to $\Delta\sigma \propto (\phi t)^{0.5}$, and thus can represent data where Eq. (6.81) might also be used.

After irradiation, it has been found that the flow stress of FCC metals is strongly temperature and strain rate dependent, showing that the barriers to movement can be overcome by thermal activation. Makin *et al.* [*127, 128*] have studied the deformation of irradiated and annealed copper, and correlated the density and size of defect clusters with mechanical properties: it was demonstrated that the grains initially contain a spectrum of obstacles of varying size in the form of vacancy clusters, all of which have sufficient strength at 4.2 K to cause dislocations to bow between them, but which become effectively transparent to dislocations at various temperatures by thermal activation. Hence the deformation mechanisms for irradiated copper have been clarified [*129*]. Irradiation produced obstacles are destroyed by the deformation, making it easier for a dislocation source to emit a second dislocation after the first has moved out. Consequently, dislocations from an active source catch up on the first dislocation loop formed, with an increase in the effective stress at the top of the pile-up [Eq. (4.25)], and slip bands form very rapidly. This leads to a jerky stress-strain curve, with deformation by Lüder's bands whenever the stress is thermally dependent.

The interaction of dislocations with small voids has been examined by Coulomb [*130*], who found that the athermal increase in strength was given by an equation of the Orowan type,

$$\Delta\sigma = 2Gb/L \qquad (6.85)$$

where L is the average distance between cavities along the dislocation line. This cavity hardening theory has been shown to describe the strengthening due to voids produced by irradiation in an austenitic stainless steel [*131*].

---

[15] This relationship is derived in Section 6.3.1 for dislocations held up at dispersed barriers, which they can bypass by thermal activation [see Eq. (6.12)].

The hardening mechanisms in BCC and HCP metals are considered generally similar to those occurring in FCC metals and as discussed above for copper, but they are complicated by the interstitial atoms normally present [*129*]. These impurity atoms can strengthen the obstacles to dislocation movement by diffusing to them either during irradiation or in subsequent annealing. Vacancy clusters strengthened by interacting interstitials may be sufficiently strong that they are not destroyed by moving dislocations, causing a change in slip mode (to fine slip as seen in dispersion hardened alloys), and in strain hardening characteristics. While not of great significance to the strengthening of FCC alloys, interstitial loops are considered likely to contribute significantly to the strengthening of BCC metals, such as α-iron, as they are in the appropriate size range [*124*]. Another factor distinguishing BCC from FCC metals is that in the former the Peierls barrier to dislocation motion is high, which produces a strong dependence of the yield stress on temperature in the absence of radiation. The effect dominates at low temperatures, hence radiation induced hardening is comparatively athermal under such conditions [*132*].

REFERENCES

1. Kelly, A., and Nicholson, R. B., (eds.), *Strengthening Methods in Crystals*, Elsevier, London (1971).
2. Stoloff, N. S., *Strengthening Methods in Crystals*, p. 193 (A. Kelly and R. B. Nicholson, eds.), Elsevier, London (1971).
3. Wilm, A., *Metallurgie, Zeitschrift für die gesamte Hüttenkunde: Aufbereitung—Eisen und Metallhüttenkunde—Metallographie*, (W. Borchers and F. Wüst, eds.), Vol. 8 (1911).
4. Martin, J. W., *Precipitation Hardening*, Pergamon Press, Oxford (1968).
5. Krishnadev, M. R., and Le May, I., *Journal of the Iron and Steel Institute*, **208**, 458 (1970).
6. Orowan, E., *Symposium on Internal Stresses in Metals*, p. 451, Institute of Metals, London (1948).
7. Tegart, W. J. McG., *Elements of Mechanical Metallurgy*, Macmillan, New York (1966).
8. Greetham, G., and Honeycombe, R. W. K., *J. Inst. Metals*, **89**, 13 (1960-61).
9. Nicholson, R. B., Thomas, G., and Nutting, J., *J. Inst. Metals*, **87**, 429 (1958-59).
10. Honeycombe, R. W. K., *The Plastic Deformation of Metals*, Ed. Arnold, London (1968).
11. Felbeck, D. K., *Introduction to Strengthening Mechanisms*, Prentice-Hall, Englewood Cliffs, New Jersey (1968).
12. Friedel, J., *Dislocations*, Pergamon Press, Oxford (1964).
13. Diehl, J., *Vacancies and Interstitials in Metals*, p. 739, (A. Seeger *et al.*, eds.), North-Holland, Amsterdam (1969).

14. Cochardt, A. W., Schoek, G., and Wiedersich, H., *Acta. Met.*, **3**, 533 (1955).
15. Saxl, I., *Czech. J. Phys.*, **B14**, 381 (1964).
16. Kelly, A., *Strong Solids*, Clarendon Press, Oxford (1966).
17. Fisher, J. C., *Acta. Met.*, **2**, 9 (1954).
18. Flinn, P. A., *Acta. Met.*, **6**, 631 (1958).
19. Flinn, P. A., *Phys. Rev.*, **104**, 350 (1956).
20. Stoloff, N. S., and Davies, R. G., *Acta. Met.*, **12**, 473 (1964).
21. Brown, L. M., and Ham, R. K., *Strengthening Methods in Crystals*, p. 9 (A. Kelly and R. B. Nicholson, eds.), Elsevier, London (1971).
22. De Wit, R., and Koehler, J. S., *Phys. Rev.*, **116**, 1113 (1959).
23. Kelly, P. M., *Int. Metall. Reviews*, **18**, 18 (1973).
24. Foreman, A. J. E., and Makin, M. J., *Phil. Mag.*, **14**, 911 (1966).
25. Ashby, M. R., *Acta. Met.*, **14**, 679 (1966).
26. Lui, M.-W., and Le May, I., *Metal Sci.*, **8**, 435 (1974).
27. Scarlin, R. B., and Edington, J. W., *Metal Sci. J.*, **7**, 208 (1973).
28. Lui, M.-W., and Le May, I., *Scripta Met.*, **9**, 587 (1975).
29. Phillips, V. A., *Acta. Met.*, **14**, 1553 (1966).
30. Ashby, M. F., *Z. Metallkunde*, **55**, 5 (1964).
31. Lui, M.-W., and Le May, I., *Metal Sci.*, **11**, 54 (1977).
32. Pattanaik, S., Tromans, D., and Lund, J. A., *Proc. Int. Conf. on the Strength of Metals and Alloys*, p. 381, Japan Inst. of Metals, Tokyo (1967).
33. Fujii, A., Nemoto, M., Suto, H., and Monma, K., *Proc. Int. Conf. on the Strength of Metals and Alloys*, p. 374, Japan Inst. of Metals, Tokyo (1967).
34. Goodman, S. R., Brenner, S. S., and Low, J. R., *Met. Trans.*, **4**, 2363 and 2371 (1973).
35. Russell, K. C., and Brown, L. M., *Acta. Met.*, **20**, 969 (1972).
36. Munjal, V., and Ardell, A. J., *Acta. Met.*, **23**, 513 (1975).
37. Raynor, D., and Silcock, J. M., *Metal Sci. J.*, **4**, 121 (1970).
38. Martens, V., and Nembach, E., *Acta. Met.*, **23**, 149 (1975).
39. Jack, D. H., and Guiu, F., *Proc. 2nd Int. Conf. on the Strength of Metals and Alloys*, Vol. II, p. 606, ASM, Metals Park, Ohio (1970). Jack, D. H., and Honeycombe, R. W. K., *Acta Met.*, **20**, 787 (1972).
40. Tyson, W. R., *Acta Met.*, **11**, 61 (1963).
41. Fukui, S., and Uehara, N., *Proc. 2nd Int. Conf. on the Strength of Metals and Alloys*, Vol. II, p. 616, ASM, Metals Park, Ohio (1970).
42. Liu, C. T., and Gurland, J., *TMS-AIME*, **242**, 1535 (1968).
43. Lewis, M. H., and Martin, J. W., *Acta. Met.*, **11**, 1207 (1963).
44. Kupcis, A., Ramaswami, B., and Woo, O. T., *Acta. Met.*, **21**, 1131 (1973).
45. Rezek, J., *Can. Metall. Quarterly*, **13**, 545 (1974).
46. Brown, L. M., *Scripta Met.*, **9**, 591 (1975).
47. Gerold, V., and Haberkorn, H., *Phys. Stat. Solidi*, **16**, 675 (1966).
48. Gleiter, H., *Acta. Met.*, **16**, 829 (1968).
49. Kelly, A., and Nicholson, R. B., *Prog. Materials Sci.*, **10**, 151 (1963).
50. Harkness, S. D., and Hren, J. J., *Met. Trans.*, **1**, 43 (1970).
51. Hirsch, P. B., and Kelly, A., *Phil. Mag.*, **12**, 881 (1965).
52. Gerold, V., and Hartmann, K., *Phil. Mag.*, **12**, 509 (1965).

53. KNOWLES, G., and KELLY, P. M. Quoted in Ref. [*23*].
54. WEEKS, R. W., PATE, S. R., ASHBY, M. F., and BARRAND, P., *Acta Met.*, **17**, 1403 (1969).
55. GLEITER, H., and HORNBOGEN, E., *Phys. Stat. Solidi*, **12**, 251 (1965).
56. HAM, R. K., *Ordered Alloys: Structural Applications and Physical Metallurgy*, p. 365, Claitor's, Baton Rouge, Louisiana (1970).
57. CHOU, Y. T., and ESHELBY, J. D., *J. Mech. Phys. Sol.*, **10**, 27 (1962).
58. COTTRELL, A. H., *Dislocations and Plastic Flow in Crystals*, Clarendon Press, Oxford (1953).
59. FOREMAN, A. J. E., *Acta Met.*, **3**, 322 (1955).
60. STOLOFF, N. S., *The Superalloys*, p. 79, (C. T. Sims and W. C. Hagel, eds.), Wiley, New York (1972).
61. ASHBY, M. F., *Physics of Strength and Plasticity*, p. 113, (A. Argon, ed.), MIT Press, Cambridge, Mass. (1969).
62. WILCOX, B. A., and CLAUER, A. H., *The Superalloys*, p. 197, (C. T. Sims and W. C. Hagel, eds.), Wiley, New York (1972).
63. ANSELL, G. S., *Oxide Dispersion Strengthening*, p. 61, (G. S. Ansell *et al.*, eds.), Gordon and Breach, New York (1968).
64. WILCOX, B. A., and CLAUER, A. H., *Oxide Disperson Strengthening*, p. 323 (G. S. Ansell *et al.*, eds.), Gordon and Breach, New York (1968).
65. ANSELL, G. S., and WEERTMAN, J., *Trans. AIME*, **215**, 838 (1959).
66. CLAUER, A. H., and WILCOX, B. A., *Mat. Sci. J.*, **1**, 86 (1967).
67. LUI, M.-W., and LE MAY, I., *Met. Trans.*, **6**, 583 (1975).
68. LUI, M.-W., and LE MAY, I., *Trans. ASME, J. Eng. Matls. and Technol.*, **98**, 173 (1976).
69. HALL, E. O., *Proc. Phys. Soc.*, **B64**, 747 (1951).
70. PETCH, N. J., *JISI*, **174**, 25 (1953).
71. ARMSTRONG, R. W., CODD, I., DOUTHWAITE, R. M., and PETCH, N. J., *Phil. Mag.*, **7**, 45 (1962).
72. ESHELBY, J. D., FRANK, F. C., and NABARRO, F. R. N., *Phil. Mag.*, **42**, 351 (1951).
73. TAYLOR, G. I., *J. Inst. Metals*, **62**, 307 (1938).
74. PRASAD, Y. V. R. K., MADHAVA, N. M., and ARMSTRONG, R. W., *Grain Boundaries in Engineering Materials*, p. 67, (J. L. Walter, J. H., Westbrook, and D. A. Woodford, eds.), Claitor's, Baton Rouge, Louisiana (1975).
75. JOHNSON, A. A., *Can. Met. Quarterly*, **13**, 215 (1974).
76. WARRINGTON, D. H., *JISI*, **201**, 610 (1963).
77. REZEK, J., and CRAIG, G. B., *TMS-AIME*, **221**, 715 (1961).
78. EMBURY, J. D., *Strengthening Methods in Crystals*, p. 331, (A. Kelly and R. B. Nicholson, eds.), Elsevier, London (1971).
79. KELLY, A., and NICHOLSON, R. B., *Strengthening Methods in Crystals*, p. 615 (A. Kelly and R. B. Nicholson, eds.), Elsevier, London (1971).
80. WONG, W. A., MCQUEEN, H. J., and JONAS, J. J., *J. Inst. Metals*, **95**, 129 (1967).
81. MCQUEEN, H. J., WONG, W. A., and JONAS, J. J., *Can. J. Phys.*, **45**, 1225 (1967).
82. ASHBY, M. F., *Strengthening Methods in Crystals*, p. 137 (A. Kelly and R. B. Nicholson, eds.), Elsevier, London (1971).

83. PEACH, M., and KOEHLER, J. S., *Phys. Rev.*, **80**, 436 (1950).
84. NABARRO, F. R. N., BASINSKI, Z. C., and HOLT, D. B., *Adv. Phys.*, **13**, 193 (1964).
85. UNKEL, H., *J. Inst. Metals*, **61**, 171 (1937).
86. HONEYCOMBE, R. W. K., and BOAS, W., *Aust. J. Scient, Res.*, **1**, 70 (1948).
87. GENSAMER, M., *Trans. ASM*, **36**, 30 (1946).
88. LUI, M.-W., and LE MAY, I., *Can. Met. Quarterly*, **15**, 37 (1976).
89. CHRISTIAN, J. W., *Strengthening Methods in Crystals*, p. 261, (A. Kelly and R. B. Nicholson, eds.), Elsevier, London (1971).
90. BARRETT, C. R., NIX, W. D., and TETELMAN, A. S., *The Principles of Engineering Materials*, Prentice-Hall, New Jersey (1973).
91. DAVIES, G. J., *Strengthening Methods in Crystals*, p. 485, (A. Kelly and R. B. Nicholson, eds.), Elsevier, London (1971).
92. MOLLARD, F. R., and FLEMINGS, M. C., *TMS-AIME*, **239**, 1526 and 1534 (1967).
93. LEMKEY, F. D., and SALKIND, M. J., *Crystal Growth*, p. 171 (H. S. Peiser, ed.), Pergamon Press, Oxford (1967).
94. DE SILVA, A. R. T., and CHADWICK, G. A., *Metal. Sci. J.*, **3**, 168 (1969).
95. KLEMENT, W., Jr., WILLENS, R. H., and DUWEZ, P., *Nature*, **187**, 869 (1960).
96. GRANGE, R. A., *Trans. ASM*, **59**, 26 (1966).
97. IRVINE, K. J., PICKERING, F. B., and GLADMAN, T., *JISI*, **205**, 161 (1967).
98. GLADMAN, T., DULIEU, D., and McIVOR, I. D., *Micro-Alloying 75 Proceedings*, p. 32, Union Carbide, New York (1977).
99. KORCHYNSKY, M., and STUART, H., *Proceedings of the Symposium on High-Strength Steels*, p. 17, The Metallurg Companies, Nuremberg (1970).
100. BRAMFITT, R. L., and MARDER, A. R., *Proceedings of the Conference on Processing and Properties of Low-Carbon Steel*, p. 191, AIME (1973).
101. IRANI, J. J., and LATHAM, D. J., *Low Alloy Steels*, p. 55, Publn. No. 114, Iron and Steel Institute, London (1968).
102. ZACKAY, V. F., PARKER, E. R., FAHR, D., and BUSCH, R., *Trans. ASM*, **60**, 252 (1967).
103. GRANGE, R. A., *Proc. 2nd Int. Conf. on the Strength of Metals and Alloys*, Vol. III, p. 861, ASM, Metals Park, Ohio (1970).
104. BROWN, C., *The Physics of Irradiation Produced Voids*, AERE, Harwell (1974).
105. HUDSON, J. A., *J. Br. Nucl. Energy Soc.*, **14**, 127 (1975).
106. COTTRELL, A. H., 46th Thomas Hawksley lecture, *Proc. I. Mech E.*, **174**, 16 (1960).
107. GITTUS, J. H., *Creep, Viscoelasticity and Creep Fracture in Solids*, Halsted Press, John Wiley, New York (1975).
108. STIEGLER, J. O., and WEIR, J. R., *Ductility*, p. 311, ASM, Metals Park, Ohio (1968).
109. KINCHIN, G. H., and PEASE, R. S., *Rept. Progr. Phys.*, **18**, 1 (1955).
110. SILCOX, J., and HIRSCH, P. B., *Phil. Mag.*, **4**, 1356 (1959).
111. MAKIN, M. J., WHAPMAM, A. D., and MINTER, F. J., *Phil. Mag.*, **7**, 285 (1962).

112. MAKIN, M. J., and MANTHORPE, S. A., *Phil. Mag.*, **8**, 1725 (1963).
113. URBAN, K., and JÄGER, W., *Phys. Stat. Solidi*, (b) **68**, K1 (1975).
114. URBAN, K., and SEEGER, A., *Phil. Mag.*, **30**, 1395 (1974).
115. HUDSON, J. A., MAZEY, D. J., and NELSON, R. S., *Voids Formed by Irradiation of Reactor Materials*, p. 213 (S. F. Pugh, M. H. Loretto and D. I. R. Norris, eds.), BNES, London (1971).
116. CAWTHORNE, C., and FULTON, E. J., *Nature*, **216**, 575 (1967).
117. PUGH, S. F., LORETTO, M. H., and NORRIS, D. I. R., (eds.), *Voids Formed by Irradiation of Reactor Materials*, BNES, London (1971).
118. BAGLEY, K. Q., BRAMMAN, J. I., and CAWTHORNE, C., *Voids Formed by Irradiation of Reactor Materials*, p. 1 (S. F. Pugh, M. H. Loretto, and D. I. R. Norris, eds.), BNES, London (1971).
119. MAKIN, M. J., *Radiation Effects*, p. 627 (W. F. Sheely, ed.), Gordon and Breach, New York (1967).
120. FOSTER, A. R., and WRIGHT, R. L., *Basic Nuclear Engineering*, 2nd. Ed., Allyn and Bacon, Boston (1973).
121. BLEWITT, T. H., COLTMAN, R. R., JAMISON, R. E., and REDMAN, J. K., *J. Nuc. Mat.*, **2**, 277 (1960).
122. DIEHL, J., *Radiation Damage in Solids*, p. 129, IAEA, Vienna (1962).
123. BLEWITT, T. H., and ARENBERG, C. A., *Suppl. Trans. Japan Inst. Metals*, **9**, 226 (1968).
124. LITTLE, F. A., *Int. Metals Reviews*, **21**, 25 (1976).
125. MAKIN, M. J., and MINTER, F. J., *Acta Met.*, **8**, 691 (1960).
126. BEMENT, A. L., *Radiation Effects*, p. 671 (W. F. Sheely, ed.), Gordon and Breach, New York (1967).
127. MAKIN, M. J., MINTER, F. J., and MANTHORPE, S. A., *Phil. Mag.*, **13**, 729 (1966).
128. MAKIN, M. J., *Phil. Mag.*, **18**, 1245 (1968).
129. MAKIN, M. J., *Irradiation Embrittlement and Creep in Fuel Cladding and Core Components*, p. 35, BNES, London (1973).
130. COULOMB, P., *Acta Met.*, **7**, 556 (1959).
131. HOLMES, J. J., ROBBINS, R. E., BRIMHALL, J. L., and MASTEL, B., *Acta Met.*, **16**, 955 (1968).
132. EYRE, B. L., *Int. Metall. Reviews*, **19**, 240 (1974).

7

COMPOSITE MATERIALS

7.1 INTRODUCTION

It is an unfortunate fact of life that the higher a material's tensile strength, in general the lower are its ductility and toughness. Also, while we can produce solids having very high tensile strengths, normally they can be made only in the form of thin fibers or else significantly large defects may be produced, providing high local stress concentrations which lower their strength appreciably. Such ultrastrong materials should also have an absolute minimum number of dislocations present in them if they are to be able to approach their theoretical strength for slip: again, this means that they must be grown in a very controlled and perfect manner, and it is extremely difficult, if not impossible, to achieve this in other than very thin fibers. In addition, when such strong fibers have been produced, their surfaces are easily damaged, and surface scratches will reduce their strength drastically: hence some means must be found to keep them separate and to prevent mutally damaging contact when they are to be used in an engineering application.

The foregoing indicates something of the rationale for the production of composite materials consisting of strong fibers set in a more ductile matrix which serves to transmit the load between them, impart toughness to the whole by providing a crack-arresting medium, and protect the surfaces of the fibers. The fibers may also be aligned along the direction or directions of maximum stress, to make most efficient use of the material. In addition to strength, weight and stiffness are frequently important factors in design.

Specific strength (UTS/specific gravity = σ_u/ρ) and specific stiffness (modulus/specific gravity = E/ρ) are important design quantities for tensile members, while the avoidance of buckling in compression with minimum weight requires E/ρ^2 to be a maximum. These quantities can be controlled to a considerable degree in composite materials, and stiff, strong components can be produced readily. Probably the most familiar and widely used example of a composite material is glass fiber reinforced resin, which has been in use for many years.

In this chapter we shall look first at the criteria required of materials for them to have high intrinsic strength, which suggests the most suitable materials for use as the strong reinforcing fibers, before discussing the governing parameters relating to composite materials and the relations governing their strength. The use of laminates is discussed briefly, and finally some examples of composite materials systems in practice are given, including a brief discussion of composites produced by controlled solidification.

7.2 STRONG MATERIALS FOR REINFORCEMENT

7.2.1 Theoretical Stress for Cleavage

The *theoretical strength* or *ideal strength* of a solid, σ_{max}, can be estimated in a simple manner using the approach due to Polanyi [*1*] and Orowan [*2, 3*]. In this approach, following Kelly [*4*], we consider the stress required to pull apart two planes in a perfect solid at absolute zero, which avoids the complexities involved in considering the oscillatory motion of the atoms, these being assumed to be at their equilibrium positions. The separation of the two planes with no external force applied is a_o, and the form of the stress-displacement curve will be as shown in Fig. 7.1, the stress rising to some maximum value, and then falling to zero as the planes are pulled infinitely far apart.

Although the exact form of the curve will depend on the interatomic forces, it may be represented approximately by a sine wave. Hence,

$$\sigma \simeq K \sin [(\pi/a)(x - a_o)] \tag{7.1}$$

where a_o is the equilibrium spacing between planes, a is the half wavelength of the sine wave, x the interplanar spacing, and K a constant. For small displacements (x-a_o small),

$$a_o\, d\sigma/dx = K\pi\,(a_o/a) \cos [(\pi/a)(x - a_o)] = E$$

where E is Young's modulus for the solid in the appropriate orientation, and noting that dx/a_o represents strain. Thus,

$$K = (E/\pi)(a/a_o) \tag{7.2}$$

The value of a is not known, and is a measure of the interatomic forces and their range, but we can evaluate it from a First Law balance, since the work

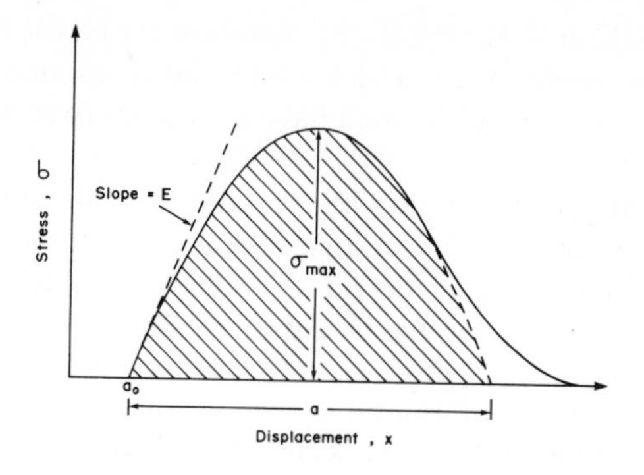

Fig. 7.1 Stress-displacement curve for separation of two planes of atoms under normal stress

done in fracture must equal the energy of the two surfaces created, in the absence of any plastic flow in which energy would be dissipated. Hence,

$$\int_{a_o}^{a_o+a} \sigma \, dx = 2\gamma \tag{7.3}$$

where γ is the specific surface energy. Substituting for σ from Eq. (7.1), we have

$$\int_{a_o}^{a_o+a} K \sin [(\pi/a)(x - a_o)] \, dx = 2\gamma$$

and, thus,

$$a = \pi\gamma/K \tag{7.4}$$

The theoretical strength, σ_{max}, is given by the maximum value of σ from Eq. (7.1) (= K) and hence from Eq. (7.2), we have

$$\sigma_{max} = K = (E/\pi)(a/a_o) \tag{7.5}$$

Substituting for a from Eq. (7.4),

$$\sigma_{max} = (E\gamma/a_o)^{1/2} \tag{7.6}$$

and the strain at fracture is

$$e_{max} = a/2a_o = (\pi/2)(\gamma/Ea_o)^{1/2} \tag{7.7}$$

The surface energy of a solid is then given approximately from Eq. (7.4) as

$$\gamma = Ka/\pi = (E/a_o)(a/\pi)^2 \tag{7.8}$$

Taking a as approximately equal to a_o (a reasonable assumption), we have a first estimate for a solid's surface energy as

$$\gamma \simeq Ea_o/10 \tag{7.9}$$

The analysis indicates that for strong materials the elastic modulus must be large, the surface energy high, and the interplanar spacing small. This last point implies that such solids should have a large number of atoms per unit volume.

The physical properties of crystalline material are anisotropic, and we must use appropriate values in Eq. (7.6). In a crystal, E is replaced by $1/S_{11}$, where S_{11} is the elastic compliance for the direction being considered, rather than by C_{11}, the elastic constant for the orientation, since we are computing the breaking strength of a rod whose sides are not constrained.

From Eq. (7.6), values of σ_{max} have been calculated for many solids, and Table 7.1 gives values derived by Kelly [*4*], using the best (although still somewhat uncertain) values of γ available. These data of Table 7.1 are for 293 K, and have not been corrected to 0 K, since the approach used in deriving Eq. (7.6) is an approximate one and the uncertainty in the values of γ does not warrant such refinements.

For FCC and BCC metals, γ does not vary greatly with orientation, and variations in the value of σ_{max} are determined largely by variation in the elastic modulus. In Table 7.1 it may be seen that such metals have relatively large

Table 7.1 Theoretical Cleavage Stress (σ_{max}) for a Number of Materials From Eq. (7.6). (From Kelly [*4*])

Material	Direction	E ($GN\ m^{-2}$)	γ ($J\ m^{-2}$)	σ_{max} ($GN\ m^{-2}$)	σ_{max}/E
Ag	<111>	121	1.13	24	0.198
Au	<111>	110	1.35	27	0.245
Cu	<111>	192	1.65	39	0.203
Cu	<100>	67	1.65	25	0.373
W	<100>	390	3.00	86	0.221
α-Fe	<100>	132	2.00	30	0.227
α-Fe	<111>	260	2.00	46	0.177
Zn	<0001>	35	0.10	3.8	0.109
C (graphite)	<0001>	10	0.07	1.4	0.140
Si	<111>	188	1.20	32	0.170
C (diamond)	<111>	1210	5.40	205	0.169
SiO_2 (glass)	-	73	0.56	16	0.219
NaCl	<100>	44	0.12	4.3	0.098
MgO	<100>	245	1.20	37	0.151
Al_2O_3	<0001>	460	1.00	46	0.100

surface energies as compared to the low values for the cleavage planes of layered structures such as zinc and graphite. Diamond is seen to have a high surface energy, but the other non-metallic crystals listed have cleavage planes of relatively low surface energy.

More exact methods of calculating σ_{max} have been formulated, based on various models of the interatomic potential function. These are discussed by Kelly [*4*] and in a review of ultimate strength by Macmillan [*5*]. These methods reduce the predicted values to some extent, use of a Morse function for interatomic energy leading to the conclusion that the simple Polanyi-Orowan estimate is too large by a factor of about 2 for covalent and ionic solids. In the present context, however, we shall use the simple approach only, merely noting that it represents an upper limit on possible strength.

7.2.2 Failure by Slip

The simple method of estimating the theoretical stress for slip, originally due to Frenkel [*6*], was discussed in Section 4.2.3, and we may note here that this was given approximately by Eq. (4.4) as

$$\tau_{max} = Gb/2\pi a$$

Values of τ_{max}, using a more refined estimation method as discussed in Section 4.2.3, are listed in Table 7.2, being taken from Kelly [*4*]. These data are estimates for room temperature, except where specifically designated otherwise.

7.2.3 Experimental Data

There are only very limited data of a reliable nature for the strength of smooth tensile specimens free from internal cracks, defects or dislocations: the experimental difficulties make such tests all but impossible. However, some data for fine *whiskers*, which are thin single crystal fibers having virtually no crystalline defects, are shown in Table 7.3.

It is found that both crystalline and non-crystalline materials of high strength have strength values which are comparable [*5*], and this is because glassy materials retain a considerable degree of short range order, and thus contain nearest neighbor bond networks similar to those in related crystalline materials. It is also clear from the published work that the presence of very small cracks or of dislocations greatly reduces a material's strength, effects of dislocations having been discussed previously in Chapter 4.

From comparison of the data in Table 7.3 with those in Tables 7.1 and 7.2, it is seen that the measured values of σ_u/G (or τ_u/G) are in all cases lower than the theoretical values predicted for perfect crystals.

7.2.4 Requirements for a Strong Solid

On the basis of the foregoing, we can now set down the requirements for a strong solid.

Table 7.2 Theoretical Shear Stress (τ_{max}) for a Number of Materials. (From Kelly [*4*])

Material	Slip Plane and Direction	G (GN m^{-2})	τ_{max} (GN m^{-2})	τ_{max}/G
C (diamond)	$\{111\}\ \langle 1\bar{1}0\rangle$	505	121.0	0.24
Cu (10 K)	$\{111\}\ \langle 11\bar{2}\rangle$	33.2	1.29	0.039
Cu (20° C)	$\{111\}\ \langle 11\bar{2}\rangle$	30.8	1.2	0.039
Au	$\{111\}\ \langle 11\bar{2}\rangle$	19.0	0.74	0.039
Ag	$\{111\}\ \langle 11\bar{2}\rangle$	19.7	0.77	0.039
Al	$\{111\}\ \langle 11\bar{2}\rangle$	23	0.9	0.039
Al	$\{111\}\ \langle 1\bar{1}0\rangle$	23	2.62	0.114
Si*	$\{111\}\ \langle 1\bar{1}0\rangle$	57	13.7	0.24
Fe	$\{111\}\ \langle 1\bar{1}1\rangle$	60	6.6	0.11
W	$\{111\}\ \langle 1\bar{1}1\rangle$	150	16.5	0.11
NaCl (0 K)	$\{111\}\ \langle 1\bar{1}1\rangle$	23.7	2.84	0.120
Al_2O_3	$\{0001\}\langle 11\bar{2}0\rangle$	147	16.9	0.115
Zn	$\{0001\}\langle 10\bar{1}0\rangle$	38	2.3	0.034
C (graphite)	$\{0001\}\langle 10\bar{1}0\rangle$	2.3	0.115	0.05

*τ_{max}/G for Si is taken as that of diamond, probably overestimating the value of τ_{max} for Si.

First, to obtain a high cleavage strength, the elastic modulus and surface energy must be high, and the number of atoms per unit volume large [see Eq. (7.6)]. σ_{max} is highest in covalent and metallic solids, as ionic solids contain planes of low surface energy. Among the metals, the transition elements have higher values of σ_{max} than do aluminum or the noble metals, because they have higher E values and small values of a_o.

Normally, the theoretical stress for cleavage is greater than that for shear, and we may see from Tables 7.1 and 7.2 that metals have a low ratio of τ_{max}/σ_{max}. The covalent solids have the largest ratios, this being close to unity for diamond; them come the ionic solids and the BCC transition metals, with the noble metals at the end of this ranking.

Large values of both σ_{max} and τ_{max} are required of a strong solid. For τ_{max} to be large, G must be large, together with the ratio τ_{max}/G. Such is found in solids where the interatomic forces are directional, as in covalent solids and those with strongly polarized bonds. The density of these bonds must also be large, and we require a three dimensional network of bonds; thus a valence of at least two is required of the constituent atoms, and the bond lengths should be as short as possible. Hence the strongest materials can be expected to

Table 7.3 Experimental Data for High Strength Whiskers at Room Temperature. (From Kelly [*4*])

Material	Maximum Tensile Strength, σ_u ($GN\ m^{-2}$)	E ($GN\ m^{-2}$)	σ_u/E	Specific Gravity, ρ	Specific Strength, σ_u/ρ	Specific Modulus, E/ρ	Melting Temperature (°C)
C (graphite)	19.6	686	0.03	2.2	8.91	312	<3000
Al_2O_3	15.4	532	0.03	4.0	3.85	133	2050
Fe	12.6	196	0.06	7.8	1.62	25	1540
Si_3N_4	14.0	385	0.04	3.1	4.52	124	1900
SiC	21.0	700	0.03	3.2	6.56	219	2600
Si	7.0	182	0.04	2.3	3.04	79	1450
BeO	7.0	357	0.02	3.0	2.33	119	2520
AlN	7.0	350	0.02	3.3	2.12	106	2000
NaCl	0.98	-	-	-	-	-	-

contain beryllium, boron, carbon, nitrogen, oxygen, aluminum and silicon.

The requirement of small atoms implies that only the lightest elements will be present in a strong solid, while directional bonding implies that the coordination number be low. Thus, we expect the density of strong materials to be low. Because of the requirement of a large value of E, strong solids will have high melting points.

Some of the properties of commercially produced high strength fibers and wires are given in Table 7.4, and these materials form the reinforcing element of many of the high strength composites used today.

7.3 STRENGTH OF COMPOSITE MATERIALS

In this discussion of the relations governing the strength of composites, we shall be concerned largely with fiber composites. Other composites include two phase alloys, including pearlite-ferrite or similar mixtures, which were discussed in Chapter 6: the same principles apply in calculating the strength of such materials as will be elucidated for fiber composites. Laminated materials pose particular problems of analysis which are difficult to deal with fully other than at an advanced level, and the main differences in treatment from that for fiber composites will be outlined (only) in Section 7.3.4.

In a fiber reinforced composite, strong fibers (as discussed in the preceding section) are dispersed in a softer, more ductile matrix. In general, the fibers are brittle, or have little ductility prior to fracture, but when combined with a ductile matrix, the resulting composite may have excellent strength and toughness. It is important to note that for considerable reinforcement to occur and for the fibers to carry the major part of the load, the matrix must have a lower elastic modulus than do the fibers to ensure a transfer of stress from matrix to fibers at their interface. The composite material will have an elastic modulus lying between those of its two constituents. When loaded in tension along the fiber axes (assuming the fibers to be aligned), the matrix stress will increase gradually until it reaches its yield point, σ_{mo}, at which time plastic flow may occur. This will initiate at the ends of fibers where the shear stress is a maximum[1]. Further increase of load will cause the proportion carried by the fibers to increase as the matrix continues to deform plastically, until the fibers reach their ultimate strength, σ_{fu}. Figure 7.2 shows the load-extension and stress-strain relations for a fiber composite with a ductile metallic matrix loaded in this way.

---

[1] The interfacial shear stress is a maximum at the ends of fibers and will be zero half way along their length. Because of the stress concentrating effects of the fiber ends, its value can be considerably greater than simple analysis would indicate. In a metallic matrix, plastic flow normally initiates here: in a resin matrix shear may take place.

Table 7.4 Tensile Strength of Some Commercial Fibers and Wires at Room Temperature. (From Kelly [4])

Material	E (GN m^{-2})	Tensile Strength, σ_u (GN m^{-2})	Specific Gravity, ρ	σ_u/ρ	E/ρ	Melting Temperature (°C)
Fibers						
E glass	72	3.5	2.55	1.37	28.2	700
S glass	84	4.6	2.50	1.84	33.6	840
SiO_2	72	6.0	2.19	2.74	32.9	1660
Al_2O_3	470	2.0	3.96	0.51	119	2072
C, type I	390	2.0	1.90	1.05	205	<3000
C, type II	240	2.6	1.90	1.37	126	<3000
Boron nitride	90	1.4	1.90	0.74	47.4	2980
Nylon 66	4.9	1.05	1.1	0.95	4.5	-
PRD-49 (Kevlar)	133	2.8	1.5	1.87	88.7	-
Wires						
Patented 0.9%C steel	210	4.2	7.9	0.53	26.6	1300
Stainless steel	203	2.1	7.9	0.27	25.7	
W	350	3.9	19.3	0.20	18.1	3390
Mo	343	2.1	10.3	0.20	33.3	2610
Be	315	1.3	1.8	0.72	175	1284

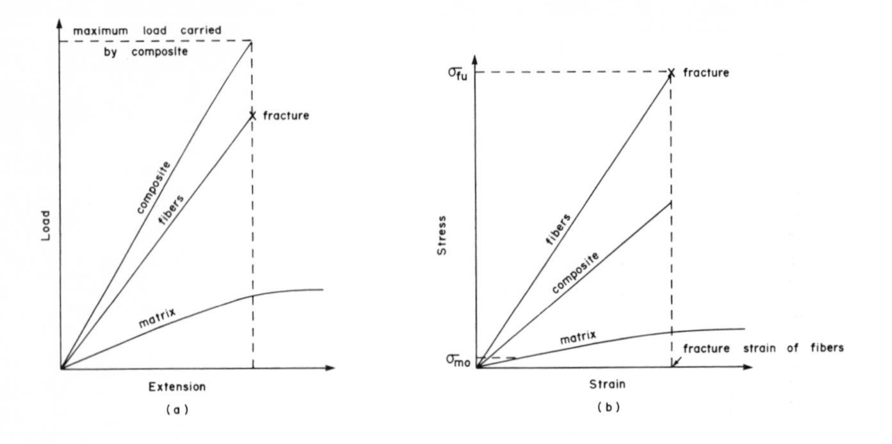

Fig. 7.2 (a) Load-extension curves and (b) corresponding stress-strain curves for an idealized fiber composite consisting of continuous brittle fibers in a ductile metallic matrix

With a polymeric matrix, the stress-strain relations are expected to be non-linear, and there is not any sudden change from elastic to plastic deformation. Also in this case the limiting factor in transferring load to the fibers is **not** the shear yield strength as in the case of a metallic matrix, rather it is the frictional forces acting between fiber and matrix. The resin of the matrix shrinks when curing and sets up a normal stress at the interface with the fiber. The load transfer is then accomplished by μn, where μ is the frictional coefficient and n the pressure at the interface [7].

7.3.1 Continuous Fibers

Let us consider a composite consisting of a series of strong, dispersed, continuous fibers set in a ductile matrix and loaded axially as in Fig. 7.3. The tensile strength of the composite, σ_{cu}, can be represented using the *Rule of Mixtures* (see Section 6.3.6) as

$$\sigma_{cu} = \sigma_{fu} V_f + \sigma_m (1 - V_f) \tag{7.10}$$

where σ_{fu} is the UTS of the fibers, σ_m the stress in the matrix when the fibers fail, and V_f the volume fraction of the fibers. Similarly, we can represent the elastic modulus of the composite, E_c, by the relation

$$E_c = E_f V_f + (d\sigma_m / de_m)_e (1 - V_f) \tag{7.11}$$

where E_f is the elastic modulus of the fibers and $(d\sigma_m / de_m)$ is the slope of the stress-strain curve at a strain e. For a metallic matrix in the elastic region,

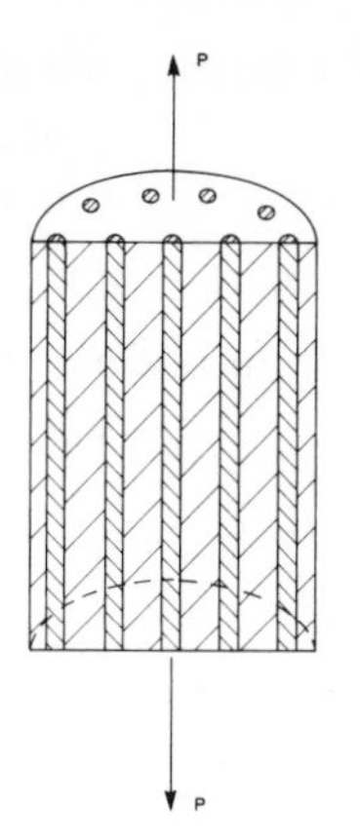

Fig. 7.3 Section through element of composite containing continuous aligned fibers and loaded axially

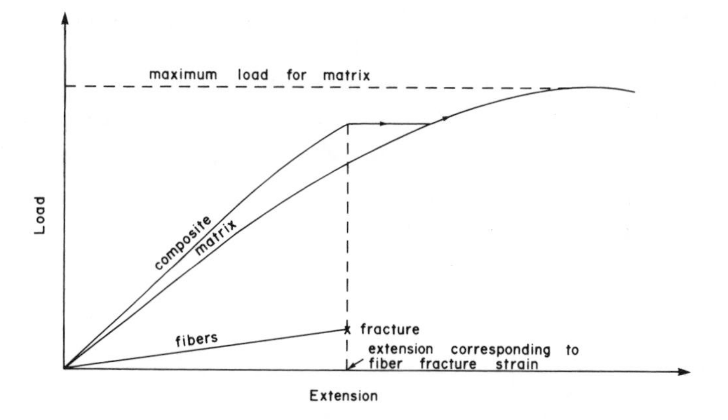

Fig. 7.4 Load-extension curves for composite and its constituents where the volume fraction of fibers is less than the minimum for reinforcement. The fracture load of the composite would in fact be less than for the matrix alone because of the volume of broken fibers

$d\sigma_m/de_m$ may be replaced by E_m.

If the volume fraction of fibers is insufficiently great, they may fracture before the matrix has strain hardened completely and no strengthening of the matrix will result. This situation is illustrated in Fig. 7.4, the tensile strength of the composite being that of the matrix, and in fact its load carrying capacity will have been reduced on account of the proportion of the cross-section given over to (broken) fibers. Hence Eq. (7.10) is valid provided

$$\sigma_{cu} \geqslant \sigma_{mu} (1 - V_f) \tag{7.12}$$

and the minimum volume fraction of fibers for Eq. (7.10) to hold is

$$V_{f\,(min)} = (\sigma_{mu} - \sigma_m)/(\sigma_{fu} + \sigma_{mu} - \sigma_m)$$

or

$$V_{f\,(min)} = 1/[1 + \sigma_{fu}/(\sigma_{mu} - \sigma_m)] \tag{7.13}$$

Where the matrix material has a greater strain at failure than do the fibers and can also strain harden, the strength of the composite will be greater than that of the matrix by itself only if $\sigma_{cu} > \sigma_{mu}$, which defines a critical volume fraction of fibers from Eq. (7.10) as

$$V_{f\,(crit)} = (\sigma_{mu} - \sigma_m)/(\sigma_{fu} - \sigma_m) \tag{7.14}$$

Figure 7.5 illustrates the variation in composite strength with volume fraction of continuous brittle fibers, and the situation is also shown where the fibers are continuous and ductile. In the latter case they may fail by necking down, but this will be prevented when the fibers have a strong adhesion to the matrix. Deformation of the fibers will then continue so that they elongate uniformly to strains greater than that corresponding to their uniaxial ultimate tensile strength [*8*], provided the matrix can support an increase in nominal stress greater than the corresponding decrease due to fiber elongation.

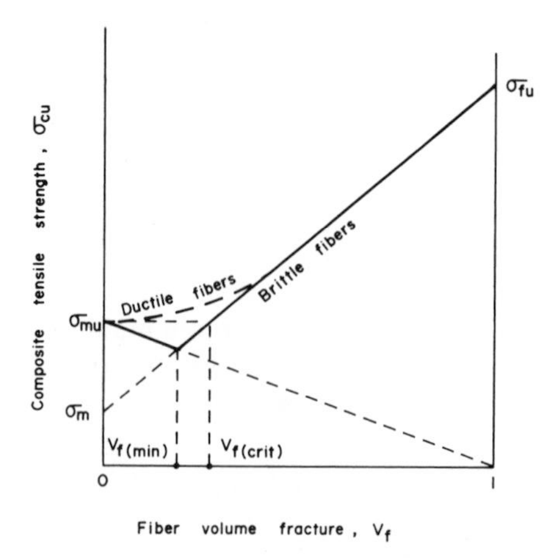

Fig. 7.5 Idealized variation in tensile strength of composite with volume fraction of brittle fibers. The case of ductile fibers is also illustrated

7.3.2 Discontinuous Fibers

In the more common situation, the fibers are discontinuous[2], and the strains in fibers and matrix can no longer be considered to be identical: thus the simple Rule of Mixtures will no longer apply. In such a situation, providing there is no loading on the ends of the fibers and that no shear occurs between them and the matrix, we can represent the distribution of tensile stress along a fiber as in Fig. 7.6, the load being transferred to the fiber by a shear stress τ at the interface. In the central portion of the fiber as shown, the tensile stress within it is constant, and the fiber and matrix strains will be equal. The maximum fiber stress, $\sigma_{f(max)}$, is given by $E_f\, e_c$, where e_c is the strain in the composite. For a fiber of circular cross-section and diameter = $2r$,

$$\sigma_{f(max)} = E_f\, e_{cu} = 2\pi r(l_c/2)\tau/\pi r^2 \tag{7.15}$$

where l_c is the *minimum fiber length* to ensure that it is strained to the same extent as is the composite, and e_{cu} is the fracture strain in the fibers, at which point the composite will also fracture. Hence,

$$l_c = E_f\, e_{cu}\, r/\tau = \sigma_{fu}\, r/\tau \tag{7.16}$$

For a metallic matrix, τ is limited to the shear yield stress of the matrix [*4*], or $\sim \sigma_{my}/2$, while for a resin matrix it is limited to μn, as noted previously.

It may be seen from Fig. 7.6, that the *average* stress in a discontinuous fiber at fracture, $\bar{\sigma}_f$, will always be less than σ_{fu}, and hence the strength of a composite containing a given volume of fibers will always be less when these are discontinuous rather than continuous. Accordingly, for the case of discontinuous fibers, Eq. (7.10) must be modified to

$$\sigma_{cu} = \bar{\sigma}_f\, V_f + \sigma_m\,(1 - V_f) \tag{7.17}$$

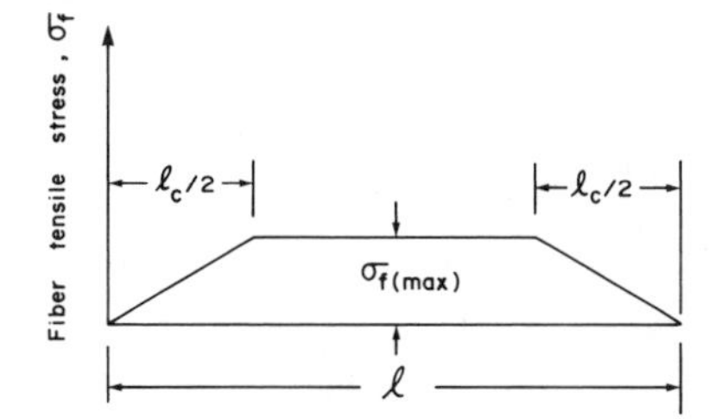

Fig. 7.6 Variation of tensile stress along a discontinuous fiber

---

[2] Indeed, continuous fibers may be considered as a special case of discontinuous fibers, where the *aspect ratio* (length/diameter) is extremely large.

This holds for $V_f > V_{f(min)}$ and for a *random distribution* of aligned fibers, i. e., all cross-sections normal to the tensile (and fiber) axis are identical.

For a linear increase in fiber stress as in Fig. 7.6,

$$\bar{\sigma}_f = [\sigma_{fu}\,(l - l_c)/2 + \sigma_{fu}\, l_c/4]/(l/2)$$

Hence,

$$\bar{\sigma}_f = \sigma_{fu}\,(1 - l_c/2l) \tag{7.18}$$

If the fiber stress increases in a nonlinear manner, we may write

$$\bar{\sigma}_f = \sigma_{fu}\,[1 - (1 - \beta)\, l_c/l] \tag{7.19}$$

where $\beta\sigma_{fu}$ is the average stress in a fiber over the distance $l_c/2$ from each end. Thus, the UTS of a composite material containing aligned, randomly distributed, discontinuous fibers may be written as

$$\sigma_{cu} = \sigma_{fu}\, V_f\,[1 - (1 - \beta)/\alpha] + \sigma_m\,(1 - V_f) \tag{7.20}$$

where $\alpha = l/l_c$ and $V_f > V_{f(min)}$. For the linear situation of Fig. 7.6, $\beta = 1/2$.

The value of the critical volume fraction of fibers for strengthening to take place with discontinuous fibers having $\alpha > 1$, is obtained by substituting for $\sigma_{cu} = \sigma_{mu}$ in Eq. (7.20). Hence

$$V_{f(crit)} = (\sigma_{mu} - \sigma_m)/\{\sigma_{fu}\,[1 - (1 - \beta)/\alpha] - \sigma_m\} \tag{7.21}$$

The validity of Eq. (7.20) has been established experimentally for a number of composite materials for values of V_f up to ~0.6 [*8, 9*].

It may be noted that when a composite containing discontinuous fibers fractures, fibers whose ends are less than $l_c/2$ from the fracture surface will have their ends pulled out rather than breaking, the proportion being $1/\alpha$, this being verified by the experiments of Kelly and Tyson [*10*]. The minimum value of fiber volume fraction for Eq. (7.20) to hold is then found from the relation

$$\sigma_{cu} \geqslant \sigma_{mu}\,(1 - V_f) + (\beta/\alpha)\,\sigma_{fu}\, V_f$$

Substituting from Eq. (7.20), we obtain

$$V_{f(min)} = (\sigma_{mu} - \sigma_m)/[\sigma_{fu}\,(1 - 1/\alpha) + \sigma_{mu} - \sigma_m] \tag{7.22}$$

We may also determine the *critical aspect ratio*, l_c/d, where d is the fiber diameter, to give reinforcement. For a metallic matrix where l_c is determined by the matrix rather than interface shear strength, the critical fiber length is given by Eq. (7.16), and the critical aspect ratio by

$$l_c/d = \sigma_{fu}/2\tau \tag{7.23}$$

where τ is taken as the *ultimate* shear strength of the matrix.

When the variation in the strength of individual fibers is small, l_c/d can be determined from the proportion of fibers pulled out of the matrix at the fracture surface, provided $V_f > V_{f(crit)}$, and the fibers are of uniform length and aspect ratio. If the number pulled out is n_p and the number fractured is n_f, then

$$n_f/n_p = (l/l_c) - 1 \tag{7.24}$$

and l_c can be determined as a fraction of l [*10*].

Alternatively, l_c/d can be determined experimentally by pulling out a single fiber embedded in the matrix, and measuring the load required. The stress required to pull out a fiber of length $l < l_c/2$ is

$$\sigma = \sigma_o + k\ l/d \tag{7.25}$$

where σ_o and k are constants. From a plot of σ versus l/d, k may be found and is a constant for a given fiber and matrix. Hence, l_c/d is found from the relation [*4*]

$$l_c/d = 2\sigma_u/k \tag{7.26}$$

To conclude this discussion, Fig. 7.7 shows the fracture surface of a fiber composite (although the fibers were continuous in this particular case) and pulled out and fractured fibers may be clearly seen.

Fig. 7.7 SEM view of fracture surface of carbon fiber, polyester matrix composite, X250. (Courtesy of K. K. Chawla, Inst. Militar de Engenharia, Rio de Janeiro)

7.3.3 Effects of Orientation

Fiber composites may be loaded other than in tension along the fiber axes. The case of compressive loading in the direction of the fibers was discussed by Kelly [*4*] and it was demonstrated that for a metallic matrix composite compressive strength is at least as great as tensile strength. The more general matter of oriented fibers will now be discussed.

Krenchel [*11*] developed two approximate theories to deal with fibers whose orientation does not correspond with the tensile stress. In the first, and simpler theory, the transverse deformations are neglected, while in the second they are taken into account. The two theories give similar results for a number of orientations but in specific cases they may differ considerably. *In general*, the approximate method gives results which correspond closely with experiment, and we shall describe it briefly here.

In Krenchel's simple theory, an element of a plane containing fibers is considered as shown in Fig. 7.8, with half the fibers lying at an angle ϕ to the x-axis and the other half at the angle $-\phi$. The element is subjected to a load in the direction of the x-axis, and it is assumed that deformation takes place *only* in this direction (i. e., Poisson effects are neglected). A fiber of unit length and extending to section *a-a* requires a force P_f for an elongation of e_f. The transverse component of this is neglected as being balanced by the corresponding force from a fiber oriented symmetrically at $-\phi$. The elongation of the composite in the x-direction is the result of elongation of the fibers and matrix in the direction of the fibers and of their rotation. Thus $e_f = e_{fc} \cos^2\phi$, where e_{fc} is the elongation of the composite along the fiber direction, and it is easily shown that the horizontal component of the fiber's contribution to the total internal force, P_x, is given by

$$P_x = E_f\, e_{fc}\, \Delta S_f \cos^4\phi \qquad (7.27)$$

where E_f is the elastic modulus of the fiber, and ΔS_f is its effective cross-sectional area in the horizontal direction. The quantity $\Delta S_f \cos^4\phi$ is termed the *equivalent area*, $\Delta S_f'$, and summing over the entire section,

$$S_f' = \Sigma \Delta S_f' = \Sigma\, \Delta S_f \cos^4\phi \qquad (7.28)$$

For fibers aligned in the x-direction, the reinforcement would have maximum efficiency, and the equivalent area would be S_f. Thus we may write the *efficiency factor of reinforcement*, η_o, as

$$\eta_o = S_f'/S_f = \Sigma\, (\Delta S_f/S_f) \cos^4\phi \qquad (7.29)$$

For fibers of different orientations, η_o is determined by summing the various groups of parallel fibers in proportion to their portion of the total groups' reinforcement per unit volume of material. Thus,

$$\eta_o = \Sigma a_n \cos^4\phi \qquad (7.30)$$

where a_n is the proportion of the nth group to the total reinforcement.

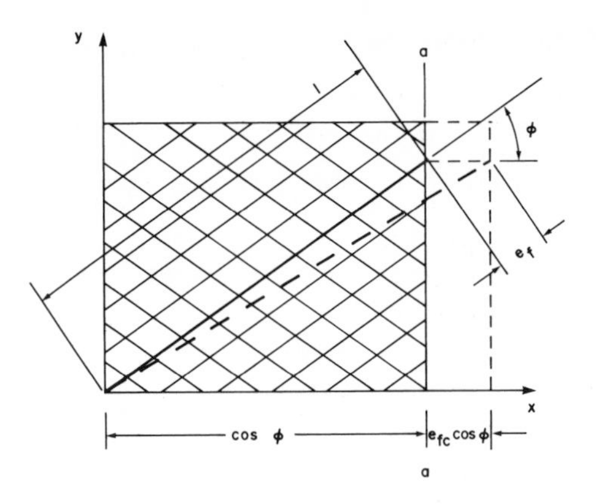

Fig. 7.8 Element of a plane containing fibers oriented at +ϕ and -ϕ to the *x*-axis and loaded in tension along this axis. (After Krenchel [*11*])

Krenchel has evaluated η_o for a number of configurations. For the fibers oriented parallel to the stress axis, $\eta_o = 1$, while for them at 90° to the axis, $\eta_o = 0$. For two groups aligned at $\pm \pi/4$, $\eta_o = 1/4$, while if they are uniformly distributed in a plane, $\eta_o = 3/8$. For fibers randomly distributed in three dimensions, $\eta_o = 1/5$.

Hence, we can approximate the strengthening of discontinuous fibers distributed non-uniformly by the relation [cf. Eq. (7.17)]

$$\sigma_{cu} = \bar{\sigma}_f V_f \eta_o + \sigma_m (1 - V_f) \tag{7.31}$$

Alternatively, we may write $\eta_l \sigma_f$ in place of $\bar{\sigma}_f$, where η_l is an *efficiency factor* relating to the transfer of load to a fiber, and σ_f is the fiber fracture stress, as before.

A useful equation for the correlation of experimental data or for commercial application is, therefore,

$$\sigma_{cu} = \sigma_f V_f \eta_l \eta_o + \sigma_m (1 - V_f) \tag{7.32}$$

A work of caution must be sounded, however, as the composite may fail by shear or yielding for some fiber orientations before the estimated value of σ_{cu} is reached. This is likely when the fibers are all oriented at a considerable angle to the applied stress axis: in the case of Krenchel's tests, the premature failures were in specimens with two groups of parallel fibers arranged symmetrically about the loading direction [*11*].

Kelly [*4*] has discussed the effect of orientation of continuous fibers on a more rigorous basis, considering the possible failure modes. These are as follows:

(1) *failure by flow parallel to the fibers*, with the required applied stress being given by

$$\sigma = \sigma_{cu} \sec^2\phi \tag{7.33}$$

where σ_{cu} is defined by Eq. (7.10), and ϕ is the angle between fibers and applied stress axis;

(2) *failure by shear of the matrix or at the fiber-matrix interface,* with the required applied stress given by

$$\sigma = 2\ \tau_u \operatorname{cosec} 2\phi \tag{7.34}$$

where τ_u is the shear stress for failure of the matrix, or of the fiber-matrix interface[3];

(3) *failure normal to the fiber direction*, which will require an applied stress

$$\sigma = \sigma_u \operatorname{cosec}^2\phi \tag{7.35}$$

where, for failure by plastic flow in the matrix under plane strain

$$\sigma_u = \sigma_{ult}\ (1 - V_f) \tag{7.36a}$$

where σ_{ult} is the matrix UTS in plane strain, or for failure at the interface, σ_u is given by

$$\sigma_u = \sigma_i\ V_f \tag{7.36b}$$

where σ_i is the tensile strength of the interface in plane strain.

Figure 7.9 shows schematically the predicted failure stress as a function of fiber orientation, and it may be seen that there appears to be a critical angle, ϕ_{crit}, above which the strength falls off. From Eqs. (7.33) and (7.34),

$$\phi_{crit} = \tan^{-1}\ (\tau_u/\sigma_{cu}) \tag{7.37}$$

7.3.4 Laminates

Laminates are widely used today, the most familiar, perhaps, being plywood. With fiber composites, planar isotropy can be achieved by bonding together sheets of fiber reinforced material, each sheet containing aligned fibers, and the orientation of the sheets being varied. The advantage over distributing fibers within a single sheet in a planar isotropic fashion is that a much higher volume fraction of fibers can be achieved.

The analysis of structural elements composed of fiber reinforced laminates is discussed at some length by several authors [*12-14*]. Here we shall note two

---

[3] When shear takes place in the matrix, the value of τ_u will be higher than that measured for the matrix alone owing to constraints introduced by the fibers. Also, when the matrix strain is large, as in many metallic matrices, rotation of the fibers will take place and ϕ should be defined on the basis of the angle when the matrix fails by shear. For shear at the interface, the change in ϕ will be small.

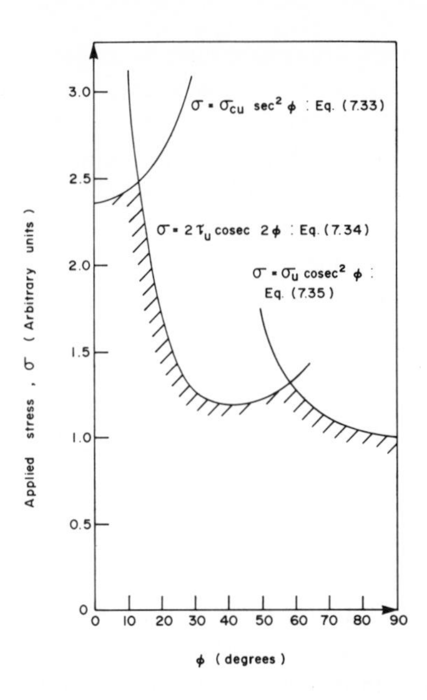

Fig. 7.9 Predicted stress for failure as a function of fiber orientation for aligned continuous fibers

complexities which arise, without discussing details of their analysis. First, when a laminate consisting of orthotropic sheets of fiber reinforced material is stretched in tension, it may not behave in a simple manner, but bending of the plate may take place. Analysis shows that twisting-stretching, stretching-shearing, or bending-twisting coupling may take place, depending on the number of laminae (even or odd), the angular relationships between them and the loading [*14*].

The other problem relates to transverse shear deformation which must be considered in structural analysis of many laminates. For an isotropic material [see Eq. (1.35)]

$$2 \leqslant E/G = 2\,(1 + \nu) \leqslant 3 \tag{7.38}$$

However, for a fibrous composite the elastic modulus in the direction of the fibers (E_{11}) is large, while the transverse shear modulus (G_{13}) approximates to that of the matrix. Hence, it is common to find that

$$20 \leqslant E_{11}/G_{13} \leqslant 50 \tag{7.39}$$

in such materials, and transverse shear effects become significant [*14*].

Another type of laminate is the metal sandwich composed of sheets of dissimilar metals bonded together by rolling, explosive bonding or some other means. Generally the individual sheets are isotropic, simplifying the analysis. The modulus of elasticity can be obtained directly from the Rule of Mixtures, but this is not strictly true for prediction of the pronounced yield point of the laminate when plastic yielding begins in the second component which has the greater tensile yield strain. Beese and Bram [*15*] have discussed this matter and, in addition to developing a theoretical equation for tensile yield, have shown that an empirical Rule of Mixtures does, in fact, hold satisfactorily, and may be preferred in view of uncertainties concerning the stress at which yield actually commences.

7.4 COMPOSITE MATERIAL SYSTEMS

As will have been apparent during the foregoing discussion, many different composite materials have been developed and are in commercial use today in a wide range of applications extending from household items in glass fiber reinforced resin to aircraft turbine components of superalloys reinforced by refractory metal wires. Carbon fiber, epoxy matrix composites have been developed for some time, while more recently composites having a matrix of pyrolitic graphite and reinforcement with carbon fibers, have been developed. These various systems, and the production of the high strength fibers and fabrication of the composite, are described clearly in Chapter 2 of Reference [*14*]. Here we shall describe briefly one specific method of producing composite materials, namely from the melt by means of controlled eutectic or dendritic growth.

7.4.1 Controlled Solidification and Growth

This is an attractive method of producing components because the "fibers" may be grown in specified directions in an overall complex shape. It also simplifies production because of its directness. However, these merits are balanced by the limitations on the choice of binary systems suitable for such growth, and on the variation in volume fraction of reinforcing phase possible.

The first requirement is that the reinforcing phase, which we wish to have grow in a lamellar or rod-like fashion, should be hard and strong. The former morphology is developed when the volume fractions of the two phases are fairly equal, while the latter occurs when the reinforcing phase has a relatively low volume fraction. In general, the systems of interest are binary alloys with a eutectic reaction which produces a solid solution and an intermetallic compound, solidification being directional to develop the correctly oriented structure.

When eutectic solidification is employed, the volume fractions of the two phases are specific, and this limits the strength and other properties obtainable: frequently the volume fraction of reinforcement is too low for satisfactory strength levels to be achieved. This problem may be overcome by

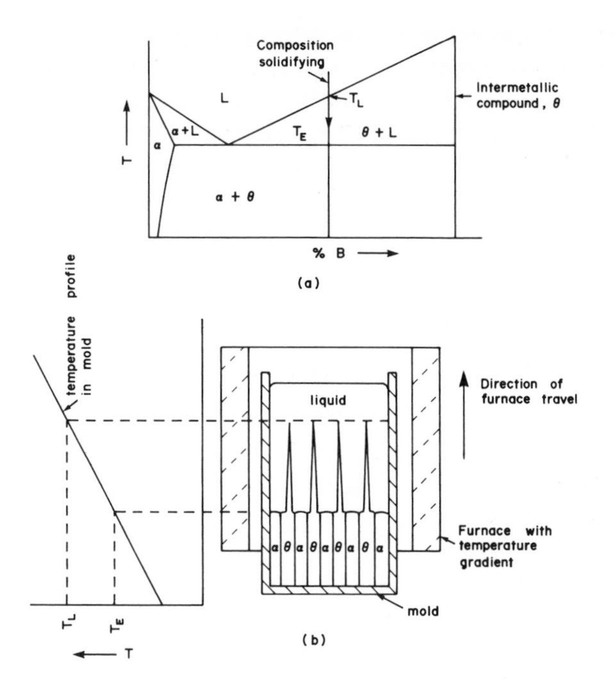

Fig. 7.10 Production of aligned fibers during directional solidification of non-eutectic alloy. (a) Phase diagram showing composition solidifying. (b) Section through gradient furnace and mold showing correspondence between solidified microstructure and temperature profile

directional solidification of an alloy of other than eutectic composition and which develops primary dendrites of the intermetallic compound. The process is illustrated in Fig. 7.10, and a range of volume fractions of reinforcement can be produced in this way, depending on the composition of the melt.

A considerable number of binary alloy systems have been studied, including Al-$CuAl_2$, Al-Al_3Ni, Nb-Nb_2C, Ta-Ta_2C and Fe-Fe_2B eutectics. Several ternary and pseudo-ternary systems have also been studied in an attempt to overcome the limitations on fiber volume fraction imposed by eutectic solidification in a binary system: however, much work remains to be done in this area.

In general, the strength of such *in situ* composites with rod-like reinforcement can be estimated using the Rule of Mixtures and the equations developed in Section 7.3.1, but for lamellar eutectics interface effects may be significant, and these are reviewed by Vinson and Chou [*14*]. The high temperature strength and isothermal stability of a number of the eutectic composites is considered to be excellent, and this emphasizes their applicability to such components as turbine blades. However, some doubts have been cast on their stability under temperature gradients [*16*].

For a detailed discussion of directionally solidified eutectic and dendritic alloy systems, the reader is referred to the reviews of Davies [*17, 18*].

REFERENCES

1. Polanyi, M., *Z. Phys.*, **7**, 323 (1921).
2. Orowan, E., *Z. Krist.*, **A89**, 327 (1934).
3. Orowan, E., *Rep. Prog. Phys.*, **12**, 185 (1949).
4. Kelly, A., *Strong Solids*, 2nd Ed., Clarendon Press, Oxford (1973).
5. MacMillan, N. H., *Can. Metall. Quarterly*, **13**, 555 (1974).
6. Frenkel, J., *Z. Phys.*, **37**, 572 (1926).
7. Outwater, J. O., Jr., *Mod. Plast.*, **33**, 156 (1956).
8. Kelly, A., and Tyson, W. R., *J. Mech. Phys. Solids,* **13**, 329 (1965).
9. Kelly, A., and Davies, G. J., *Metall. Rev.,* **10**, 1 (1965).
10. Kelly, A., and Tyson, W. R., *High Strength Materials*, p. 578 (V. F. Zackay, ed.), Wiley, New York (1965).
11. Krenchel, H., *Fibre Reinforcement*, Akademisk Forlag, Copenhagen (1964).
12. Ashton, J. E., Halpin, J. C., and Petit, P. H., *Primer on Composite Materials: Analysis*, Technomic, Stanford, Conn. (1969).
13. Calcote, L. R., *The Analysis of Laminated Composite Structures*, Van Nostrand Reinhold, New York (1969).
14. Vinson, J. R., and Chou, T. W., *Composite Materials and Their Use in Structures*, Halsted Press, John Wiley, New York (1975).
15. Beese, J. G., and Bram, G., *Trans. ASME, J. Eng. Matls. and Technology*, **97**, 10 (1975).
16. Jones, D. R. H., *Mat. Sci. and Eng.*, **15**, 203 (1974).
17. Davies, G. J., *High Strength Materials*, p. 603 (V. F. Zackay, ed.), Wiley, New York (1965).
18. Davies, G. J., *Strengthening Methods in Crystals*, p. 485 (A. Kelly and R. B. Nicholson, eds.), Elsevier, London (1971).

8

FRACTURE FUNDAMENTALS

8.1 INTRODUCTION

During the latter part of the nineteenth century in particular, and up to World War II, many engineering structures and components made from steel failed suddenly and catastrophically, in many cases with considerable loss of life [*1*]. The incidence of such failures was reduced by improvements in design, but more importantly by improvements effected in steel quality and fabrication techniques so that the number and size of defects present were reduced. The subject of sudden and brittle fracture of engineering structures was brought into particular prominence in the public domain by the large number of sudden fractures which occurred during the 1940s in all-welded Liberty ships and T2 tankers, including the spectacular failure of the tanker *Schenectady* in calm cool weather, while at her fitting-out dock moorings [*2*]. As a consequence of these events, extensive study was made of brittle fracture in steel and of conditions of temperature and local stress concentration which led to its occurrence. It became clear that the steel employed in the all-welded ships had a transition temperature for the change from ductile to brittle fracture which corresponded closely to cold weather service conditions, that there were local stress concentrations present due to poor structural design and welding procedures, and that these corresponded with the fracture initiation points, and finally that the continuous nature of the all-welded structure did not provide for the arrest of propagating cracks as do rivets and riveted joints in plates.

The ductile-brittle transition temperature in steel and its "toughness" have commonly been determined in the past using Charpy V-notch impact specimens, a typical curve being shown in Fig. 8.1. However, such data are not of significant value in designing thicker-section components because of a size effect, nor do they provide quantitative information relating to the stress required to propagate a crack rapidly through a component. In an attempt to overcome these deficiencies, a number of other test methods have been devised. These include the Robertson crack arrest test, defining the temperature and stress at which a large running crack will arrest (the CAT curve) [*3*]; the U.S. Navy Bulge Explosion Test, defining the temperature above which full plastic bulging would occur without fracture (the FTP temperature) and the temperature above which cracks would not propagate when stressed up to the yield stress (the FTE temperature); and the standardized drop weight test [*4*], defining the temperature below which a small crack will propagate across the plate (the NDT temperature). However, none of these provides good design data independent of size effect, although Pellini and Puzak [*5*] have developed the *fracture analysis diagram* (FAD) to provide useful guidance for design in low-strength steel. The FAD is shown schematically in Fig. 8.2.

With the post-World War II trend to higher strength alloys, the problem of essentially brittle failures became more acute, as such materials generally have a reduced resistance to crack propagation or reduced toughness. Thus, a number of spectacular failures of structures, pressure vessels, and heavy components have occurred during the past two decades, and it became clear that design procedures for high strength materials or for thick sections which were based on a classical approach using yield stress or ultimate tensile

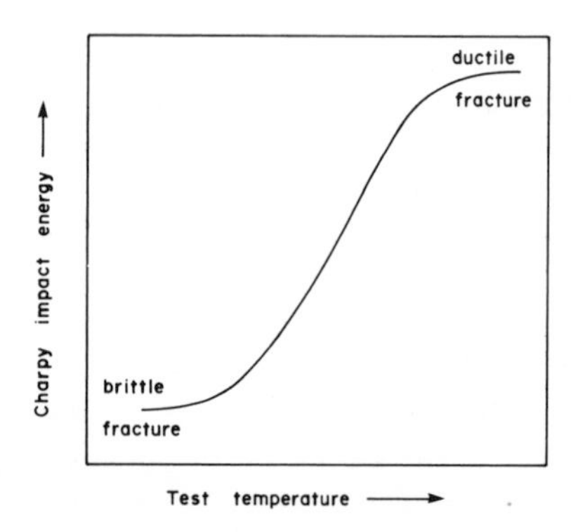

Fig. 8.1 Variation in fracture energy with test temperature in Charpy V-notch impact testing of a typical steel

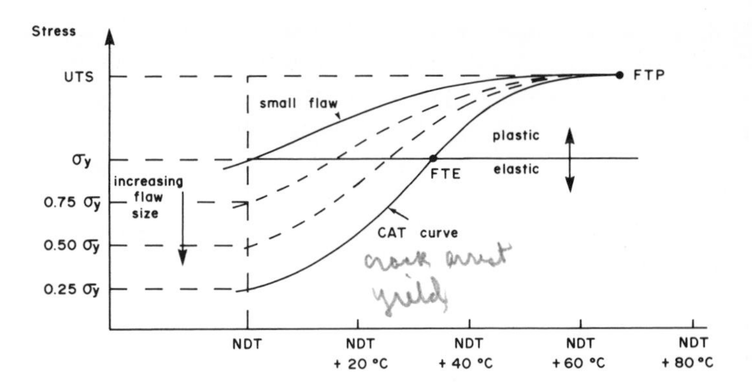

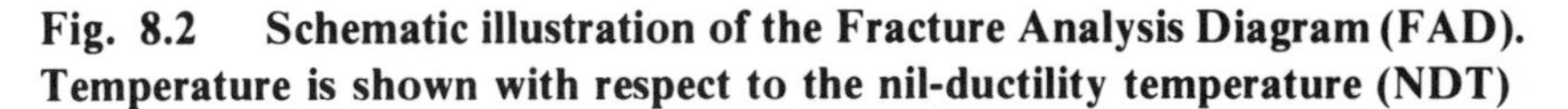

Fig. 8.2 Schematic illustration of the Fracture Analysis Diagram (FAD). Temperature is shown with respect to the nil-ductility temperature (NDT)

strength, together with such factor of safety (or of ignorance) as was deemed appropriate, were inadequate and might be inherently unsafe.

In the 1920s Griffith had published his classic papers which proposed that the propagation of a crack in a brittle solid required an adequate release of elastic strain energy to provide for the additional energy associated with the new surface created [*6, 7*]. He also demonstrated the interrelation between fracture stress and defect size in glass, corresponding to his theory. This work had been considered to be, to a large extent, an historical scientific curiosity rather than of practical engineering value, until Orowan [*8*] and Irwin [*9*] (who can truly be said to be the father of engineering fracture mechanics) independently modified the Griffith expression for fracture stress to allow for the energy expended at the crack tip in *local* plastic deformation, so making the concept applicable to many engineering materials. Thus, the concept of *fracture mechanics* as an approach to safe design has been developed over the past twenty years or so, the approach being to establish the value of stress which may cause the largest defect likely to be present to propagate, and subsequently to establish a safe working stress on this basis.

In the foregoing, it has been indicated that fracture may take place in either a brittle or a ductile manner, the criterion being primarily the energy which is absorbed during crack propagation. In an idealized situation, brittle fracture may be viewed as separation of atomic planes without any plastic flow occurring, i.e., as *cleavage*, and such a condition is *approached* (only) in the fracture of ceramics and in BCC and some HCP metals under appropriate conditions. However, cracks in ductile FCC and other metals may propagate under tensile stress with *local* plastic deformation only, depending on the geometry of the crack and the loading. Additionally, cracks may propagate under low stress owing to the synergistic action of a gaseous or liquid medium, while grain boundary (intercrystalline) cracks may develop either from deformation at high temperature (creep) or by reason of local embrittlement caused by the segregation of specific elements to grain boundaries (e.g., temper embrittlement in steels). Finally, cracks may propagate under cyclic stress, giving rise to a fatigue failure.

It is important that mechanical engineers engaged in design have a clear understanding of the basic mechanisms of fracture and of the mechanics of the fracture process. Equally, it is crucial that metallurgists involved in the development and the selection of materials have an understanding of the service conditions (in terms of stress, environment and the loading geometry) which may lead to premature failure. In the present chapter, the basic mechanisms of crack initiation and of fracture are first reviewed with a view both to providing a background to the design engineer and to introducing the concepts relevant to the analysis of failures based on fracture surface appearance, then the basic Griffith and subsequent Irwin-Orowan criteria for crack growth are introduced in a straightforward manner, while the criteria for cleavage crack growth and the transition to ductile fracture under the appropriate conditions of temperature and microstructure are discussed in the light of these ideas. Finally, the analytical methods used to describe the stress fields at the tips of cracks or defects are described, these leading to quantitative methods for the determination of the energy absorbed during crack extension. In Chapter 9 the application of these basic concepts to fracture mechanics and design is discussed briefly.

Of necessity, the material covered in this and the subsequent chapter is limited, and it has not been possible to do much more than introduce the concepts. Nevertheless, the main principles have been covered, which should enable the interested reader to turn to more specialized texts. For detailed discussion, the reader should consult the seven volume treatise edited by Liebowitz [*10*], while excellent introductions to fracture mechanics are provided in the works of Broek [*11*] and Knott [*12*], and a good overview of fracture is given by Tetelman and McEvily [*13*].

8.2 FRACTURE MECHANISMS

The two basic types of fracture are *cleavage* and *ductile fracture*. The former occurs under tensile stress and involves separation along crystallographic planes without significant plastic flow occurring: the fracture is brittle and a small amount of energy is expended during the process. Ductile fracture involves plastic flow by slip and, depending on the extent of plastic deformation, the energy absorbed may vary widely. It will be large where extensive shear, necking or void formation takes place, but when plastic deformation is localized and confined to the tip of a propagating crack, the fracture may be brittle in an engineering sense, but ductile on a strictly physical or mechanistic basis.

In addition to these basic types of fracture, and as noted in the introduction, cracks may be initiated and may propagate as a result of other factors until they reach a sufficient length that final fast fracture takes place either by cleavage or plastic flow, or some combination of the two. Specifically, we may identify *environmentally assisted cracking, intergranular cracking*, and *fatigue crack growth*. The first of these refers to cracking produced under the *combined* action of stress and an environmental effect such as corrosion, the presence of embrittling hydrogen, or the adsorption of some chemical species on the fracture surface. The second may occur during high temperature creep deformation as a result of grain boundary sliding or the deposition of vacancies at grain boundary sinks, so producing voids, or it may be induced through segregation of specific elements to grain boundaries, so causing embrittlement. While we shall designate this an essentially distinct fracture type, it does involve cleavage or ductile fracture in the final analysis. Finally, fatigue cracking occurs under cyclic stress and, although the micromechanisms are ductile in nature, the fracture process can be considered to be specific, with the resulting fracture surface having a distinctive morphology.

The study of fracture mechanisms involves detailed study of fracture surfaces by optical and electron microscopy, together with related examinations of microstructure and crack paths. The techniques involved and the methods of interpretation are detailed in a number of sources [*14-16*] to which the interested reader is referred. In addition to providing information concerning the micromechanisms of fracture, fractographic techniques may, in some cases, provide direct information concerning the microstructure itself [*17*]. A general and more detailed review by the author of fracture mechanisms and their relation to fracture morphology [*18*] may serve to complement the discussions here.

8.2.1 Cleavage Fracture

Because of its crystallographic nature, cleavage leads to a fracture surface consisting of flat, shiny facets. The essentially flat crack, whose surfaces lie roughly normal to the tensile stress axis, changes its orientation somewhat when it crosses a grain boundary because of the differing orientations of the grains and the fact that a cleavage crack propagates on specific families of planes (e.g., on {100} in α-iron). Where the boundary is a tilt or high angle one, the crack may re-initiate in the next grain as a single one, but in crossing a twist boundary, many small parallel cracks may be formed as shown in Fig. 8.3, these having *cleavage steps* between them. The small steps run together to

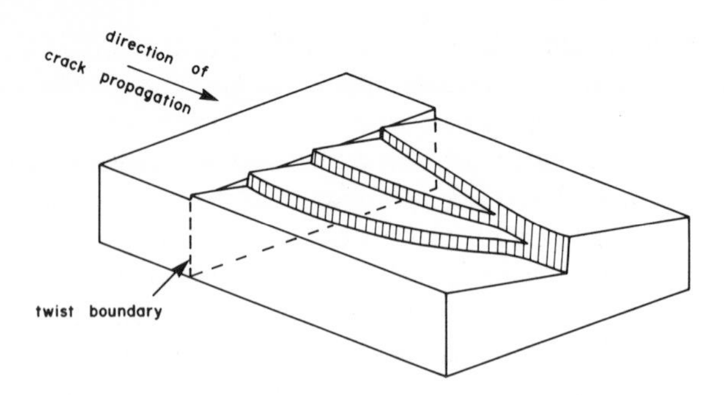

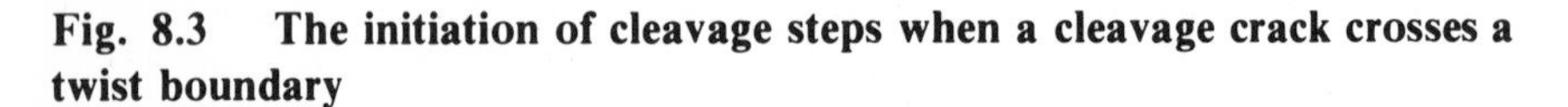

Fig. 8.3 The initiation of cleavage steps when a cleavage crack crosses a twist boundary

produce larger ones downstream, (see Fig. 8.3) and this leads to the characteristic features known as *river patterns* which are observed on fractographs prepared from cleaved polycrystalline metals and may be seen in Fig. 8.4. Cleavage steps may also be produced when a cleavage crack intersects a screw dislocation [*19*] as indicated in Fig. 8.5.

Another characteristic feature of cleavage in iron is the formation of *tongues* [*20*] which are formed by local fracture along the interfaces between the matrix and a twin. The crack, which is growing on a {100} plane in the

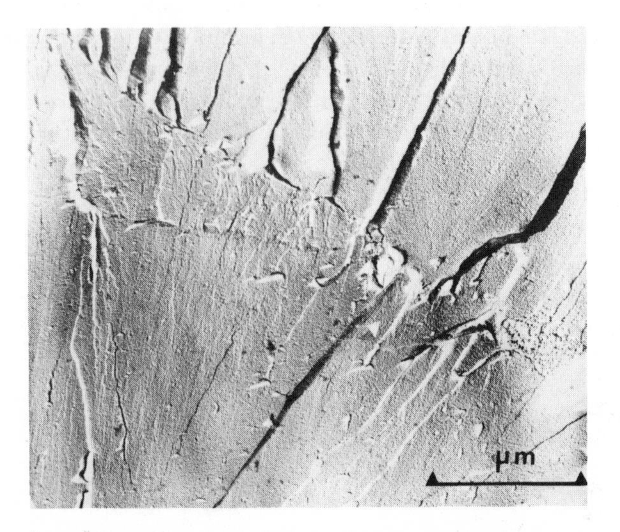

Fig. 8.4 River patterns on the cleaved surface of mild steel

matrix, propagates some distance along a {112} twin interface before linking up once more with the main crack which has propagated around the twin on the original {100} plane. This is shown schematically in Fig. 8.6, while Fig. 8.7 shows the appearance of tongues on a microfractograph. Because a cleavage crack propagates at high velocity, normally a considerable number of twins are formed in α-iron just in advance of the moving crack front, the local strain rate being too high for slip processes to provide all the accommodation required.

Although cleavage is crystallographic in nature, the initiation of a cleavage crack requires that the local tensile stress at the tip of a flaw or crack reach the theoretical tensile stress for fracture, and cracks will propagate from sharp notches with virtually no plastic flow in covalent and ionic solids. However, in

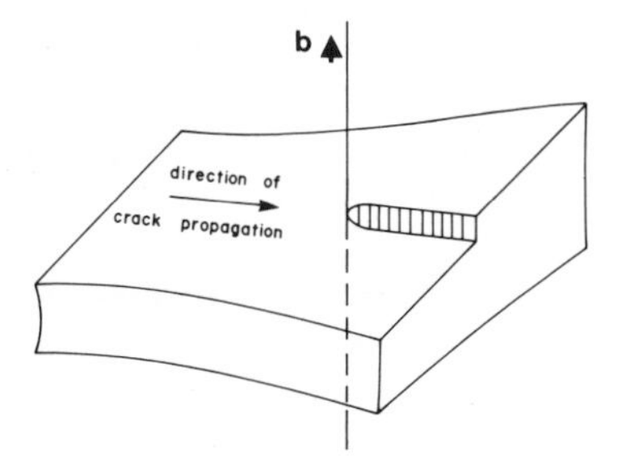

Fig. 8.5 Cleavage step produced at intersection of a screw dislocation by the propagating cleavage crack

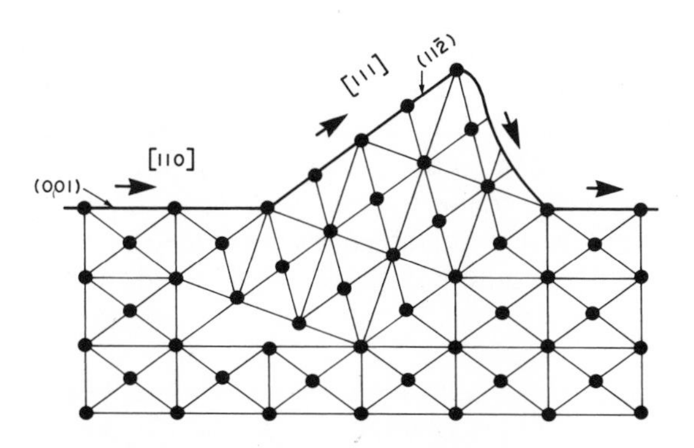

Fig. 8.6 Schematic illustration of tongue formation in α-iron at a twin-matrix interface

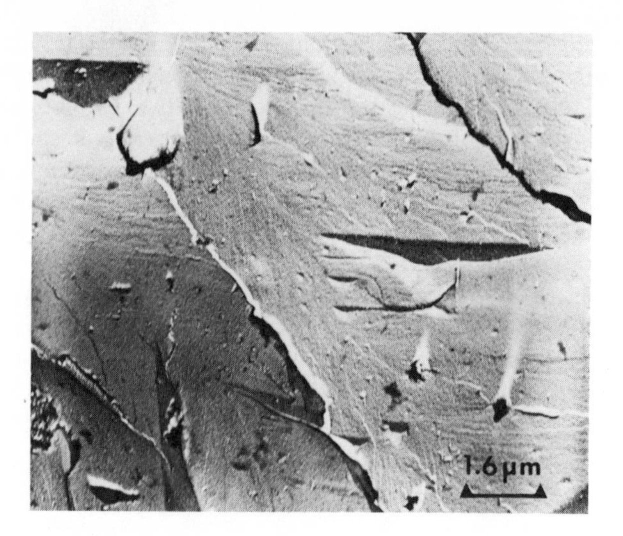

Fig. 8.7 Microfractograph showing tongues on cleaved mild steel

metals local plastic flow will begin when the local stress reaches τ_Y, the yield stress in shear[1], and a cleavage crack will be initiated at this stress level after some critical displacement has been produced at the crack tip. In the absence of defects in the material, yielding will commence and, if the temperature is below the ductile-brittle transition value[2], cleavage will then initiate. During the propagation of a cleavage crack in a metal, some plastic flow always takes place at the crack tip where a high stress concentration exists. Thus, the energy required to create the new surfaces is raised above that strictly for separation of atomic planes. In addition, in some cases ductile tear ridges form between cleavage facets rather than cleavage steps [*21*]: again, this raises the quantity of plastic work which is done. The interrelation between fracture stress and yield stress for low-carbon steel tested in tension at 77 K is shown in Fig. 8.8 [*22*] as a function of grain size, the data being taken from the work of Low [*23*]. Below some critical grain size, whose value is discussed in Section 8.3.3, yield takes place before fracture; above this the extrapolated line for yield stress corresponds with the data points for fracture stress, the resultant fracture being brittle in nature, as indicated by the ductility curve.

---

[1] Because of local triaxial stresses, the local tensile yield stress may be greatly in excess of the uniaxial tensile stress, following the discussion of Chapter 2 concerning yield criteria.

---

[2] Note that this is a function of the local stress state and specimen dimensions.

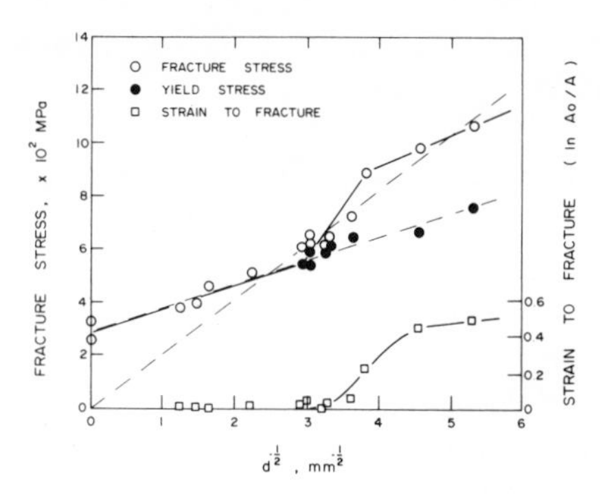

Fig. 8.8 The variation in fracture stress, yield stress and fracture strain at 77 K for low carbon steel tested in tension. The data are from the work of Low [*23*]. (From Ref. [*22*])

A number of theories have been formulated to explain the mechanisms of nucleation of cleavage cracks on the basis of dislocation movement, and it is instructive to examine some of them.

The simplest model envisages dislocations at the head of a pile-up (see Fig. 4.37), coalescing to form a wedge-shaped crack nucleus [*24*]. The resulting stress concentration may cause a crack to initiate on a plane roughly normal to the slip plane, when the stress on this plane reaches a value equal to the theoretical strength. Stroh [*25*] analyzed this situation, and his model is represented in Fig. 8.9. The value of the normal stress, $\sigma_{\theta\theta}$, is

$$\sigma_{\theta\theta} = (3/2)\,(d/2r)^{1/2}\,(\tau - \tau_o)\sin\theta\cos(\theta/2) \tag{8.1}$$

where τ is the applied shear stress and τ_o is the internal friction stress (see Section 4.3.10). The maximum value of $\sigma_{\theta\theta}$ occurs at $\theta = 70.5°$ and we may set this equal to the theoretical strength given by Eq. (7.5). Hence, for cleavage crack nucleation,

$$(\sigma_{\theta\theta})_{max} = (2/\sqrt{3})\,(d/2r)^{1/2}\,(\tau - \tau_o) \geqslant (E\gamma/a_o)^{1/2} \tag{8.2}$$

and the applied stress required to initiate cleavage is

$$\tau = \tau_c = \tau_o + (2E\gamma/a_o)^{1/2} d^{-1/2} \tag{8.3}$$

For $r = a_o$, the lattice spacing, which is the smallest value possible, and taking $E = 2G(1 + \nu)$ from Eq. (1.35), we have

$$\tau_c = \tau_o + [4G\gamma\,(1 + \nu)]^{1/2} d^{-1/2} \tag{8.4}$$

The number of dislocations in a pile-up required for initiation of cleavage fracture in this manner is given approximately from Eq. (4.28) as

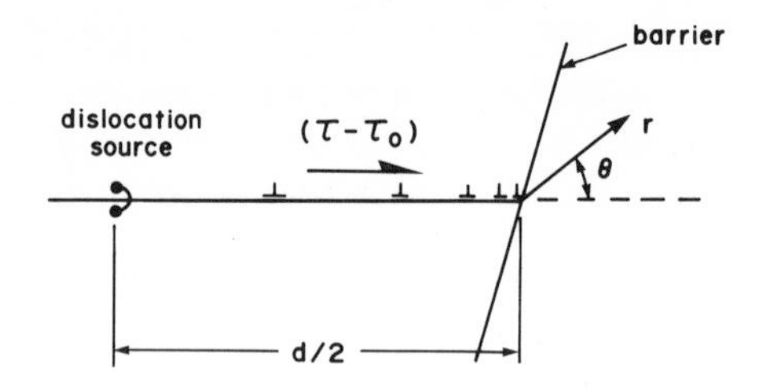

Fig. 8.9 Stroh's model for dislocation pile-up leading to cleavage fracture

$$n = d\pi(\tau_c - \tau_o)/4Gb \tag{8.5}$$

Eq. (8.4) indicates that the stress for cleavage crack nucleation should obey a Hall-Petch-type dependence on grain size, similar to that observed for yield stress [cf. Eq. (6.57) and Section 6.3.3]. Such a relationship is illustrated in Fig. 8.8 for fracture stress, and on the basis that the observed stresses for yield and fracture correspond at larger grain sizes and at low temperature, we may rewrite Eq. (8.4) as

$$\tau_c = \tau_o + (k_Y\, d^{-1/2})/2 \tag{8.6}$$

where k_Y is the Hall-Petch constant from Eq. (6.57).

An alternative dislocation mechanism for cleavage crack nucleation in BCC crystals was proposed by Cottrell [*26*], being shown in Fig. 4.25 and represented by Eq. (4.21). In this, intersecting dislocations on {110} slip planes coalesce to form an immobile dislocation on a {001} cleavage plane, the crack nucleus extending as more dislocations pass down the slip planes.

Cleavage cracks may also be nucleated at twins. At low temperatures and high strain rates twinning becomes the preferred deformation mode in BCC metals, taking place by the movement of $a/6\,[111]$ dislocations on successive {112} planes. If a twin is blocked, for example at a grain boundary or at another twin, a large pile-up of dislocations is produced, and a cleavage crack may be nucleated. Such twin-initiated cleavage has been observed in practice [*27, 28*]. The stress for twinning follows a Hall-Petch relationship but its dependence on grain size is greater than that for slip, although the friction stress component is lower [*29*]. Hence, twinning is only preferred over slip in larger grain sizes, and cleavage is initiated by twinning at larger grain sizes only.

Another important factor in the nucleation of cleavage (and other) cracks is the presence and size of second phase particles, such as carbides in steel. Smith [*30*] has discussed the case of cleavage being initiated in a carbide at a grain boundary by plastic deformation in the neighboring ferrite grain as shown in

Fig. 8.10. This, in turn, may initiate a cleavage fracture in the next adjacent grain if the local stress at the grain boundary is sufficiently high and the formation of such a crack becomes energetically favorable. The theory is of particular value as it allows us to examine the important effect of carbide thickness on the brittle fracture of steels on a mechanistic and semi-quantitative basis.

Our discussion of detailed dislocation mechanisms has been concerned primarily with the initiation of cleavage cracks, but the fact that such cracks may or do form does not necessarily mean that they will propagate, resulting in fracture of the component. For example, cleavage cracks have been observed to nucleate at twin/grain boundary intersections under compression [*31*], but they require a change to tensile stress in order to propagate further. We shall return to the subject of cleavage crack propagation in Section 8.3.3, after examining other fracture mechanisms and dealing with the criteria and energy requirements for continued crack growth.

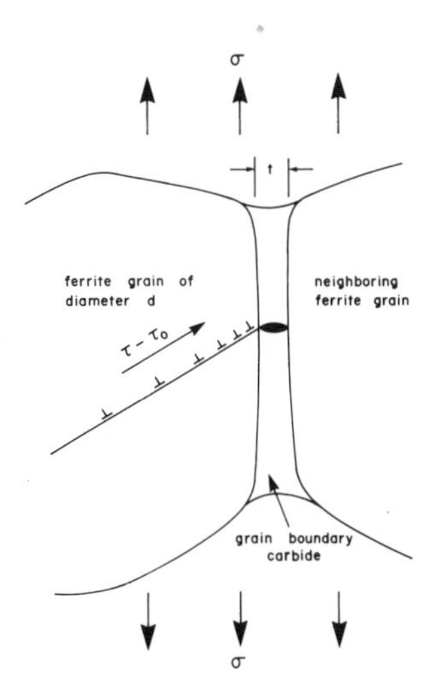

Fig. 8.10 The initiation of cleavage in a grain boundary carbide by slip within a neighboring ferrite grain

8.2.2 Ductile Fracture

Ductile fracture involves plastic flow, and the general term covers a range of failure modes. Under monotonic loading, and in the absence of cleavage, ductile fracture may take place with a large variation in the extent of plastic deformation: when this is small, the fracture may be considered "brittle" in an engineering sense, with the crack propagating in a plane normal to the applied tensile stress. In such a case, however, the localized crack tip plastic deformation precludes the formation of cleavage facets and steps.

Figure 8.11 shows schematically various types of ductile fracture ranging from tensile fracture in pure, ductile metals which can result in necking down to a point, to creep fracture by intergranular void formation. Here, the discussion will be concerned primarily with rupture in materials containing second phase particles, and intergranular creep fracture by cavitation will be dealt with briefly in Section 8.2.4. Concentration on materials containing second phase particles is hardly a severe restriction, as engineering alloys always contain large quantities of second phase particles, ranging from fine precipitates detectable only using thin foil electron microscopy, to large particles or inclusions which are readily visible under the optical microscope.

The mechanism of ductile fracture is generally related to the initiation of microvoids at second phase particles, and their subsequent growth and coalescence. Resulting from this, the fracture surfaces contain dimples which

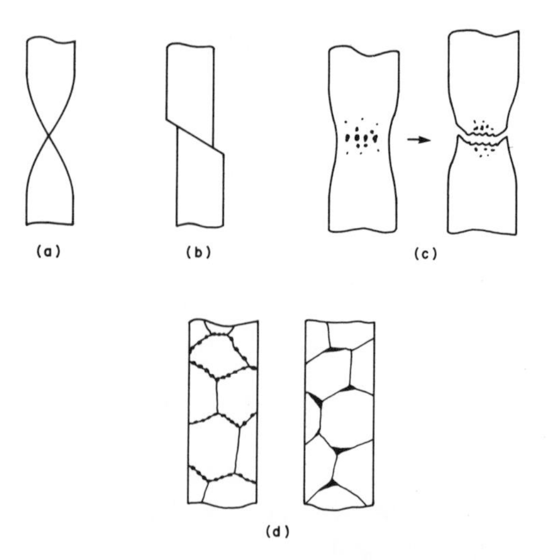

Fig. 8.11 Schematic illustration of different types of ductile fracture in tension: (a) necking to a point in very pure metal; (b) failure by shearing off; (c) cup and cone fracture caused by void nucleation and coalescence with final shear; (d) intergranular creep fracture showing different void nucleation points

frequently show evidence of their initiation at the second phase particles. Figure 8.12 shows schematically the dependence of dimple morphology on the fracture mode, while Fig. 8.13 shows typical equiaxed dimples on an AISI 4140 steel fractured in tension. In the case of large particles, microvoid initiation may result from cleavage fracture of the particles, rather than decohesion at the particle-matrix interface as more commonly occurs.

Gurland and Plateau [32] proposed a simple model for microvoid initiation by interfacial decohesion on the basis that the strain energy relieved by void formation must be sufficient to produce the energy required for the newly created surfaces. Their treatment leads to an expression for the uniaxial tensile stress, σ, required for particle-matrix decohesion of the form

$$\sigma = (1/q)\,(E\gamma/a)^{1/2} \tag{8.7}$$

where q is the stress concentration factor at the particle, γ the specific surface energy of the crack, E a weighted average of the elastic moduli of particle and matrix, and a the particle size (diameter). The model has been extended to take into account the plastic work being dissipated around the particle [33], and the following expression derived:

$$\sigma = (1/q)\,(E\gamma/a)^{1/2} + (\sigma_Y/q)\,(\Delta V/V)^{1/2} \tag{8.8}$$

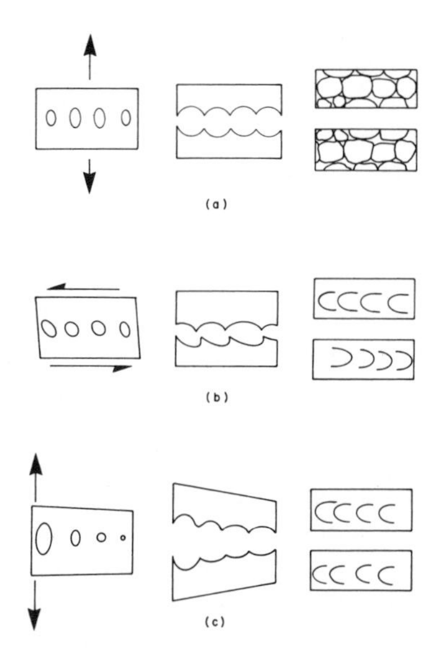

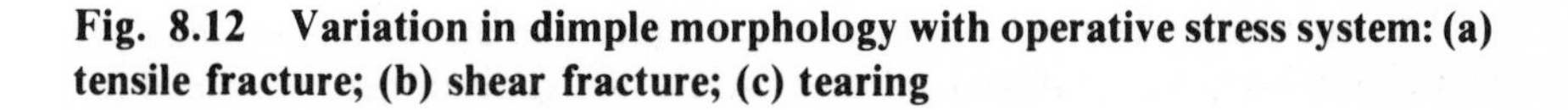
Fig. 8.12 Variation in dimple morphology with operative stress system: (a) tensile fracture; (b) shear fracture; (c) tearing

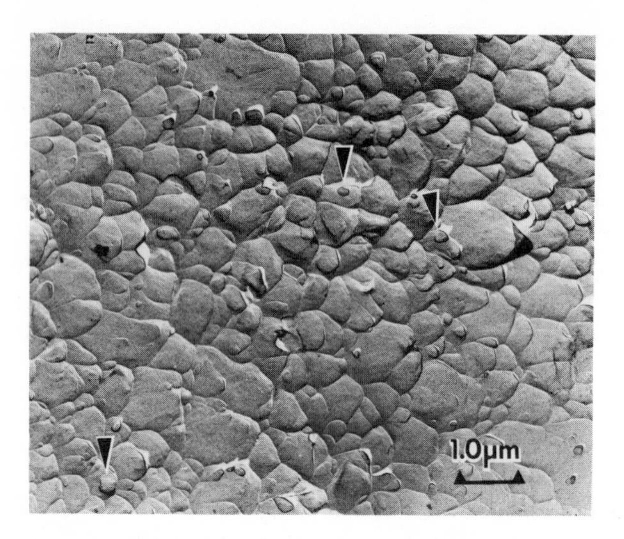

Fig. 8.13 Equiaxed dimples on fracture surface of quenched and tempered 4140 steel. The arrows indicate carbides at which local fracture initiated

in which σ_Y is the average yield stress of the matrix for a given particle shape and volume fraction of precipitates, and is assumed to be independent of particle size and strain, V is the volume of a particle, and $\Delta V / V$ indicates the volume deformed around the particle in relation to the particle volume, this being regarded as approximately constant for given particle shape.

Equation (8.8) indicates a linear dependence of the tensile stress to initiate ductile fracture on $(1/a)^{1/2}$, and it has been shown [*33, 34*] that UTS has a similar dependence for some high strength quenched and tempered steels and for spheroidized carbon steels, for a given carbide shape, such plots being shown in Fig. 8.14. Thus, it has been suggested that the controlling step in the process of ductile fracture in such materials may be microvoid initiation, this occurring at a load close to the maximum on the tensile load-extension curve.

Where extensive work hardening takes place, voids grow under the influence of a tensile stress as well as of a lateral stress, as occurs during necking, and the fracture ductility may be calculated on the basis of the voids coalescing, being a function of the volume fraction of the voids [*36*]. In high strength materials of low work hardening capacity, microvoids formed by interfacial decohesion may link up by separation along slip bands [*37*], as indicated in Fig. 8.15, giving rise to a jagged fracture surface.

8.2.3 Environment Assisted Fracture

This is a large and complex subject and a number of different cases may be identified where the combined action of stress and environment may lead to

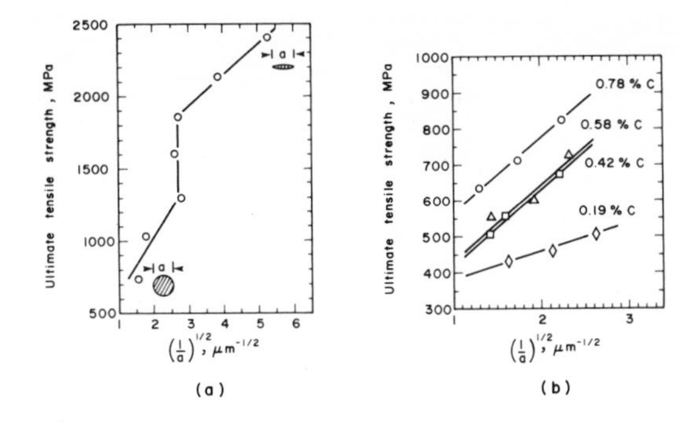

Fig. 8.14 Variation of ultimate tensile stress with carbide size in steels: (a) for AISI 4140, quenched and tempered; (b) for spheroidized carbon steels using the data of Fukui and Uehara [*35*]. (From Lui and Le May [*34*])

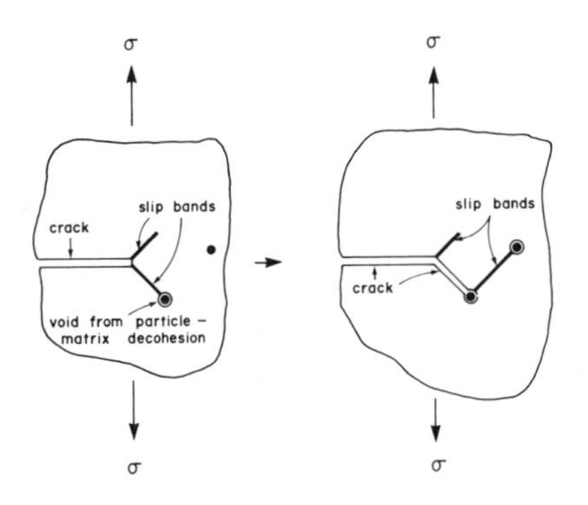

Fig. 8.15 Microvoids linking by shear along slip bands in high strength material having limited work hardening capacity

fracture. These include stress corrosion cracking (SCC), hydrogen embrittlement and liquid metal embrittlement. In addition, we may include corrosion fatigue and embrittlement under radiation under the heading, the former being referred to when discussing fatigue fracture mechanisms in Section 8.2.5, and discussed at greater length in Section 10.8, while the latter has been dealt with to some extent in Section 6.5.

Stress corrosion cracking can occur by several mechanisms, a simple model which will serve as a basis for discussion being shown in Fig. 8.16. Here, corrosion is considered to take place at the tip of a propagating crack when

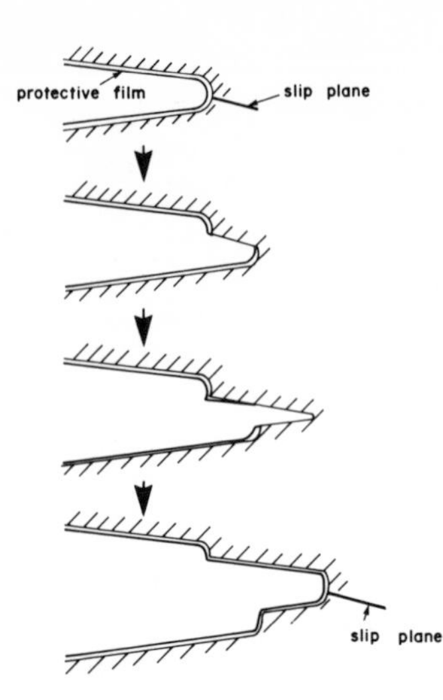

Fig. 8.16 Schematic view of a stress corrosion crack propagating by rupture of the protective film following plastic deformation at the crack tip, subsequent corrosion, repassivation, and build up of local stress to repeat the cycle

bare metal is exposed upon rupture on the protective film or passivated layer, consequent upon slip taking place at the tip. The corroding metal becomes passivated or otherwise protected in its turn, and the process is repeated, with the crack progressing in a stepwise manner. The resulting crack path may be either intergranular or transgranular depending on the material and environmental conditions, while the fracture is characterized by little ductility, and an oxide coating is frequently to be seen on the crack surfaces.

An additional mechanism causing SCC is thought to be the evolution of hydrogen at the crack tip in some metal-environment combinations. This may be produced during corrosion in an aqueous medium or it may arise from dissociation of water vapor on newly exposed bare metal at the tip. In either case, the hydrogen produced may diffuse into the metal ahead of the crack, causing local embrittlement.

It should also be noted that it is the local environmental conditions at the tip of the crack which are of importance, rather than those of the bulk fluid, as these local conditions can differ greatly in pH and in solute concentration.

Hydrogen embrittlement can be a particularly severe problem in ferrous materials because of the very high mobility of hydrogen in the BCC lattice at ambient temperatures, as well as its reduced solubility in ferrite as compared with austenite. When present in sufficient quantity, hydrogen lowers the cohesive strength of the lattice [38], and it also tends to accumulate at defects, so producing high internal pressure which may lead to rupture, particularly in high strength or martensitic alloys. Hydrogen pickup may occur in several ways, e.g., from water in the scrap used in steelmaking, from pickling or plating operations, from the tip of an SCC crack as mentioned above (or, similarly, at a corrosion fatigue crack), or when welding using electrodes having a moist coating. The last mentioned source leads to underbead cracks developing in the heat affected zone which is the last material to transform from austenite and into which hydrogen is concentrated during cooling.

Hydrogen embrittlement is not confined to ferrous materials, and can occur in Ti, Zr and the refractory metals. In Ti and Zr, brittle hydrides are formed on specific crystallographic planes, while the refractory metals are thought to be embrittled from supersaturation of hydrogen in solution in the lattice [39].

Liquid metal embrittlement is thought to be caused by a reduction in the bond strength of the atoms in a solid by chemisorption of a liquid atom at the tip of a crack or at the end of a pile-up of dislocations near a surface obstacle. Attachment of an atom in this way reduces the stress required for cleavage, the whole stress-displacement curve of Fig. 7.1 being moved downwards. Thus, normally ductile metals may fail in a brittle manner by cleavage or, more commonly for polycrystalline metals, by rapid intergranular crack propagation. A review of the mechanisms of liquid metal embrittlement and fracture has been made by Kamdar [40].

8.2.4 Intergranular Fracture

At higher temperature and increased strain rates, polycrystalline metals which normally fail in a transcrystalline ductile manner undergo fracture of an intergranular nature. Thus, the majority of creep service failures are intergranular, as illustrated by the typical fracture of Fig. 8.17.

The development of intergranular creep fracture depends on the nucleation, growth and subsequent linking of voids on the grain boundaries. The two distinct types of cavity formed are illustrated in Fig. 8.11, being termed "r" or rounded-type and "w" or wedge-type, respectively. The nucleation and growth of r-type cavities at high temperature is strongly dependent on the stress state. The local effective stress may lead to nucleation, while subsequent growth depends on the maximum tensile stress which produces the flow of vacancies to voids which grow preferentially on boundaries having high tensile stress acting on them [41]. Their formation can be suppressed by superposition of hydrostatic compression [42], while at lower levels of tensile stress, fracture may occur by shear of material between them, rather than by void coalescence, so producing a mixed intergranular-transgranular fracture.

The w-type cavities initiate at triple points consequent upon grain boundary sliding, and their formation may be promoted by decohesion at the interfaces between grain boundary precipitates and matrix.

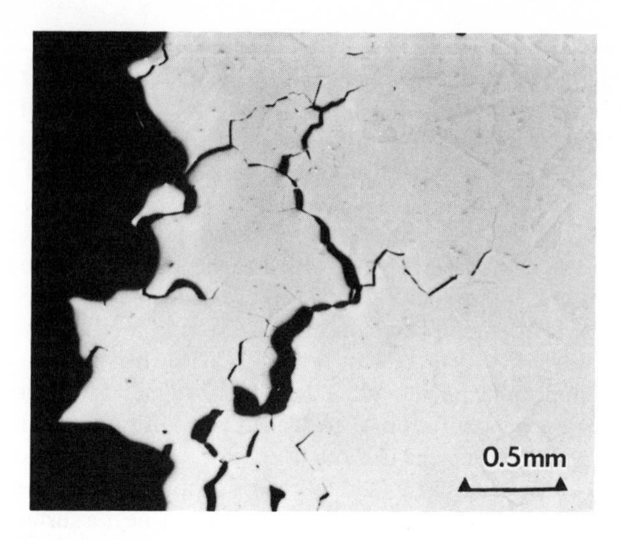

Fig. 8.17 Intergranular creep fracture in a nickel-base superalloy gas turbine blade. (Courtesy of Dr. W.E. White)

Embrittlement, which gives rise to intergranular fracture at ambient temperatures, may arise from segregation of specific elements to grain boundaries, from precipitation of brittle phases (such as carbides) on boundaries, of from environmental sources, such as SCC or liquid metal ions, as discussed already. In all these cases a fracture surface with a faceted appearance results.

8.2.5 Fatigue Fracture

In discussing the mechanisms of fatigue failure, we must distinguish between the processes of crack initiation, crack propagation and final rupture. Fatigue is not in itself a mechanism of fracture, the term applying rather to the phenomenon by which materials fail under cyclic or repeated loads, which if applied as steady state values, would not cause failure.

Fatigue cracks generally initiate at surface discontinuities or stress raisers, and their formation is aided by the presence of a notch or inclusion, although they may form in some cases at a subsurface defect. Fatigue damage arises from local cyclic plastic flow, which generates high local dislocation densities on slip bands and can lead to the formation of small surface intrusions and extrusions which may be detected by optical or replica electron microscopy. They are shown schematically in Fig. 8.18. The nominal stress levels at which fatigue cracks initiate may be far below the elastic limit of the material, but on a micro-scale the local stresses may be much higher as a result of local stress concentration at discontinuities.

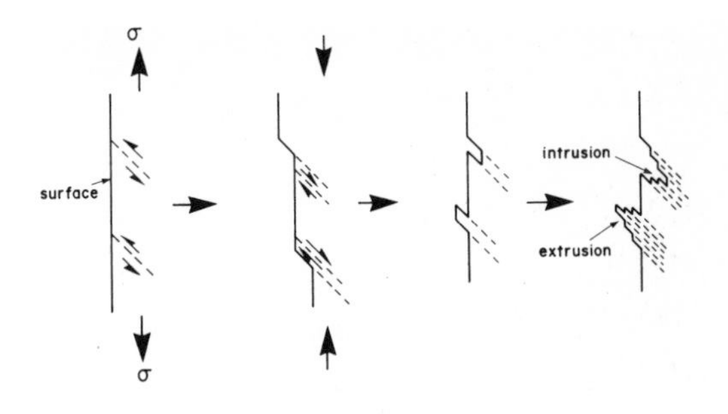

Fig. 8.18 The formation of intrusions and extrusions at the surface during cyclic stressing

A fatigue crack, once initiated, grows as a result of local plastic deformation at the crack tip during each tensile load application: fatigue cracks will not grow under compressive loading conditions. Initially, the crack grows slowly along a slip band in which the crack nucleated in the highly dislocated substructure, and after some time it changes from this *Stage I* crack propagation mode to *Stage II* mode, during which it grows roughly normal to the tensile axis. Figure 8.19 shows this schematically, and also indicates the region of rapid fracture by final shear to leave a shear lip.

The length of the Stage I crack is generally small and its surfaces are often essentially featureless; however, during Stage II growth distinctive features termed *fatigue striations* are frequently formed. Figure 8.20 shows a typical example of such striations, although their morphology varies widely with material and environment.

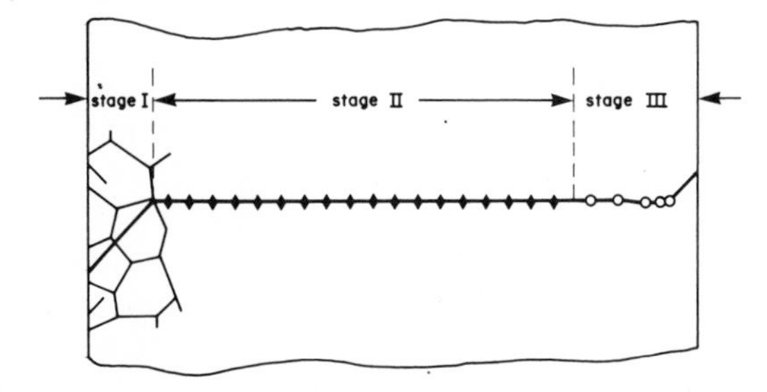

Fig. 8.19 The different stages of fatigue crack propagation

Fig. 8.20 Fatigue striations on high strength steel

The "plastic blunting model" of Laird and Smith [*43*], illustrated in Fig. 8.21, provided a reasonable first model for striation formation, and more detailed models have been formulated subsequently both on a continuum mechanics approach and from consideration of slip processes at the crack tip. The coarse slip model [*44-46*] is shown in Fig. 8.22, in which slip occurs on

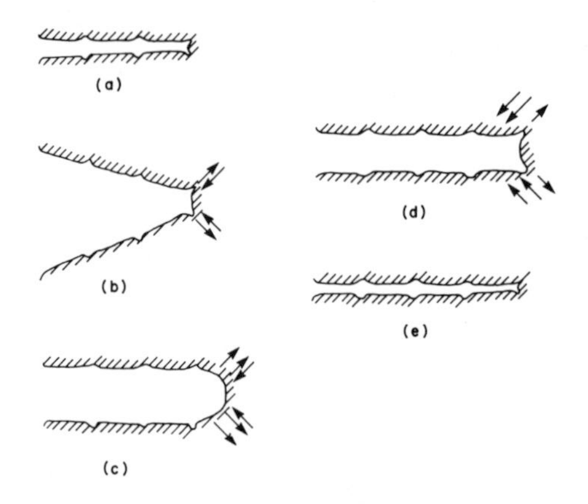

Fig. 8.21 The plastic blunting model of Laird and Smith [*43*] for the formation of fatigue striations. One complete stress cycle is shown

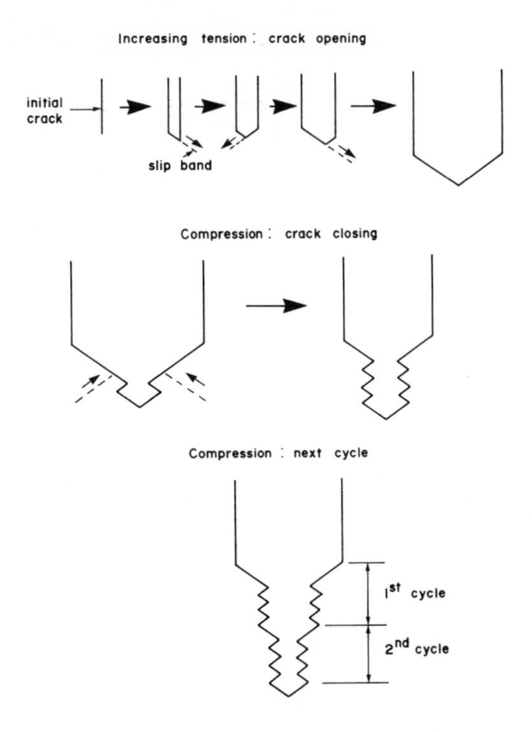

Fig. 8.22 The coarse slip model for fatigue crack growth

intersecting slip planes meeting at the crack tip during tensile loading to produce crack blunting. On load reversal, reverse slip takes place, but rewelding does not occur because of surface oxidation, and crack tip resharpening takes place. The result is a series of fine steps on the fracture surface and a larger step or striation is produced during each load cycle. However, slip will not always be possible on slip planes passing exactly through the crack tip because of dislocations or second phase particles which restrain it. Thus, models have been derived in which slip occurs on planes which do not pass through the crack tip [*47, 48*], and Lui's model is shown in Fig. 8.23. Depending on the number of active forward and reverse slip planes, a wide range of striation shapes can be predicted. Although drawn for FCC metals, the model can be applied to BCC metals by modifying the slip geometry.

We may note that Stage II fatigue fracture surfaces formed in vacuum do not have striations on them, and that the crack growth rate is much reduced. This can be explained because reverse slip may occur on the same planes as forward slip and crack rewelding may take place, as no surface oxide film will

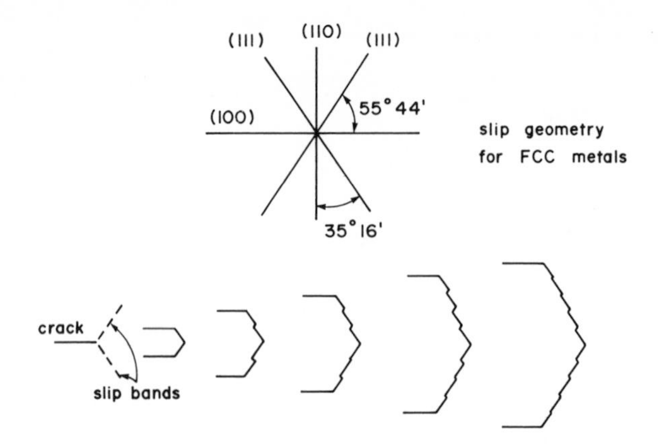

Fig. 8.23 The model of Lui [*47*] for fatigue crack growth on multiple slip planes

have been formed. It is postulated that a crack opens first as in air, rewelds to a decreasing extent over some 5 to 10 cycles, and then moves forward another step [*49*]. This would produce a series of fine irregular slip steps with a spacing similar to that occurring in air, and this has been observed in practice [*49*].

Corrosion fatigue has already been referred to, and the effect of a corrosive environment is to reduce greatly a material's resistance to cyclic loading. The corrosive fluid may attack the freshly exposed metal at the crack tip upon loading, or water vapor may dissociate causing hydrogen penetration and embrittlement ahead of the crack. The subject is a complex one, and much remains to be done: in particular, it is very difficult to predict fatigue strengths and rates of crack propagation in corrosive media on the basis of data determined in air, which is usually taken to be "inert". Schijve [*50*] has prepared an good review of this subject, and the proceedings of two conferences provide good discussion of the state of knowledge [*51, 52*].

This discussion of fatigue fracture mechanisms has been brief and qualitative, but the broader subject of design for conditions of cyclic or repeated loading is considered in more detail in Chapter 10.

8.3 CRACK GROWTH CRITERIA

Having examined the criteria and mechanisms for crack initiation, we shall now examine the criteria for propagation of a crack through a crystal by cleavage or ductile fracture. The special case of fatigue crack growth, which is not unstable but occurs in cyclic steps, is discussed in Chapter 10.

8.3.1 The Griffith Criterion

The theoretical strength of a perfectly brittle solid containing no flaws was given by Eq. (7.6) as

$$\sigma_{max} = (E\gamma/a_o)^{1/2} \tag{8.9}$$

and at failure a large quantity of elastic strain energy is released, as indicated by the strain at fracture [see Eq. (7.7)]. In practice, the observed strength is generally much lower, either because slip occurs at a lower stress level or because a crack propagates from the small cracks, defects or notches present in a real material, unless it is a very perfect fiber or crystal (see Section 7.2.3). Such defects cause a local high stress concentration, but this alone is not a sufficient factor to explain the propagation of a crack from a defect at a low value of applied stress.

Consider an infinitely wide plate of unit thickness as shown in Fig. 8.24, containing an elliptical crack of length $2a$ and tip radius, ρ, and subjected to a uniaxial stress σ in a plane perpendicular to the crack. The Inglis [53] solution for the stress at the crack tip was given in Eq. (1.53) and is

$$\sigma_{tip} = \sigma\,[1 + 2(a/\rho)^{1/2}] \tag{8.10}$$

or

$$\sigma_{tip} \simeq 2\sigma\,(a/\rho)^{1/2} \tag{8.11}$$

for $2(a/\rho)^{1/2} >> 1$. In the limiting case of a very sharp crack we may consider $\rho \rightarrow a_o$, the lattice parameter. Thus, setting σ_{tip} equal to the theoretical stress for

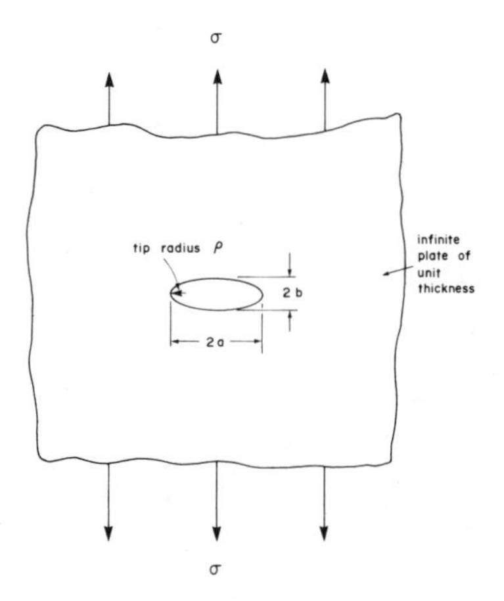

Fig. 8.24 Elliptical crack in infinite plate and subject to uniaxial stress

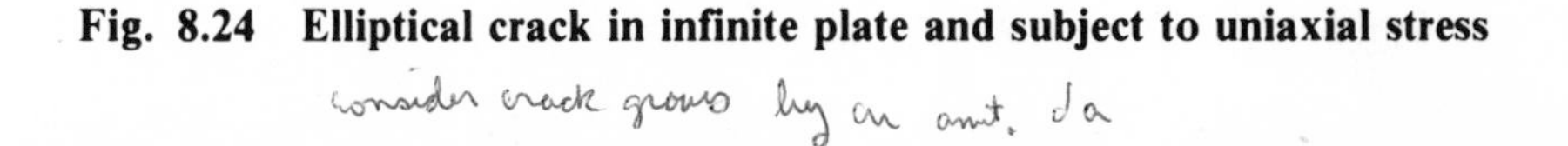

fracture [Eq. (8.9)], the criterion for crack propagation or failure ($\sigma = \sigma_F$) is

$$\sigma_F = (E\gamma/4a)^{1/2} \tag{8.12}$$

However, this derivation ignores the fact that the equation for stress concentration [Eq. (8.10)] was derived on the basis of linear elastic behavior holding far beyond what is known to be the elastic limit, while the theoretical stress for fracture [Eq. (8.9)] is based on a sinusoidal relationship obtaining between force and atomic spacing. This inconsistency is avoided by the approach taken by Griffith [*6, 7*] in determining a thermodynamically-based criterion for crack propagation.

Consider the crack growing by an amount da. There will be a release of elastic stress distributed around the elliptical crack, and Griffith calculated the elastic strain energy released in the plate per unit thickness for a crack length $2a$ to be

$$U_E = \pi\sigma^2 a^2/E \tag{8.13a}$$

for plane stress conditions, and

$$U_E = [\pi\, \sigma^2\, a^2/E]\,(1 - \nu^2) \tag{8.13b}$$

for plane strain. The surface energy due to the presence of the crack is

$$U_S = 4a\gamma \tag{8.14}$$

where γ is the specific surface energy. Hence the criterion for crack propagation,

$$\partial U_E/\partial a \geqslant \partial U_S/\partial a \tag{8.15}$$

leads to the stress to cause failure, as

$$\sigma_F \geqslant (2E\gamma/\pi a)^{1/2}, \qquad \text{plane stress} \tag{8.16a}$$

or

$$\sigma_F \geqslant [2\, E\gamma/\pi\,(1 - \nu^2)a]^{1/2}, \qquad \text{plane strain} \tag{8.16b}$$

Alternatively, we may write the length of the "Griffith crack" for fracture to take place under a specified stress, σ, as[3]

$$a = 2\, E\gamma/\pi\, \sigma^2, \qquad \text{plane stress} \tag{8.17a}$$

and

$$a = 2\, E\gamma/\pi\,(1 - \nu^2)\sigma^2, \qquad \text{plane strain} \tag{8.17b}$$

The stress at a crack tip was given by Eq. (8.10), and putting this equal to the theoretical strength [Eq. (8.9)] as a necessary condition for fracture to occur at the crack tip, we obtain the fracture stress as

---

[3] It should be noted that a surface elliptical crack of length a and an interior one of length $2a$ give rise to identical fracture stresses.

$$\sigma_F = [(E\gamma/4a)\,(\rho/a_o)]^{1/2} \tag{8.18}$$

Equating this to the fracture stress found using Griffith's energy approach [Eq. (8.16a)], we obtain the condition that

$$\rho = (8/\pi)a_o \simeq 3\ a_o \tag{8.19}$$

This represents the lower limit of the effective radius of an elastic crack, and the Griffith criterion based on energy considerations is applicable, independent of the sharpness of any crack whose tip radius is less than $\sim 3\ a_o$. At larger values of ρ, the requirement for crack propagation is given by Eq. (8.18), which is the more severe condition.

From the foregoing it is clear that the presence of a defect in a brittle solid will have a strong effect in reducing the fracture stress from its theoretical value in the absence of defects, and that increasing size of a defect causes further substantial reduction in strength. It is extremely difficult to produce defect-free materials, except in the form of vapor-deposited fibers or very carefully grown single crystals as discussed in Chapter 7, when reviewing the highest observed strengths. It is also clear that the Griffith theory, being based on thermodynamic principles, provides a necessary criterion for fracture to take place, but does not say that it will in fact occur. As noted above, if the radius, ρ, at the end of the notch is at all large, there may be insufficient concentration of stress for fracture to commence, even though the energy requirement would permit crack growth.

Griffith's assumption that the energy requirement was the thermodynamic value of surface energy is incorrect in that dissipative processes at the crack tip absorb much larger amounts of energy; for example, in Al_2O_3 at liquid nitrogen temperature, these processes absorb some fifteen times the value of the surface energy [*54*]. In engineering materials such as steel, the plastic deformation at the crack tip may absorb at least a thousand times the surface energy requirement. Thus, while Griffith's work allowed a reasonable prediction of the effect of defect size on the fracture strength of a brittle material such as glass, a modified criterion for fracture is required for practical engineering materials.

8.3.2 The Irwin-Orowan Criterion

Orowan [*8*] and Irwin [*9*] independently modified the Griffith criterion to take into account the plastic work at the crack tip. Orowan rewrote Griffith's equation for fracture stress [Eq. (8.16a)] in the form

$$\sigma_F = [E(2\gamma + \gamma_p)/\pi a]^{1/2} \tag{8.20}$$

where γ_p represents the energy expended in plastic work per unit area of crack surface. As indicated above, experiment has shown that $(2\gamma + \gamma_p)$ is much greater that 2γ, thus Eq. (8.20) can be rewritten as

$$\sigma_F = (E\gamma_p/\pi a)^{1/2} \tag{8.21}$$

Irwin [9] postulated that a critical amount of energy, $\mathcal{G}_c$, was required to create additional area of crack surface, and that this quantity could be measured for a specific material in a fracture test. Thus, by comparing the quantity of energy which would be released when a given crack extended under a specified stress level with the known value of $\mathcal{G}_c$, prediction could be made as to whether or not crack growth would take place.

The energy equation for crack propagation [Eq. (8.15)] can be rewritten in the form

$$\partial U_E / \partial a = 2\mathcal{G} \geqslant \partial W / \partial a \tag{8.22}$$

where $\mathcal{G}$ is the *strain energy release rate* for each crack tip, and W is the energy required for crack growth. Assuming that the energy requirement $\partial W / \partial a$ is, to a first approximation, constant, and considering the elliptical crack of Fig. 8.24, we can write the criterion for fracture stress as

$$\sigma_F = (E\mathcal{G}_c / \pi a)^{1/2}, \quad \text{plane stress} \tag{8.23a}$$

$$\sigma_F = [E\mathcal{G}_c / \pi (1 - \nu^2) a]^{1/2}, \quad \text{plane strain} \tag{8.23b}$$

Thus,

$$\mathcal{G} = \pi \sigma^2 a / E, \quad \text{plane stress} \tag{8.24a}$$

or

$$\mathcal{G} = (\pi \sigma^2 a / E)(1 - \nu^2), \quad \text{plane strain} \tag{8.24b}$$

and $\mathcal{G}$ reaches the critical value, $\mathcal{G}_c$, when crack propagation commences. $\mathcal{G}$ can be measured by means of compliance tests on a specimen containing a notch, loaded as in Fig. 8.25(a)[4]. The energy release for an increase in crack length to $a + \delta a$ is obtained from the elastic loading curves for the two crack lengths shown in Fig. 8.25(b). The energy release is $\mathcal{G} t \delta a$ where t is the thickness of the plate and is equal to $P \delta u / 2$, which is the shaded area between the two loading curves.

The load and deflection are related by

$$P = u / C \tag{8.25}$$

where C is the *compliance* of the system at a specified value of a. For constant load,

$$\mathcal{G} t \delta a = P^2 \, \delta C / 2$$

and as $\delta a \rightarrow 0$, we have

$$\mathcal{G} = P^2 (\partial C / \partial a) / 2t \tag{8.26}$$

[4] In order to produce as sharp a notch as possible, a fatigue crack is produced at the end of the machined notch. The length of the crack, a, is the total crack length: machined notch plus fatigue cracks.

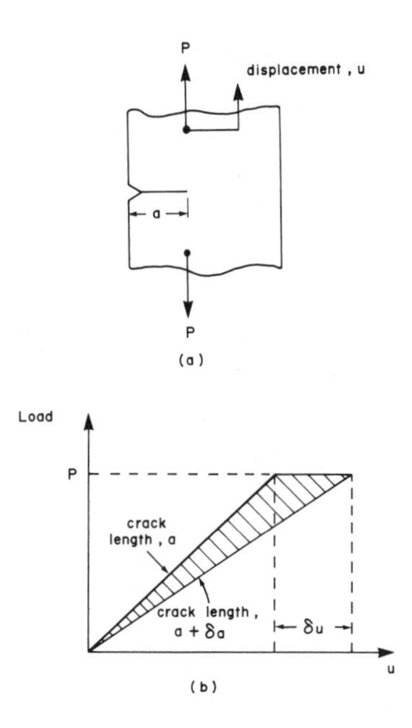

Fig. 8.25 For a cracked plate loaded as in (a), the elastic load-displacement curve is shown in (b). Two curves are shown, corresponding to crack lengths a and $a + \delta a$, both loaded to the same maximum load, P

The compliance can be measured as a function of a and its slope at the appropriate value of a (e.g., at the initial crack length) determined. The value of $\mathcal{G}_c$ is obtained by noting the load point at which the crack begins to grow, so producing a step or at least a change in the slope of the load/displacement curve.

The energy approach of Irwin was received at first with considerable scepticism, but it now forms the basis of the fracture mechanics approach to design, as discussed in the following chapter.

8.3.3 Cleavage Crack Propagation and Ductile-Brittle Transition

In Section 8.2.1 we examined the mechanism of cleavage fracture and the models postulated for cleavage crack nucleation. Here we shall briefly discuss models for the continued growth and propagation of a cleavage crack, which also show clearly the basis for the observed transition from ductile to brittle fracture as a function of temperature and grain size in BCC metals such as steel.

It should be noted that the propagation of a cleavage crack requires the presence of a tensile normal stress; also, that while the surface energy requirement may be low and close to the specific thermodynamic value for the initial separation of atoms, the value of effective surface energy during propagation is expected to be much higher because of induced dislocation movement; finally, for a crack to propagate through a crystal it must be longer than the "Griffith size", i.e., the critical size for a crack to develop by itself as given by Eq. (8.17), or a larger value still which takes into account the plastic work requirements.

Cottrell [*26*] and Petch [*55*] have analyzed the situation where cleavage crack propagation may be more difficult than nucleation. First, we consider dislocations coalescing to form a crack at an obstacle such as a grain boundary (see Fig. 8.9) or coalescing on a cleavage plane (see Fig. 4.25). The crack will be wedged open by nb, and it can itself be represented as a pile-up of n *cleavage dislocations* of Burgers vector b [*19*] as shown in Fig. 8.26, such that the distance between the sides of the crack $h = nb$. Propagation of the crack can be represented by the movement of this group of cleavage dislocations, "climbing" without diffusion in the cleavage plane perpendicular to their Burgers vectors.

For cleavage to occur, the work done by the applied stress during an increment in crack length must equal the effective energy of the crack surface produced, i.e.,

$$\sigma\, nb = 2\gamma \tag{8.27}$$

where σ is the normal stress on the crack.

If the crack is forming in a notch, transverse tensions will be developed because of localized plastic deformation, so that only a part of the applied stress is available to provide the shear stress required to coalesce the dislocations. If we designate this part as $q\sigma$, it has been shown that $q \simeq 1/3$ for a notched specimen [*26*] and it is, of course, unity for an unnotched one.

The number of dislocations in the pile-up can be computed by equating the relaxed elastic shear displacement to the plastic displacement on the crack

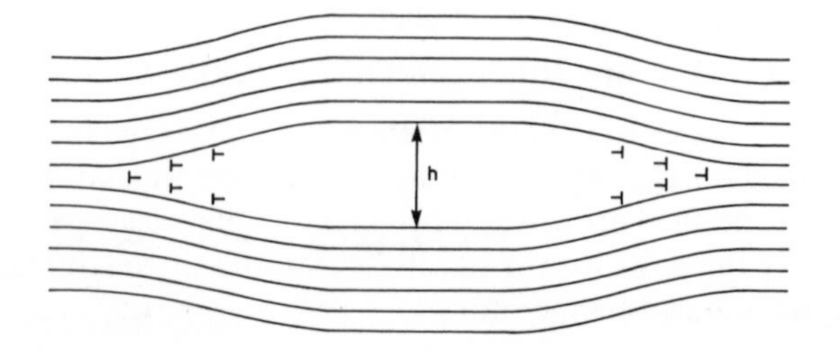

Fig. 8.26 A cleavage crack made up of cleavage dislocations

opening. The relaxed elastic displacement is $(\tau - \tau_o)\, d/G$, assuming that the slip band has formed across the grain of diameter d, and the plastic displacement is nb. Thus,

$$n = (\tau - \tau_o)\, d/Gb \tag{8.28}$$

We can rewrite this as

$$n \simeq (q\sigma - \sigma_o)\, d/2Gb \tag{8.29}$$

and the stress required for cleavage crack propagation is given by

$$\sigma\, (q\sigma - \sigma_o) = 4\, G\gamma/d \tag{8.30}$$

The dependence of yield stress on grain size is given by the Hall-Petch relation

$$\sigma_Y = \sigma_o + k_Y\, d^{-1/2}$$

Thus, we can substitute $q\sigma$ for σ_Y under conditions of yielding, and for cleavage failure to occur at yield, the criterion is

$$\sigma = \sigma_Y/q = (\sigma_o + k_Y\, d^{-1/2})/q = 4G\gamma/k_Y\, d^{1/2} \tag{8.31}$$

If cleavage does not occur at the yield stress, we can write the observed dependence of the stress for ductile fracture on grain size as

$$\sigma_F = \sigma_o + k_F\, d^{-1/2} \tag{8.32}$$

where k_F is a constant. The transition from ductile to cleavage fracture is then given by

$$\sigma = 4G\gamma/k_F\, d^{1/2} \tag{8.33}$$

Cleavage will take place at the yield stress if this has a value greater than the RHS of Eq. (8.31). The effect of increased grain size is seen to favor cleavage fracture, as also does increase in the friction stress, σ_o, and the presence of a notch which decreases the value of q.

The effect of temperature on the fracture transition may be seen from the effect of temperature on the friction stress, σ_o, as both k_Y and γ are fairly independent of temperature. For BCC metals, σ_o has a dependence on temperature of the form [56]

$$\ln \sigma_o = \ln B - \beta T \tag{8.34}$$

where B and β are constants. Thus, the transition temperature, T_c, is obtained from Eq. (8.31) as

$$T_c = [\ln B - \ln (4qG\gamma/k_Y - k_Y) - \ln (d)^{-1/2}]/\beta \tag{8.35}$$

The predicted transition temperature is seen to depend strongly on grain size, and to vary linearly with $\ln (d)^{-1/2}$, which relationship has been observed to hold in practice [57].

While the foregoing model predicts most of the observed effects of grain size, yield stress and the presence of a notch on the transition temperature, it omits consideration of the important effect of second phase particles, and we shall return to the theory of Smith [*30*] which provides an explanation for the embrittling influence of such particles, which has been discussed elsewhere [*12, 22, 58*].

Smith's model was shown in Fig. 8.10. Plastic deformation on a slip band within a ferrite grain provides a stress which may initiate a crack in a brittle grain boundary carbide of thickness t. The surface energy of the carbide (γ_c) will be low, while the neighboring ferrite grains will have a higher value of effective surface energy, γ_f. Hence, the crack in the carbide may or may not propagate through the adjacent ferrite grain by cleavage, depending on the change in energy as it extends.

The energy terms to be considered first are as follows:

Work done by the applied stress in extending the crack from $x = t$ to $x = t + c$

$$= \tfrac{1}{2}\sigma\, nb\, (t + c)$$

The change in elastic strain energy as the crack extends to $x = t + c$ [from Eq. (8.13b)]

$$= \pi\, \sigma^2\, [(t + c)/2]^2\, (1 - \nu^2)/E - \pi\, \sigma^2\, (t/2)^2\, (1 - \nu^2)/E$$

The change in surface energy = $2\, \gamma_f\, c$

The energy of the cracked edge dislocation of strength nb [from Eq. (4.11a)]

$$= [1/(1 - \nu)]\, [G\, (nb)^2/4]\, \ln\, (2R/t)$$

where R is the effective radius of the stress field of the dislocation and $t/2$ is taken as its core radius.

Solution of the resulting equation for energy balance leads to the criterion for brittle fracture as

$$\sigma\, nb/2\gamma_f + (1 - \nu^2)\, \sigma^2\, t/4E\gamma_f \geqslant 1 \tag{8.36}$$

Taking the number of dislocations in the pile-up to be given by Eq. (8.28) we obtain the fracture stress in the form

$$\sigma_F = [4(\tau - \tau_o)\, d^2/(1 - \nu)^2\, t^2 + 2\, G\gamma_f/(1 - \nu)t]^{1/2} - 2(\tau - \tau_o)\, d/(1 - \nu)t \tag{8.37}$$

where $(\tau - \tau_o)$ is the effective stress or active component of stress on the slip plane. For the condition where fracture occurs at the yield stress, we can replace $(\tau - \tau_o)$ by $k_Y d^{-1/2}/2$ from the Hall-Petch equation for yield stress, and the fracture stress is then

$$\sigma_F = [k^2_Y\, d/(1 - \nu)^2\, t^2 + 2\, G\gamma_f/(1 - \nu)\, t]^{1/2} - k_Y\, d^{1/2}/(1 - \nu)\, t \tag{8.38}$$

Similar expressions to Eqs. (8.37) and (8.38) are given by Almond *et al.* [*58*], the differences arising from different expressions used for the number of dislocations in the pile-up. Certainly, the considerable influence of carbide thickness is apparent in the expressions derived.

If there are no carbides present, i.e., $t = 0$, Eq. (8.36) simplifies to give

$$\sigma_F = 4\ G\gamma_f / k_Y\ d^{1/2}$$

for fracture at the yield stress, this being identical to Eq. (8.31). Where the grain size of a steel is small, the presence of coarse carbides will be of increased importance in determining the stress for brittle fracture, and there is a need for additional careful experimental work in this area, particularly as there is a strong tendency towards finer grain sizes in high strength, low alloy structural steels.

8.4 STRESSES AT CRACKS

As we have seen, a crack will only propagate when the elastic strain energy released by its extension from a to $(a + \delta a)$ is sufficient to provide for the energy of the additional surface and the plastic deformation produced. Assuming that this quantity is known (i.e., $\mathcal{G}_c$ is known), this being a constant for the particular material and the conditions of crack loading, our requirement is to compute the actual rate of energy release which would occur on crack growth in order to compare it with $\mathcal{G}_c$. Thus, it is necessary to compute the distribution of stress and strain around a crack.

Before we examine this matter further, we must note that there are three distinct loading modes for a crack as shown in Fig. 8.27. In Mode I loading [Fig. 8.27(a)] the crack surfaces are displaced in directions normal to the plane of the crack; in Mode II [Fig. 8.27(b)] in-plane shear or sliding takes place, while Mode III [Fig. 8.27(c)] refers to antiplane strain with all deformation occurring in the direction normal to the plate. The general case of three-dimensional cracking can be represented by superimposing the three modes; however, Mode I is by far the most important in practice.

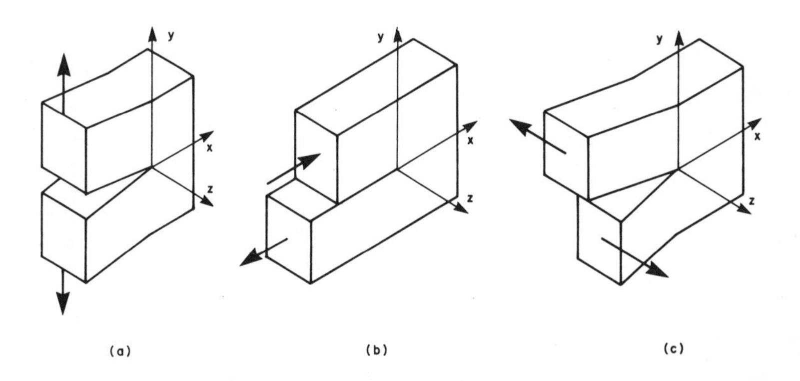

Fig. 8.27 The three loading modes for a crack: (a) Mode I, the opening mode; (b) Mode II, the edge-sliding mode; (c) Mode III, antiplane strain or tearing mode

8.4.1 Linear Elastic Crack Tip Stress Analysis

The introduction of a crack into a body causes redistribution of stress. In an elastic body, the greatest stresses occur around the crack tips, and while there may be some local plastic deformation, linear elastic methods of analysis are widely used when this small local non-linear deformation is confined within a linearly-elastic stress field.

In Chapter 1, the basic ideas concerning stress concentrations were introduced, and it was shown that a stress function could be used to obtain a solution to such a problem in linear elasticity. The stresses around a crack can be obtained conveniently by the method of Westergaard [*59*] using a complex function, and the procedure is outlined in some detail by Knott [*12*], while Eshelby [*60*] provides a useful background to such analytical methods.

For a crack of length $2a$ in an infinite plate and for the geometry shown in Fig. 8.28, the stresses in the region of the crack tip are given by

$$\left.\begin{aligned}
\sigma_{xx} &= \sigma\,(a/2r)^{1/2}\cos(\theta/2)\,[1-\sin(\theta/2)\sin(3\theta/2)]\\
\sigma_{yy} &= \sigma\,(a/2r)^{1/2}\cos(\theta/2)\,[1+\sin(\theta/2)\sin(3\theta/2)]\\
\tau_{xy} &= \sigma(a/2r)^{1/2}\sin(\theta/2)\cos(\theta/2)\cos(3\theta/2)\\
\tau_{xz} &= \tau_{yz} = 0\\
\sigma_{zz} &= 0, \qquad \text{plane stress}\\
\sigma_{zz} &= \nu\,(\sigma_{xx}+\sigma_{yy}), \quad \text{plane strain}
\end{aligned}\right\} \tag{8.39}$$

As pointed out by Irwin [9], the local stresses at the crack tip depend on the product $\sigma\sqrt{a}$, and defining the *stress intensity factor*, K_{I}, for the sharp crack in an infinitely wide plate as

$$K_{\mathrm{I}} = \sigma\,(\pi a)^{1/2} \tag{8.40}$$

Eqs. (8.39) may be written as

$$\left.\begin{aligned}
\sigma_{xx} &= [K_{\mathrm{I}}/(2\pi r)^{1/2}]\cos(\theta/2)\,[1-\sin(\theta/2)\sin(3\theta/2)]\\
\sigma_{yy} &= [K_{\mathrm{I}}/(2\pi r)^{1/2}]\cos(\theta/2)\,[1+\sin(\theta/2)\sin(3\theta/2)]\\
\tau_{xy} &= [K_{\mathrm{I}}/(2\pi r)^{1/2}]\sin(\theta/2)\cos(\theta/2)\cos(3\theta/2)\\
\tau_{xz} &= \tau_{yz} = 0\\
\sigma_{zz} &= 0, \qquad \text{plane stress}\\
\sigma_{zz} &= \nu\,(\sigma_{xx}+\sigma_{yy}), \quad \text{plane strain}
\end{aligned}\right\} \tag{8.41}$$

and the displacements in x, y and z directions are given, for plane strain conditions, by

$$\left.\begin{aligned}
u &= (K_{\mathrm{I}}/G)\,(r/2\pi)^{1/2}\cos(\theta/2)\,[1-2\nu+\sin^2(\theta/2)]\\
v &= (K_{\mathrm{I}}/G)\,(r/2\pi)^{1/2}\sin(\theta/2)\,[2-2\nu-\cos^2(\theta/2)]\\
w &= 0
\end{aligned}\right\} \tag{8.42}$$

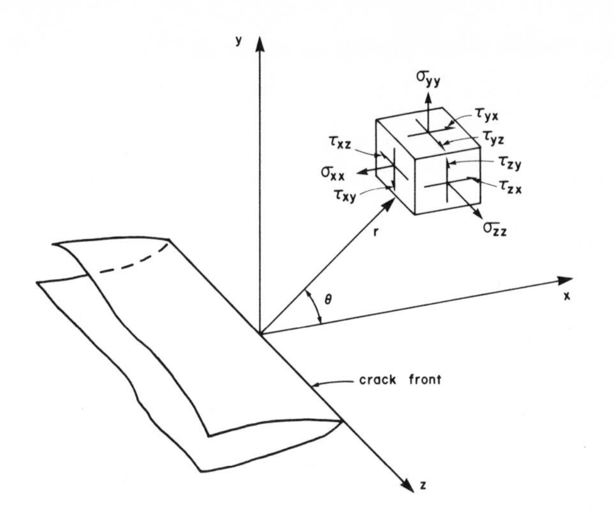

Fig. 8.28 Coordinates specified from the leading edge of a crack, and the components of stress in the crack tip stress field

From Eqs. (8.41), we see that the stress *distribution* at the tip of a crack depends only on the parameters r and θ, and that stress varies as $1/(2\pi r)^{1/2}$. This is true whatever the stress system applied, so that the stress intensity factor, K_I, defines the *magnitude* of the stress field at the crack tip. The

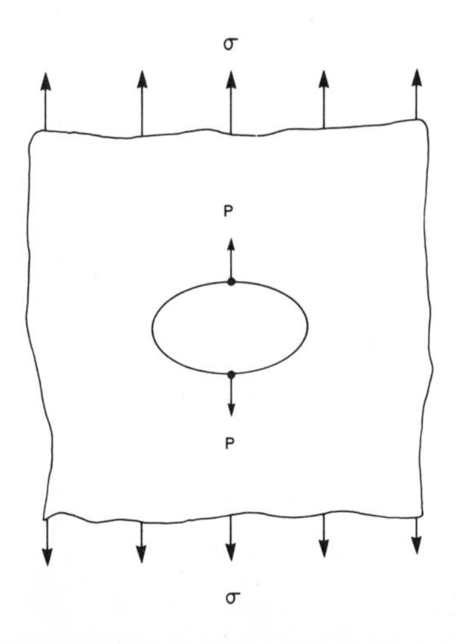

Fig. 8.29 Elliptical crack with forces applied at the crack center in addition to uniform stress on the member

subscript refers to Mode I loading as in Fig. 8.27(a).

If the loading on the crack (in Mode I) were complex, consisting of several components, for example a uniform tensile stress on the member as before, together with a force on the crack center as in Fig. 8.29, the total stress intensity factor could be obtained by summing the individual component stress intensity factors. Thus

$$K_{\mathrm{I\ resultant}} = K_{\mathrm{I\ a}} + K_{\mathrm{I\ b}} + \ldots \tag{8.43}$$

This direct summation is possible because of the identical form of the equations for crack tip stress.

In all cases K_{I} has the form

$$K_{\mathrm{I}} = \sigma\,(\alpha\pi a)^{1/2} \tag{8.44}$$

where α is a function of both specimen and crack geometry.

Similar equations for the stress and displacement fields at the tips of Mode II and Mode III cracks have been developed and are as follows.

Mode II:

$$\left.\begin{aligned}
\sigma_{xx} &= -\,[K_{\mathrm{II}}/(2\pi r)^{1/2}]\sin(\theta/2)\,[2 + \cos(\theta/2)\cos(3\theta/2)]\\
\sigma_{yy} &= [K_{\mathrm{II}}/(2\pi r)^{1/2}]\sin(\theta/2)\cos(\theta/2)\cos(3\theta/2)\\
\tau_{xy} &= [K_{\mathrm{II}}/(2\pi r)^{1/2}]\cos(\theta/2)\,[1 - \sin(\theta/2)\sin(3\theta/2)]\\
\tau_{xz} &= \tau_{yz} = 0\\
\sigma_{zz} &= 0, \qquad \text{plane stress}\\
\sigma_{zz} &= \nu\,(\sigma_{xx} + \sigma_{yy}), \qquad \text{plane strain}
\end{aligned}\right\} \tag{8.45}$$

Plane strain displacements

$$\left.\begin{aligned}
u &= (K_{\mathrm{II}}/G)\,(r/2\pi)^{1/2}\sin(\theta/2)\,[2 - 2\nu + \cos^2(\theta/2)]\\
v &= (K_{\mathrm{II}}/G)\,(r/2\pi)^{1/2}\cos(\theta/2)\,[-1 + 2\nu + \sin^2(\theta/2)]\\
w &= 0
\end{aligned}\right\} \tag{8.46}$$

Mode III:

$$\left.\begin{aligned}
\sigma_{xx} &= \sigma_{yy} = \sigma_{zz} = \tau_{xy} = 0\\
\tau_{xz} &= -\,[K_{\mathrm{III}}/(2\pi r)^{1/2}]\sin(\theta/2)\\
\tau_{zy} &= [K_{\mathrm{III}}/(2\pi r)^{1/2}]\cos(\theta/2)
\end{aligned}\right\} \tag{8.47}$$

Displacements

$$\left.\begin{aligned}
w &= (2\,K_{\mathrm{III}}/G)\,(r/2\pi)^{1/2}\sin(\theta/2)\\
u &= v = 0
\end{aligned}\right\} \tag{8.48}$$

Values of K have been determined for many different crack and loading geometries in plates of finite size, and are tabulated in a number of sources, e.g. [*61, 62*]. Some of the more commonly used stress intensity factors are listed in Table 8.1.

Table 8.1 Sample Formulae for Stress Intensity Factors

Form of Crack	Stressing	Loading Mode	Stress Intensity Factor
Crack of length $2a$ in infinite plate	Stress, σ, normal to crack	I	$K_I = \sigma(\pi a)^{1/2}$
	Shear stress, τ, parallel to crack	II	$K_{II} = \tau(\pi a)^{1/2}$
	Shear stress, τ_t, through sheet thickness	III	$K_{III} = \tau_t(\pi a)^{1/2}$
Crack of length $2a$ in plate of width w	σ, normal to crack	I	$K_I = \sigma(\pi a)^{1/2} [\sec (\pi a/w)]^{1/2}$
	τ, parallel to crack	II	$K_{II} = \tau(\pi a)^{1/2} [\sec (\pi a/w)]^{1/2}$
	τ_t, through thickness	III	$K_{III} = \tau_t(\pi a)^{1/2} [(w/\pi a) \tan (\pi a/w)]^{1/2}$
Edge crack of length a in plate of width w	Stress, σ, normal to crack	I	$K_I = \sigma(\pi a)^{1/2} [1.12 - 0.23 (a/w) + 10.6 (a/w)^2 - 21.7 (a/w)^3 + 30.4 (a/w)^4]$
Two collinear edge cracks, each of length a, in plate of width, w	Stress, σ, normal to cracks	I	$K_I = \sigma(\pi a)^{1/2} [1.12 - 0.56(a/w) - 0.015(a/w)^2 + 0.091(a/w)^3]/(1 - a/w)^{1/2}$

8.4.2 Relation Between Energy Release Rate and Stress Intensity Factor

The strain energy release rate, $\mathscr{G}$, for the crack growing under tension in an infinite plate was given by Eqs. (8.24), and from comparison with Eq. (8.44) it is seen that

$$\mathscr{G}_{\mathrm{I}} = K_{\mathrm{I}}^2/E, \qquad \text{plane stress} \qquad (8.49a)$$

$$\mathscr{G}_{\mathrm{I}} = (K_{\mathrm{I}}^2/E)(1-\nu^2), \qquad \text{plane strain} \qquad (8.49b)$$

where the subscript I is used with $\mathscr{G}$ to designate that the energy release rate is for Mode I loading.

The relationship between $\mathscr{G}$ and K represented by Eqs. (8.49) is a general one, and this is shown by the analysis due to Irwin [63] as follows.

Consider a body containing a crack to be externally loaded, but with the loading points fixed so that the energy available for crack extension is solely the strain energy released by the extension. Now, imagine the tip of the crack to be elastically pulled shut over a distance δa as shown in Fig. 8.30. The elastic strain energy will be increased by the work done in closing the crack, which is given by

$$\mathscr{G}\delta a = 2\int_0^{\delta a} (\sigma_{yy}v/2 + \tau_{yx}\,u/2 + \tau_{yx}\,w/2)dx \qquad (8.50)$$

for unit plate thickness, where σ_{yy}, τ_{yx}, τ_{yz}, and u, v, w, are the stresses and displacements at the crack surface over the portion which is pulled shut. $\mathscr{G}$ is evaluated for the limiting case where $\delta a \to 0$, substituting from Eqs. (8.41) to

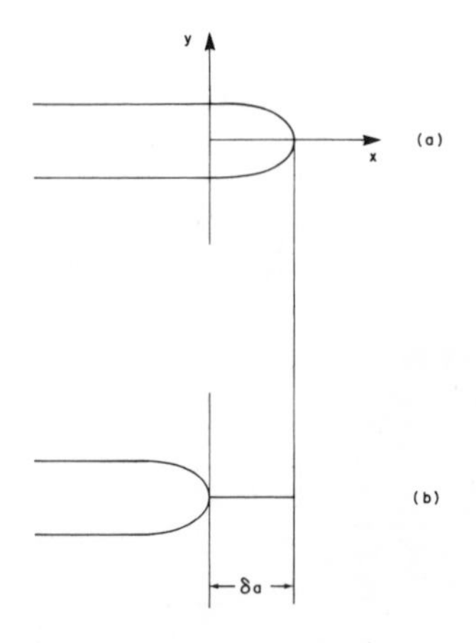

Fig. 8.30 Crack in externally loaded body, initially as at (a), and then with the tip elastically pulled shut as in (b) over distance δa

(8.48) for the stresses at $\theta = 0$ and $r = x$, and for the displacements where $r = \delta a - x$ and $\theta = \pi$. Thus, we obtain, for plane strain deformation

$$\mathscr{G} = [(1 - \nu)/2G]K_{\mathrm{I}}^2 + [(1 - \nu)/2G]K_{\mathrm{II}}^2 + (1/2G)K_{\mathrm{III}}^2 \qquad (8.51)$$

Substituting $E = 2G(1 + \nu)$ [Eq. (1.35)], and putting $E' = E/(1 - \nu^2)$, we obtain

$$\mathscr{G} = K_{\mathrm{I}}^2/E' + K_{\mathrm{II}}^2/E' + K_{\mathrm{III}}^2/2G$$

or, the total strain energy release rate is given by

$$\mathscr{G} = \mathscr{G}_{\mathrm{I}} + \mathscr{G}_{\mathrm{II}} + \mathscr{G}_{\mathrm{III}} \qquad (8.52)$$

where

$$\left.\begin{aligned} \mathscr{G}_{\mathrm{I}} &= K_{\mathrm{I}}^2/E' \\ \mathscr{G}_{\mathrm{II}} &= K_{\mathrm{II}}^2/E' \\ \mathscr{G}_{\mathrm{III}} &= K_{\mathrm{III}}^2/E' \end{aligned}\right\} \qquad (8.53)$$

and

$$\left.\begin{aligned} E' &= E/(1 - \nu^2), && \text{plane strain} \\ E' &= E, && \text{plane stress} \end{aligned}\right\} \qquad (8.54)$$

As already noted, the stress intensity factor for complex loading can be obtained by summing individual K values [Eq. (8.43)]. Hence the energy release rate for each mode can be obtained by summing the component rates in the form

$$\left.\begin{aligned} \mathscr{G}_{\mathrm{I}} &= [(\mathscr{G}_{\mathrm{Ia}})^{1/2} + (\mathscr{G}_{\mathrm{Ib}})^{1/2} + \ldots]^2 \\ \mathscr{G}_{\mathrm{II}} &= [(\mathscr{G}_{\mathrm{IIa}})^{1/2} + (\mathscr{G}_{\mathrm{IIb}})^{1/2} + \ldots]^2 \\ \mathscr{G}_{\mathrm{III}} &= [(\mathscr{G}_{\mathrm{IIIa}})^{1/2} + (\mathscr{G}_{\mathrm{IIIb}})^{1/2} + \ldots]^2 \end{aligned}\right\} \qquad (8.55)$$

The total strain energy release rate is then obtained from Eq. (8.52).

8.4.3 Crack Tip Plasticity

The foregoing discussion and stress analysis have been based on the assumption of elasticity at the crack tip but, as the stresses rise to a very high level there, this is obviously incorrect in an engineering material, and a small plastic zone is formed at the crack tip.

A first estimate of the size of the zone in plane stress can be made by assuming that it extends from the crack tip to a point where the stress falls to the value of the yield stress for the material at r_{Y} in Fig. 8.31(a). Thus, the distance is obtained by substituting $\sigma_{yy} = \sigma_{\mathrm{Y}}$ in Eq. (8.39) to give

$$r_{\mathrm{Y}} = (\sigma/\sigma_{\mathrm{Y}})^2\, a/2 = K^2/2\pi\, \sigma_{\mathrm{Y}}^2 \qquad (8.56)$$

However, the loss of load carrying capacity from limitation of σ_{yy} to σ_{Y} across r_{Y}, requires that the stress increase ahead of this. Thus the actual plastic zone size will be greater than our first estimate.

According to Irwin [*64*], we may consider an equivalent crack of length a

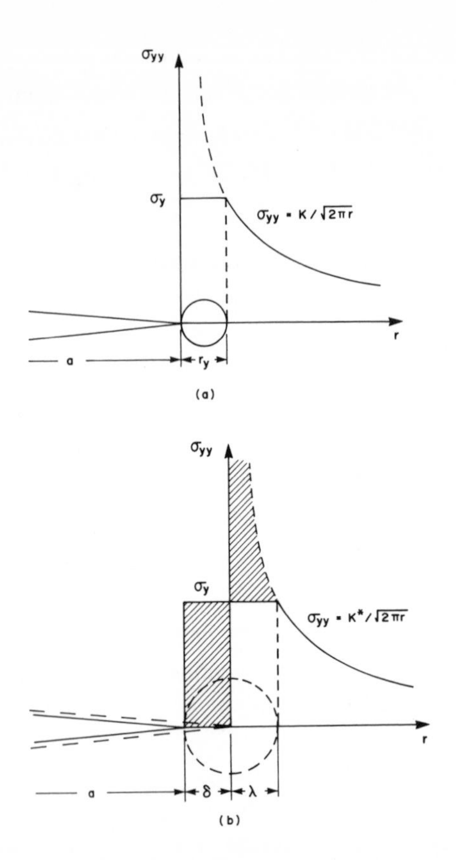

Fig. 8.31 Estimate of plastic zone size: (a) first estimate; (b) second estimate based on equivalent crack of length $a + \delta$

$+ \delta$, with stress distribution as in Fig. 8.31(b). The areas of the two hatched regions must be equal to maintain the load carrying capacity. Thus, from Eq. (8.39),

$$\sigma_{yy} = \sigma_Y = \sigma\,[(a + \delta)/2\lambda]^{1/2}$$

or

$$\lambda = (\sigma/\sigma_Y)^2\,(a + \delta)/2 \simeq r_Y \tag{8.57}$$

The plastic zone size is, therefore, approximately twice that for the first estimate, and the quantity r_Y is the plastic zone correction which is added to a to define the effective crack length.

The modified elastic stress distribution is given by

$$\sigma_{yy} = K^*/(2\pi r)^{1/2}$$

where

$$K^* = \sigma\,[\pi\,(a + r_Y)]^{1/2}$$

More accurate methods of determining the plastic zone size and shape are discussed by Broek [*11*], and in Mode I it appears that the zone forms ears sticking out on each side of the crack growth direction at angles of some 70° as shown in Fig. 8.32.

The situation is complicated under plane strain conditions because of the constraint which raises the effective yield stress to a value significantly higher than for uniaxial loading. Also, the surfaces of a plane strain specimen do not experience plane strain, and the plastic zone varies roughly as in Fig. 8.33. The degree of plane strain depends on the ratio of the plastic zone size in plane strain to the specimen thickness, and in a practical case the plastic zone

Fig. 8.32 Schematic view of actual plastic zone shape in Mode I loading

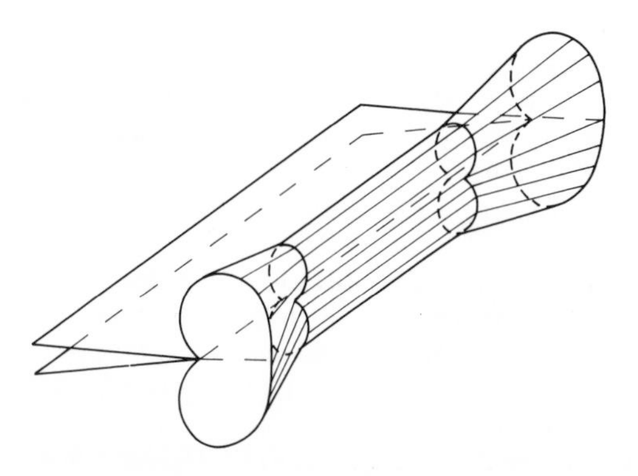

Fig. 8.33 Shape of plastic zone in Mode I loading in plate specimen, showing plane stress field at the surface, plane strain in the interior

correction in plane strain is only some one third of the plane stress case. The degree of constraint can only be found experimentally or calculated approximately using finite element methods, as a general elastic-plastic analysis has not yet been obtained.

REFERENCES

1. SHANK, M. E., in *Symposium on Effect of Temperature on the Brittle Behavior of Metals with Particular Reference to Low Temperature*, STP 158, p. 45, ASTM, Philadelphia (1954).
2. BIGGS, W. D., *The Brittle Fracture of Steel*, Macdonald and Evans, London (1960).
3. ROBERTSON, T. S., *JISI*, **175**, 361 (1953).
4. ASTM Standard E-208-69, *Standard Method for Conducting Drop-Weight Test to Determine Nil-Ductility Transition Temperature of Ferritic Steels*, ASTM, Philadelphia (1969).
5. PELLINI, W. S., and PUZAK, P. P., U. S. Naval Research Laboratory Report No. 6957, Washington, D.C. (1969).
6. GRIFFITH, A. A., *Phil. Trans. Roy. Soc.*, **221A**, 163 (1920).
7. GRIFFITH, A. A., *Proc. First. Int. Congr. App. Mech.*, p. 55, Delft (1924).
8. OROWAN, E., *Trans. Inst. Engrs. Shipbuilders Scotland*, **89**, 165 (1945).
9. IRWIN, G. R., *Fracturing of Metals*, p. 148, ASM, Cleveland (1948).
10. LIEBOWITZ, H. (ed.), *Fracture: An Advanced Treatise*, Vols. I-VII, Academic Press, New York (1968-72).
11. BROEK, D., *Elementary Engineering Fracture Mechanics*, Noordhoff, Leyden (1974).
12. KNOTT, J. F., *Fundamentals of Fracture Mechanics*, Butterworths, London (1973).
13. TETELMAN, A. S., and MCEVILY, A. J., *Fracture of Structural Materials*, Wiley, New York (1967).
14. PHILLIPS, A., KERLINS, V., and WHITESON, B. V., *Electron Fractography Handbook*, Tech. Rept. ML-TDR-64-416, AMFL, WPAFB, Ohio (1965).
15. *Metals Handbook*, Vol. 9, *Fractography and Atlas of Fractographs*, ASM, Metals Park, Ohio (1974).
16. MCCALL, J. L., in *Electron Fractography*, STP 436, p. 3, ASTM, Philadelphia (1968).
17. LE MAY, I., in *Microstructural Analysis: Tools and Techniques*, p. 153 (J. L. McCall and W. H. Mueller, eds.) Plenum Press, New York (1973).
18. LE MAY, I., *Failure Mechanisms and Metallography: A Review*, in *Metallography in Failure Analysis*, p. 1, (J. L. McCall and P. M. French, eds.), Plenum Press, New York (1978).
19. FRIEDEL, J., *Dislocations*, Pergamon Press, Oxford (1964).
20. BERRY, J. M., *Trans. ASM*, **51**, 556 (1959). See also discussion of this paper by CRUSSARD, C., PLATEAU, J., and LEAN, J. B., on p. 581.
21. BEACHEM, C. D., *Trans. ASME, J. Basic Eng.*, **87**, 299 (1965).

22. LUI, M.-W., and LE MAY, I., *Proc. II Int. Conf. on Mech. Behavior of Materials*, p. 1129, Federation of Materials Societies, Boston (1976).
23. LOW, J. R., in *Relation of Properties to Microstructure*, p. 163, ASM, Cleveland (1954).
24. ZENER, C., in *Fracturing of Metals*, p. 3, ASM, Cleveland (1948).
25. STROH, A. N., *Proc. Roy. Soc.*, **A223**, 404 (1954).
26. COTTRELL, A. H., *Trans. AIME*, **212**, 192 (1958).
27. HULL, D., *Phil. Mag.*, 8th Series, **3**, 1468 (1958).
28. REID, C. N., *J. Less Common Metals*, **9**, 105 (1965).
29. HULL, D., *Acta Met.*, **9**, 191 (1961).
30. SMITH, E., in *Physical Basis of Yield and Fracture*, p. 36, Inst. of Physics and the Physical Society, London (1966).
31. GILBERT, A., HAHN, G. T., REID, C. N., and WILCOX, B. A., *Acta Met.*, **12**, 754 (1964).
32. GURLAND, J., and PLATEAU, J., *Trans. ASM*, **56**, 442 (1963).
33. LUI, M.-W., and LE MAY, I., *Met. Trans.*, **6**, 583 (1975).
34. LUI, M.-W., and LE MAY, I., *Trans. ASME, J. of Eng. Matls. and Technology*, **98**, 173 (1976).
35. FUKUI, S., and UEHARA, N., in *Proc. Second Int. Conf. on the Strength of Metals and Alloys*, Vol. II, p. 616, ASM, Metals Park, Ohio (1970).
36. McCLINTOCK, F. A., *Int. J. Fract. Mech.*, **4**, 101 (1968).
37. KNOTT, J. F., in *Fracture 1977*, Vol. 1, p. 61 (D. M. R. Taplin, ed.) Univ. of Waterloo Press, Waterloo, Ont. (1977).
38. McMAHON, C. J., BRIANT, C. L., and BANERJI, S. K., in *Fracture 1977*, Vol. 1, p. 363 (D. M. R. Taplin, ed.) Univ. of Waterloo Press, Waterloo, Ont. (1977).
39. LAWLEY, A., LEIBMANN, W., and MADDIN, R., *Acta Met.*, **9**, 841 (1961).
40. KAMDAR, M. G., in *Fracture 1977*, Vol. 1, p. 387 (D. M. R. Taplin, ed.) Univ. of Waterloo Press, Waterloo, Ont. (1977).
41. HULL, D., and RIMMER, D. E., *Phil. Mag.*, 8th Series, **4**, 673 (1959).
42. NEEDHAM, N. G., WHEATLEY, J. E., and GREENWOOD, G. W., *Acta Met.*, **23**, 23 (1975).
43. LAIRD, C., and SMITH, G. C., *Phil. Mag.*, 8th Series, **7**, 847 (1962).
44. BROEK, D., and BOWLES, C. Q., *Int. J. Fract. Mech.*, **6**, 321 (1970).
45. NEUMANN, P., *Z. für Metallkunde*, **58**, 780 (1967).
46. PELLOUX, R. M. N., *Trans. ASM*, **62**, 281 (1969).
47. LUI, M.-W., Ph.D. Thesis, Univ. of Saskatchewan (1973).
48. NEUMANN, P., *Acta. Met.*, **22**, 1167 (1974).
49. BROEK, D., *Int. Metall. Reviews*, **19**, 135 (1974).
50. SCHIJVE, J., *Proc. Inst. Mech. Engrs.*, **191**, 107 (1977).
51. *The Influence of Environment of Fatigue*, Inst. Mech. Engrs., London (1977).
52. *Corrosion Fatigue Technology*, STP 642, ASTM, Philadelphia (1978).
53. INGLIS, C. E., *Trans. Inst. Naval Architects*, **55**, 219 (1913).
54. CONGLETON, J., and PETCH, N. J., *Acta Met.*, **14**, 1179 (1966).
55. PETCH, N. J., *Phil. Mag.*, 8th Series, **3**, 1089 (1958).
56. PETCH, N. J., in *Fracture: An Advanced Treatise*, (H. Liebowitz, ed.) Vol. 1, *Microscopic and Macroscopic Fundamentals*, p. 351, Academic Press, New York (1968).

57. Heslop, J., and Petch, N. J., *Phil. Mag.*, 8th Series, **3**, 1128 (1958).
58. Almond, E. A., Timbres, D. H., and Embury, J. D., in *Fracture 1969*, p. 253 (P. L. Pratt *et al.*, eds.) Chapman and Hall, London (1969).
59. Westergaard, H. M., *Trans. ASME, J. Appl. Mech.*, **61**, 49 (1939).
60. Eshelby, J. D., *Fracture Toughness*, Chapters 2 and 3, Publn. 121, Iron and Steel Inst., London (1968).
61. Tada, H., Paris, P. C., and Irwin, G. R., *The Stress Analysis of Cracks Handbook*, Del Research Corp., Hellertown, Pa. (1973).
62. Rooke, D. P., and Cartwright, D. J., *Compendium of Stress Intensity Factors*, HMSO, London (1976).
63. Irwin, G. R., *Trans. ASME, J. Appl. Mech.*, **24**, 361 (1957).
64. Irwin, G. R., *Appl. Mater. Res.*, **3**, 65 (1964).

9

FRACTURE MECHANICS

9.1 INTRODUCTION

Having reviewed the mechanisms and concepts of fracture in Chapter 8, we now turn to the quantitative assessment of a material's resistance to fracture, or its *toughness*, and of the application of such parameters in the design of engineering components and structures. The effect of section thickness on fracture resistance is first discussed and the testing procedures for plane strain and plane stress conditions are examined. Following this, the crack opening displacement criterion is examined, which, in addition to being available for use with high strength materials of moderate or low toughness, provides a means of assessing the fracture resistance of many structural steels and other alloys of medium strength and higher toughness than can be evaluated using the concepts of *linear elastic fracture mechanics* (LEFM). Subsequently, the assessment of toughness where extensive plastic deformation occurs before fracture is discussed using the J-integral method.

As pointed out in Chapter 8, environmental factors are important in determining a material's resistance to fracture, and the effects of temperature, corrosive environment and radiation on a material's fracture toughness are all reviewed.

The applicability of the fracture mechanics approach to design is examined in the last section of the chapter, including the selection of appropriate material for the particular operating conditions, determination of critical flaw size, and some aspects of proof testing and the leak-before-break concept which is important in the design of some pressure vessel and similar components. The important matter of cyclic loading, causing a crack to grow to a critical size at which fast fracture may take place, is discussed in Chapter 10, when examining fatigue.

9.2 FRACTURE TOUGHNESS

The critical value of the stress intensity factor to initiate fracture, K_c, is a measure of the resistance of a material to fracture, i.e., of its toughness. If we consider Mode I loading, we may term this value K_{1c}. However, because the plastic zone is much larger at the surface than the interior of a specimen, varying from plane stress conditions at the surface to plane strain at the center, as shown in Fig. 8.33, the value of K_{1c} is dependent on specimen thickness in the form shown in Fig. 9.1. Beyond a certain thickness, b_c, the value of K_{1c} often reaches a constant value, K_{Ic}, although in other cases it may continue to decrease to some extent. K_{Ic} is termed the *plane strain fracture toughness*.

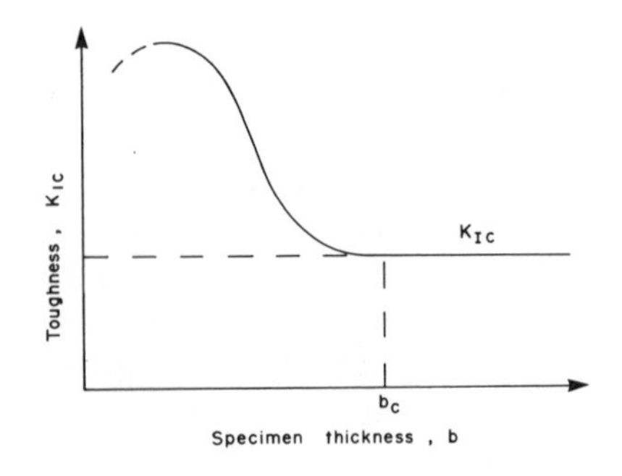

Fig. 9.1 Variation of toughness with specimen thickness

9.2.1 Plane Stress and Plane Strain

In thin sheet specimens loaded in tension, the plastic zone size will be large with respect to the specimen thickness, and for a ratio of plastic zone size to thickness approaching unity, plane stress conditions will exist. The planes of maximum shear stress at the crack tip lie at 45° to the principal stresses, as shown in Fig. 9.2(a), and the specimen will fracture in a slant mode, separation being by Mode III, as shown in Fig. 9.2(b).

The thicker the specimen the greater the proportion of the crack tip region which will be subjected to plane strain deformation. The planes of maximum

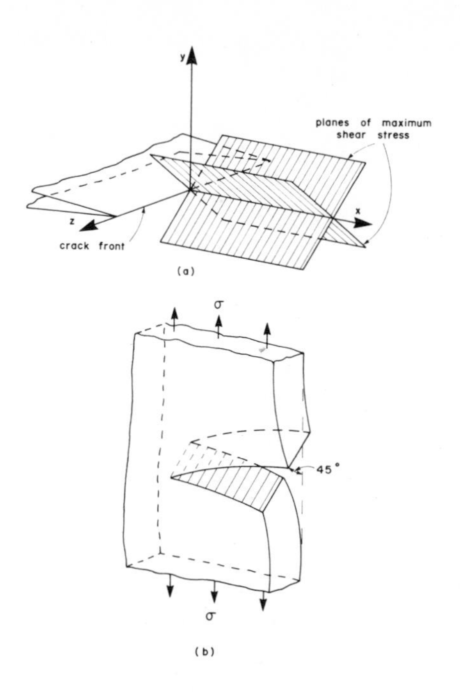

Fig. 9.2 Fracture in a thin sheet specimen under tension: (a) planes of maximum shear stress; (b) slant mode fracture

shear stress at the crack tip for plane strain deformation are shown in Fig. 9.3(a), and deformation takes place in the form of a hinge as shown in Fig. 9.3(b) [*1*].

9.2.2 Fracture Toughness Testing

The American Society for Testing and Materials has drawn up standardized procedures for fracture toughness testing which allow reproducible values of the lower limiting value of K_{1c} (K_{Ic}) to be determined [*2*]. Such test procedures provide specifications for test specimens such that the contribution of the shear lips is small, and the test is made to all intents under conditions of plane strain.

The commonly used forms of test specimen are shown in Fig. 9.4, both being of the single edge notched (SEN) type, the first loaded in three-point bending, the second under tension. Both specimens are provided with a sharp fatigue crack grown from their respective notches, this being grown under cyclic loading during which the maximum stress intensity factor does not exceed $0.6K_{Ic}$: if this is exceeded, the crack tip becomes blunted and the value of K_{Ic} determined may be higher than the correct one. The two specimens

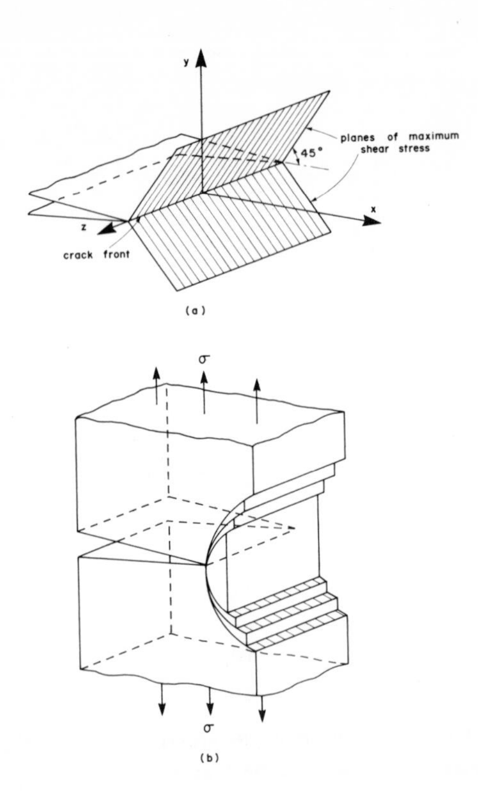

Fig. 9.3 Fracture in a thick specimen: (a) planes of maximum shear stress; (b) plastic hinge before final failure

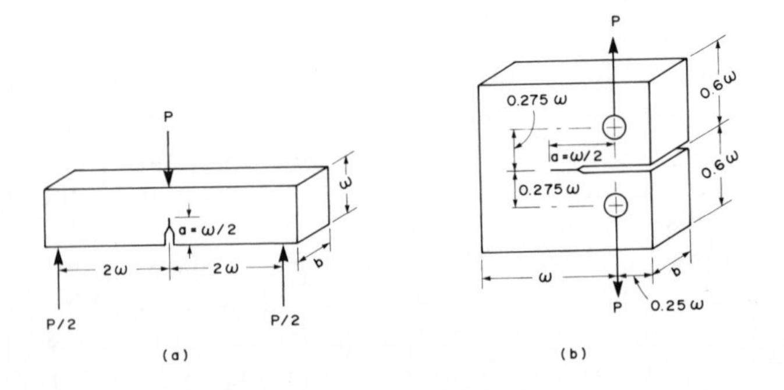

Fig. 9.4 Commonly used fracture toughness test specimens

illustrated have had their K values determined by several methods, the accepted expressions being as follows:

Bend specimen [*3*]:

$$K_I = (6P/bw)(\pi a)^{1/2} F(a/w) \tag{9.1}$$

where $F(a/w)$ is given by

$$F(a/w) = 1.090 - 1.735(a/w) + 8.20(a/w)^2 - 14.18(a/w)^3 + 14.57(a/w)^4$$

Compact tension specimen [*4*]:

$$K_I = (P/bw)\, a^{1/2} F(a/w) \tag{9.2}$$

where $F(a/w)$is given by

$$F(a/w) = 29.6 - 185.5(a/w) + 655.7(a/w)^2 - 1017.0(a/w)^3 + 638.9(a/w)^4$$

The size of the specimens must be such that plane strain conditions are essentially achieved across them, and the thickness must be large with respect to the plastic zone size. The latter is proportional to $(K_{Ic}/\sigma_Y)^2$, as may be seen in Section 8.4.3, and the criterion adopted for ASTM standard specimens is that

$$b \geqslant 2.5(K_{Ic}/\sigma_Y)^2 \tag{9.3}$$

This is found to give consistent values for K_{Ic} [*3*].

In addition, there are restrictions on the crack length. It must not be so small as to produce invalid results when the plastic zone size becomes significant with respect to the crack length, and the criterion adopted as being satisfactory is [*3*]

$$a \geqslant 2.5(K_{Ic}/\sigma_Y)^2 \tag{9.4}$$

similar to the relation for b. Similarly, there is a restriction on crack length with respect to specimen width or depth, w. If the crack length is too great, the plastic zone dimensions may become large and the quasi-elastic solution breaks down because of the proximity of the free surface. The value adopted is usually $w/a = 2$.

In plane strain fracture toughness testing, a clip gage is attached to a specimen of either type shown in Fig. 9.4, at the mouth of the crack, and records of load and displacement are made, usually on an X-Y plotter, typical records being shown in Fig. 9.5. To determine K_{Ic}, a tentative value of fracture toughness, K_Q, is determined from the plot, and if this satisfies Eqs. (9.3) and (9.4), then this is the value of K_{Ic} for the material. K_Q is determined by the following procedure [*2*]. First, a secant line OP_5 is drawn as shown in Fig. 9.5, through the origin and with slope equal to 0.95 of that of the tangent AO at the origin. If the recorded load is lower than P_5 at every point up to P_5, then $P_5 = P_Q$ as in curve (a): however, if there is a load maximum greater than P_5 before P_5 is reached on the recorded plot, then this maximum is P_Q, as shown for

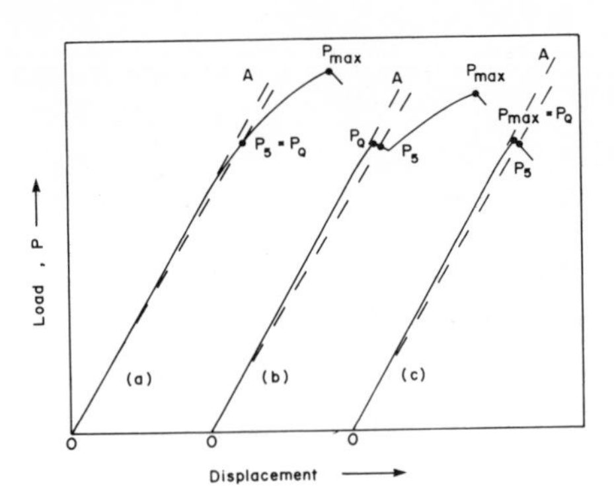

Fig. 9.5 Load-displacement records from fracture toughness specimens

curves (b) and (c) of Fig. 9.5. The ratio, P_{max}/P_Q, must also be determined, and if this does not exceed 1.10, then K_Q can be calculated using Eq. (9.1) or Eq. (9.2), as appropriate, and after determination of the crack length after fracture. This must be done at three positions; viz., at the center of the crack front, and midway between the center and the edge at each side. The allowable differences for valid testing are given in the ASTM standard [2]. It should be noted also that if $P_{max}/P_Q > 1.10$, the test does not give a valid value of K_{Ic}.

Where thin sections are employed, or in the case of tough materials where it is impossible to obtain sufficiently thick material for testing, plane strain conditions will not obtain, and it is necessary to determine the resistance to cracking in a plate. Thus, such tests may be made under plane stress conditions where thin sheet material is employed, quasi-plane stress, or conditions which deviate significantly from this state, although not providing valid plane strain conditions. A crack in a plate under these conditions while loaded in tension may propagate in a slow stable manner for some distance as load is increased, and we may characterize the stress intensity factor at failure as K_c. As would be expected on the basis of the lack of definition of the loading conditions, K_c is a function of plate thickness and of constraints such as initial crack length and plate width for a particular material, other variables such as temperature and loading rate being held constant: thus there is not a unique value for plane stress fracture toughness, as there is for plane strain (K_{Ic}). Nonetheless, a useful method does exist for prediction of the plane stress (or quasi-plane stress) fracture criterion in a material by utilizing curves which show the crack growth resistance as a function of crack extension: such curves are termed *R*-curves. At the time of writing, no standardized procedure has been adopted for their formulation, although a recommended practice has been proposed [5].

Consider a thin member containing a crack and loaded in tension. As the load is increased we may visualize the crack extending, the resistance to growth increasing with increasing load or crack length. The crack growth resistance, K_R [$= (\mathcal{G}E)^{1/2}$ from Eq. (8.49a)], is equal to the crack driving force, K_I; thus, by recording loads applied incrementally together with the corresponding values of stable crack length, the value of K_R can be computed in each case and plotted against *effective* crack length, a_{eff}. The resulting curve is the *R*-curve, which is regarded as being independent of initial crack length and specimen configuration, being a function of crack extension only [*6*]. The effective crack length is the current physical crack length, a, plus the plastic zone correction, r_Y, as discussed in Section 8.4.3.

Thus, we can predict the instability of a crack by calculating the curves for crack driving force as a function of crack length for the particular geometry of concern and for various loads, and positioning the established *R*-curve such that its origin lies at the initial crack length, a_o, as shown in Fig. 9.6. The critical value of load to cause fracture is determined from the crack driving force curve which mades a tangent with the *R* curve: this also shows the increment in crack length, Δa, at the point of instability.

The *R*-curve determined as described above provides a description of a material's resistance to fracture, independent of specimen configuration or crack length, but it does depend on specimen thickness. Hence, a number of such curves may be required to provide adequate design information for a particular material. It should be noted, however, that *predominantly* elastic behavior must obtain in the specimen: in other words, the yield stress is exceeded only in the region of the crack tip.

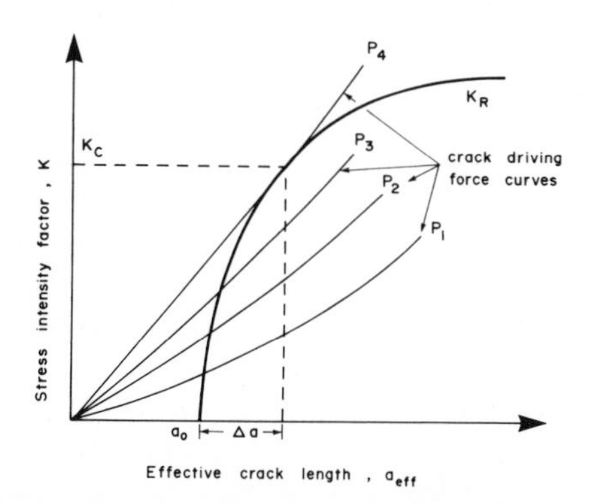

Fig. 9.6 *R*-**curve superimposed on set of curves for crack driving force as a function of crack length.** K_c **is shown together with the increment in crack length to cause instability,** Δa

9.2.3 Crack Opening Displacement

Wells [7] postulated that there is a critical and unique value of displacement at the tip of a crack in moderately stressed material at the interface between the crack tip and the plastic zone. A simplistic model is shown in Fig. 9.7 for both plane stress and plane strain conditions. The value of the tip crack opening displacement (COD) is obtained by equating elastic and plastic strains at the yield boundary. Thus,

$$\delta / 2\pi r_Y = \sigma_Y / E$$

or

$$\delta = 2\pi r_Y \sigma_Y / E, \qquad \text{plane stress} \qquad (9.5)$$

and

$$\delta / 2(2\pi r_Y) = \tau / G = \sigma_Y / 2G$$

or

$$\delta = 2\pi r_Y \sigma_Y / G, \qquad \text{plane strain} \qquad (9.6)$$

In the case of plane strain, r_Y is reduced because of triaxial stressing, as discussed in Section 8.4.3.

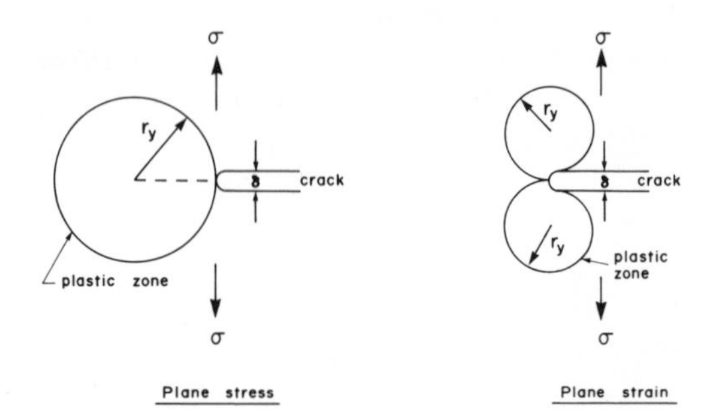

Fig. 9.7 Crack tip opening displacement in plane stress and plane strain

Alternatively, more detailed calculations can be made for δ, for example using the model of Dugdale [8] for yielding in the form of thin strips at the ends of a slit in an ideal elastic-plastic solid, Dugdale's model being shown in Fig. 9.8. Goodier and Field [9] have analyzed the model and determined the COD for plane stress conditions as

$$\delta = (8\sigma_Y a / \pi E) \ln[\sec(\sigma\pi / 2\sigma_Y)] \qquad (9.7)$$

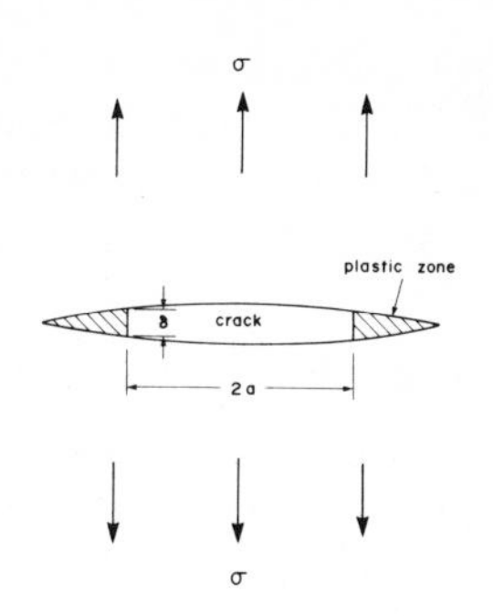

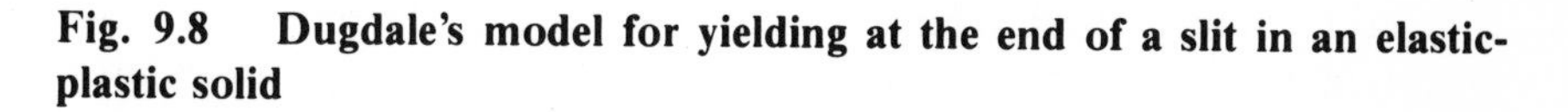

Fig. 9.8 Dugdale's model for yielding at the end of a slit in an elastic-plastic solid

which may be expanded to give

$$\delta = (8\sigma_Y a/\pi E)[(\sigma\pi/2\sigma_Y)^2/2 + (\sigma\pi/2\sigma_Y)^4/12 + ...]$$

For $\sigma << \sigma_Y$,

$$\delta \simeq \sigma^2 \pi a / E\sigma_Y$$

or

$$\delta = K^2/E\sigma_Y = \mathcal{G}/\sigma_Y \tag{9.8}$$

from Eq. (8.49a).

If we substitute for r_Y from Eq. (8.56) in Eq. (9.5), we obtain for the crude model for crack opening displacement,

$$\delta = K^2/E\sigma_Y$$

which is identical to Eq. (9.8).

Thus, we see that fracture toughness may be determined by measurement of the critical value of crack opening displacement at the crack tip, and $\mathcal{G}_c$ is found from

$$\mathcal{G}_c = \delta_c \sigma_Y \tag{9.9}$$

for plane stress conditions. For plane strain conditions, the plastic zone size must be modified, and Eq. (9.9) should be altered to [*10*]

$$\mathcal{G}_{Ic} \simeq 2\delta_c \sigma_Y \tag{9.10}$$

For materials of high toughness such that K_{Ic} cannot be determined, it is possible to measure the critical value of COD, and while the equations of this section would not apply, so that neither critical crack size nor fracture stress could be calculated explicitly, nonetheless different materials of high toughness can be compared, a higher value of COD_c indicating a higher toughness.

Because it is not normally feasible to measure COD directly at the crack tip, the displacement is normally monitored at the mouth of the crack using a clip gage as in K_{Ic} testing. The relation between this measured displacement and COD is determined by prior calibration for the specimen geometry and material of interest.

9.2.4 Fracture Toughness with Plastic Deformation

Under conditions of plane strain where the crack tip plastic zone is small, linear elastic analysis is appropriate. For materials and conditions where the plastic zone becomes larger, it is important that the effective crack length be used to calculate the effective stress intensity factor: however, as r_Y, the plastic zone correction, is itself a function of stress intensity factor, an iterative process in necessary in calculating K_{eff}, and as the ratio of r_Y/a becomes large, the computation becomes less accurate. As loading is increased towards conditions of general yielding, the allowable loads tend to a limiting value, while the crack opening displacement increases continuously until the failure condition is reached. Thus, the COD method of fracture toughness testing and assessment can extend through from the nominally fully elastic range where LEFM can be applied, to conditions where general yielding occurs across the entire cracked region.

An alternative approach to determine the fracture energy under plane strain conditions where elastic-plastic conditions obtain is the use of the *J*-integral, which is defined for two-dimensional problems and given by [*11*]

$$J = \int_{\Gamma} [W dy - T(\partial u/\partial x) ds] \tag{9.11}$$

In this, Γ is a contour surrounding the crack tip, as shown in Fig. 9.9, and evaluated in an anticlockwise direction, W is the strain energy per unit volume defined by

$$W = W(\epsilon_{mn}) = \int_0^{\epsilon_{mn}} \sigma_{ij}\, d\epsilon_{ij} \tag{9.12}$$

T is the traction vector defined by the outward normal, n, along Γ, $T_i = \sigma_{ij} n$, u is the displacement vector and ds is an element of arc length along Γ. For a closed contour, Rice [*11*] demonstrated that $J = 0$.

Consider the crack tip shown in Fig. 9.10, with two contours, Γ_1 and Γ_2, surrounding it. For the path $ABCDEFA$, $J = 0$, since the contour is closed. On the two parts of the contour comprised of crack surface, AF and CD, T and dy are both zero and make no contribution to J. Thus, $J(\Gamma_1) = -J(\Gamma_2)$, or the value of J is path independent when taken around a crack tip.

For elastic conditions, Rice [*12*] has shown that J is equivalent to the rate of release of potential energy with crack growth, or

$$J = \partial U/\partial a = \mathcal{G} \tag{9.13}$$

In the general case, Eq. (9.13) is valid even when appreciable crack tip plastic flow takes place.

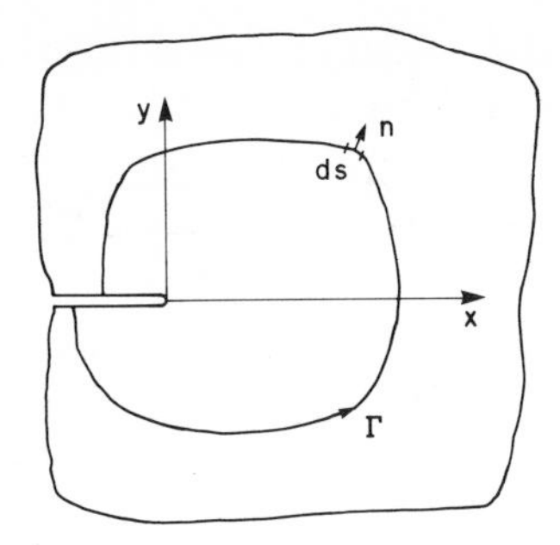

Fig. 9.9 Contour around a crack tip defining the *J*-integral

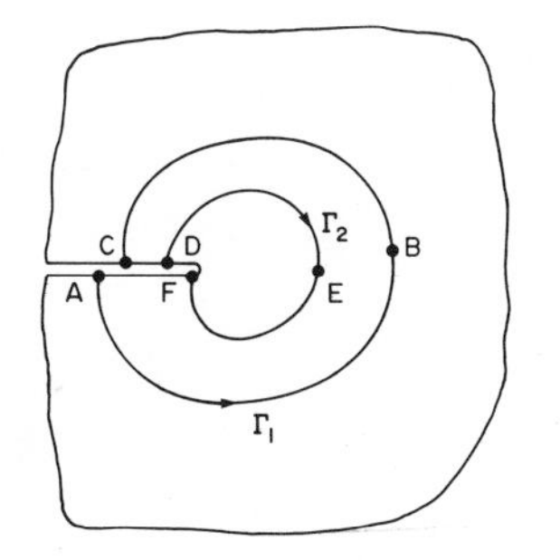

Fig. 9.10 Contours around a crack tip

Because of its independence of path, J can be evaluated along any convenient path where integration can be carried out without difficulty. Also, it may be evaluated by drawing up load-displacement curves for specimens with steadily increasing crack size. For example, Fig. 9.11 shows curves for crack lengths a and $a + \delta a$, the shaded area being equal to $J\delta ab$, where b is the specimen thickness.

The critical value of J at which a crack starts to propagate is termed J_{Ic} and this will be the same for the elastic-plastic and the elastic cases. Thus

$$J_{Ic} = \mathscr{G}_{Ic} = [(1 - \nu^2)/E]K_{Ic}^2 \tag{9.14}$$

Figure 9.12(a) illustrates a series of load versus displacement curves for tensile specimens with different crack lengths, while the energy per unit thickness is plotted against crack length for fixed values of displacement in Fig. 9.12(b). The resulting plot of the negative slope of the U versus a curves against displacement is shown in Fig. 9.12(c) for different crack lengths, the ordinate being equal to J, from Eq. (9.13). J_{Ic} can be determined from

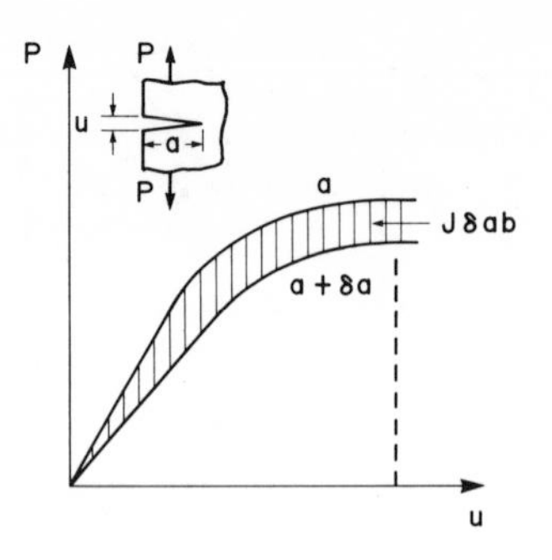

Fig. 9.11 Physical interpretation of the *J*-integral

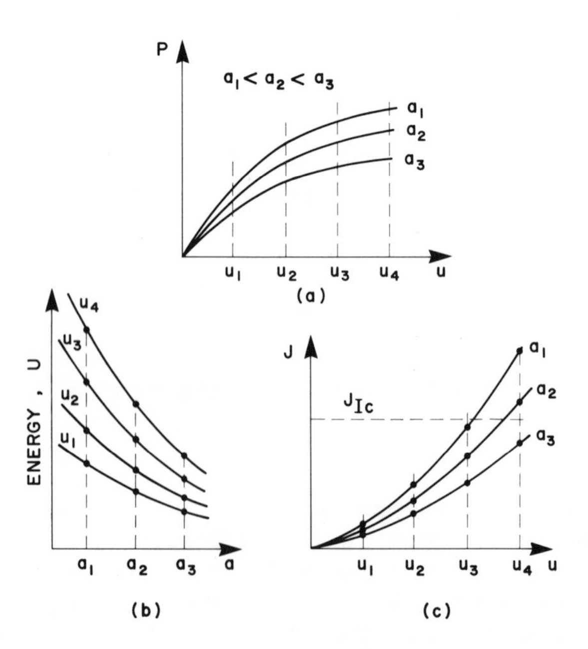

Fig. 9.12 Schematic diagram showing evaluation of *J* versus displacement

measurement of the displacement when a crack begins to propagate. Experiments have shown that J_{Ic} is a constant and of value equal to $\mathscr{G}_{Ic}$ [*13, 14*].

Thus, we see that the *J*-integral can be used where the specimen thickness and other dimensions do not allow valid K_{Ic} testing to be conducted according to the standard procedures [*2*]. There are some restrictions to the specimen size, because if the *process zone*, in which the separation processes are taking place, becomes large compared to the specimen thickness, then plane strain conditions will not apply. Standardized procedures for J_{Ic} testing, covering conditions from fully elastic to general yield, are expected to be drawn up before long.

9.3 ENVIRONMENTAL EFFECTS

The resistance to fracture of a material can be altered appreciably as a result of the environment. The effect may result from the synergistic effect of stress and environment as in stress corrosion cracking, from alteration of the material's mechanical properties owing to changes in structure as may occur under irradiation, or from changes in properties which are a direct consequence of the environment and which depend directly on it, as do temperature effects. Here we shall look briefly at each of these effects.

9.3.1 Stress Corrosion Cracking

The importance of the plane strain fracture toughness, K_{Ic}, of a material has been made clear in the preceding sections. Basically, failure in the presence of a pre-existing defect should not occur unless the applied stress is such that the stress intensity factor reaches K_{Ic}. However, it is found in practice that failure under static loading can occur when the combination of applied stress and pre-existing defect size gives rise to an initial value of stress intensity factor, K_{Ii}, very much less than K_{Ic}. This arises because subcritical crack growth may take place from the defect until there is a crack of critical size, i.e., $K_I \rightarrow K_{Ic}$. Such subcritical crack growth can occur where there is a synergistic effect between the stress and the environment, as in stress corrosion cracking (SCC), hydrogen embrittlement and liquid metal embrittlement. The general principles of all three are similar, but we shall discuss only the more common phenomenon of SCC in this section.

Stress corrosion cracking depends for its occurrence on there being a corrosive atmosphere: however, in the absence of stress, there may be little if any continuing corrosion. Early studies were made using smooth or mildly notched specimens, the time to fracture being measured. However, such tests combine crack initiation and crack growth stages and, in some cases, initiation may be virtually non-existent with resulting long or indefinite life while in others initiation may take place at pre-existing notches with subsequent rapid propagation and short overall life. In order to distinguish between initiation and propagation phenomena, pre-cracked specimens have been used.

In such SCC tests, cantilever bend specimens are commonly used, loaded as in Fig. 9.13 [*15, 16*]. The initial value of K_I ($=K_{Ii}$) is measured, as is the time to fracture, a typical plot obtained for different values of K_{Ii} being shown in Fig.

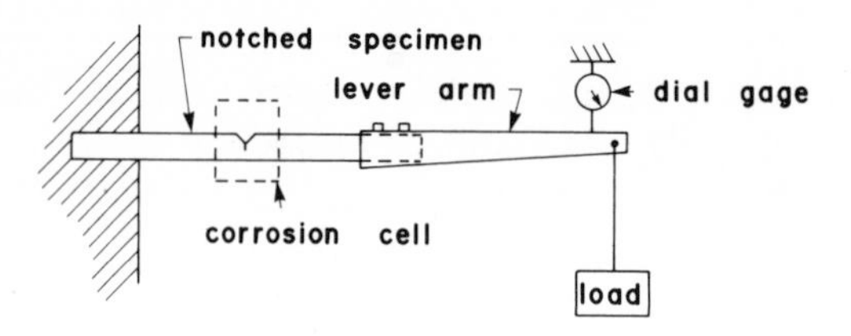

Fig. 9.13 Stress corrosion cracking cantilever bend specimen test

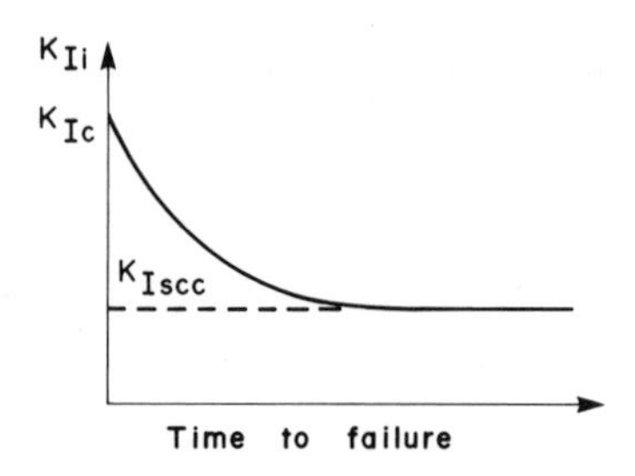

Fig. 9.14 Typical plot of K_{Ii} versus time to fracture

9.14. Commonly, a lower threshold value is found and is termed K_{Iscc}, approximately ten test specimens being required to determine it. Unfortunately, K_{Iscc} frequently does not have a unique value for a particular material/environment combination, but depends on the failure time, with lower values being determined if longer time tests are run. At low values of K_{Ii} the initiation time becomes very large, exceeding 1000 h in many cases. Another uncertainty relates to specimen size. In order to ensure plane strain conditions at the crack tip, K_{Iscc} specimens, as with K_{Ic} specimens, must satisfy certain minimum size requirements: however, these requirements have not yet been incorporated into a standard test procedure and a standard specimen design.

From crack growth studies of SCC specimens, the crack growth rate as a function of instantaneous stress intensity factor, K_I, has a sigmoidal shape as shown in Fig. 9.15, growth becoming unstable as $K_I \rightarrow K_{Ic}$. Checks on failed specimens to determine the crack length at the onset of catastrophic failure have also shown that failure occurs at $K_I = K_{Ic}$ unless crack branching has occurred. Figure 9.16 illustrates the relative magnitude of crack initiation and growth periods as a function of initial stress intensity factor, K_{Ii}.

An alternative test procedure involves the use of a modified wedge-opening-loading (WOL) specimen as shown in Fig. 9.17 [*17*]. This is given a fixed displacement using the bolt, and as the crack extends K_I decreases until it

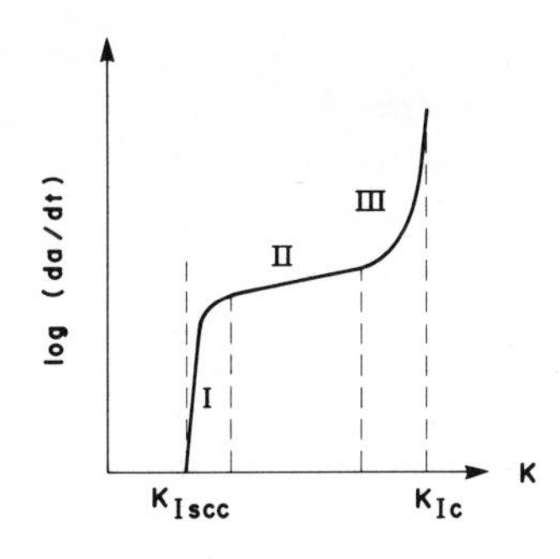

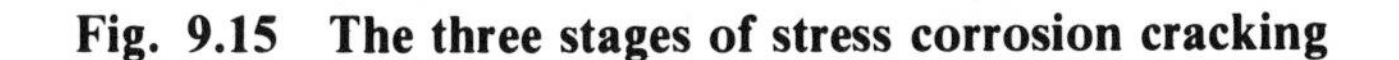
Fig. 9.15 The three stages of stress corrosion cracking

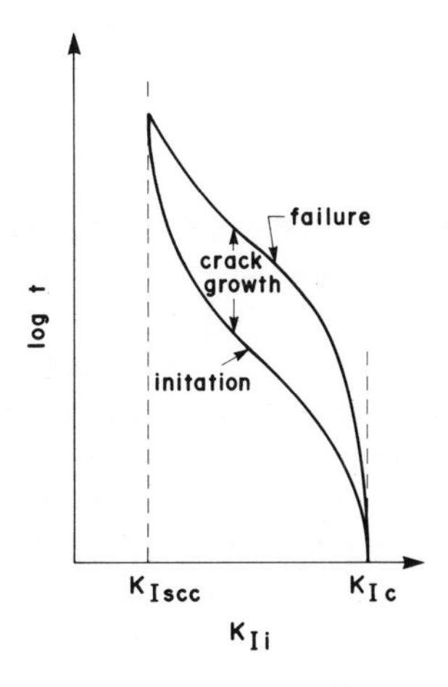

Fig. 9.16 Relative magnitude of crack initiation and crack growth periods

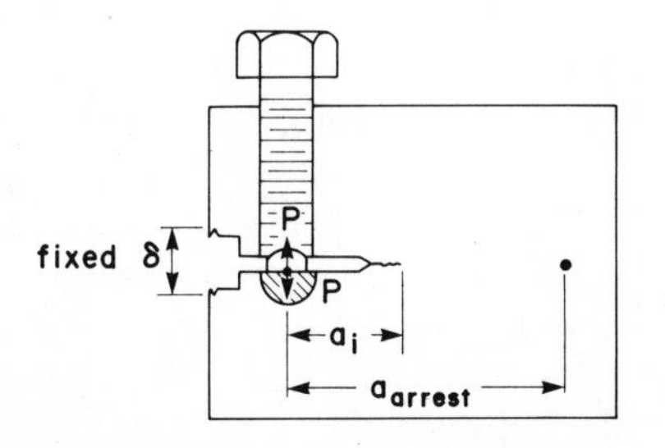

Fig. 9.17 Modified wedge-opening-loading specimen

reaches K_{Iscc}, at which time propagation ceases. A single specimen can be used to determine K_{Iscc}, the test specimen being portable and not requiring the use of a testing machine. In order to allow for material scatter, duplicate tests should be made. Figure 9.18 illustrates the change in K_I as crack growth occurs in a modified WOL specimen.

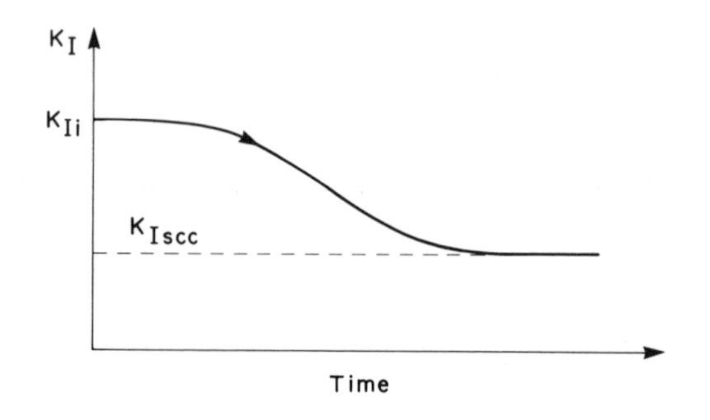

Fig. 9.18 Change in K_I as crack grows in modified WOL specimen

Unfortunately, it is necessary to determine K_{Iscc} data for specific material/environment combinations. Generally, the higher the yield stress for a particular material, the lower its K_{Iscc} value in a specific environment. To illustrate the variation in K_{Iscc} values for different material conditions and differing environments, some representative values are shown in Table 9.1. This also illustrates the extent to which K_{Iscc} data lie below corresponding K_{Ic} data.

9.3.2 Effects of Temperature

The transition temperature in steels and other BCC metals, determined from Charpy V-notch or other tests, was discussed in Chapter 8. Similarly, many structural materials display a change in fracture toughness with temperature as shown schematically in Fig. 9.19. This change is not caused by loss of constraint with increase in plastic zone size as temperature is raised, but is an inherent plane strain transition, independent of specimen geometry.

Increase of loading rate generally causes a reduction in the value obtained for fracture toughness. Hence, under impact loading, the dynamic fracture toughness is obtained, being termed K_{Id}. Figure 9.20 illustrates the variation in fracture toughness with loading rate, slow loading rates being of the order of 1 $MPa\sqrt{m}/s$, while impact loading to determine K_{Id} is around 10^5 $MPa\sqrt{m}/s$. Dynamic tests are usually made in a drop-weight machine using a pre-fatigue-cracked slow bend test specimen equipped with strain gages to

Table 9.1 Representative K_{Iscc} Data [18]

Alloy	σ_Y (MPa)	K_{Ic} (MPa$\sqrt{m}$)	Environment	K_{Iscc} (MPa$\sqrt{m}$)
Aluminum Alloys				
2024-T351	325	55*	3½%NaCl	11
7075-T6	505	25	"	21
Steels				
18 Ni (300)	1960	80	NaCl Soln.	8
4340	1150	69*	Seawater	6
4340	1205	83*	"	30
4340	860	89*	"	77
300 M	1735	70	3½%NaCl	22
Titanium Alloys				
Ti-6Al-4V	890	104	3½%NaCl	39
Ti-8Al-1Mo-1V	825	97	"	25
Ti-8Al-1Mo-1V	855	105*	Methanol	15

* K_{Ic} testing not valid in terms of Eqs. (9.3) and (9.4).

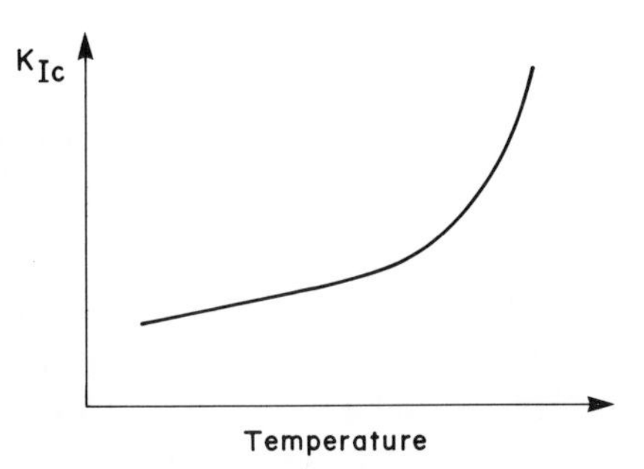

Fig. 9.19 Schematic showing effect of temperature on K_{Ic}

determine the applied stress and the initiation of fracture [19]. Both K_{Ic} and K_{Id} are affected by temperature in a similar manner, as shown in Fig. 9.21.

9.3.3 Effect of Radiation

The effect of neutron irradiation on the strength of metals was reviewed in Section 6.5. Because of the increase in strength and reduction in ductility which take place under such circumstances, a considerable decrease in

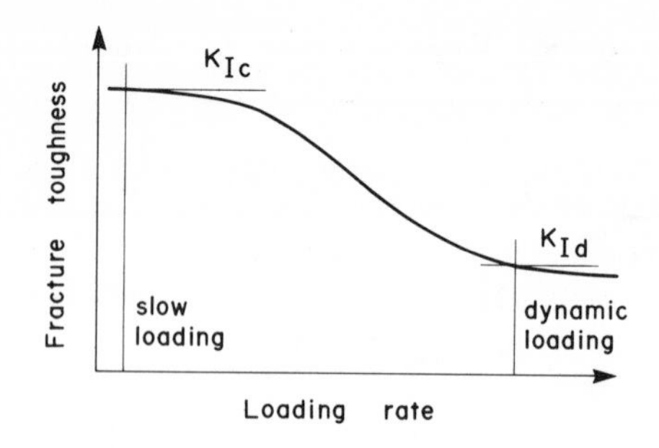

Fig. 9.20 Variation of fracture toughness with loading rate

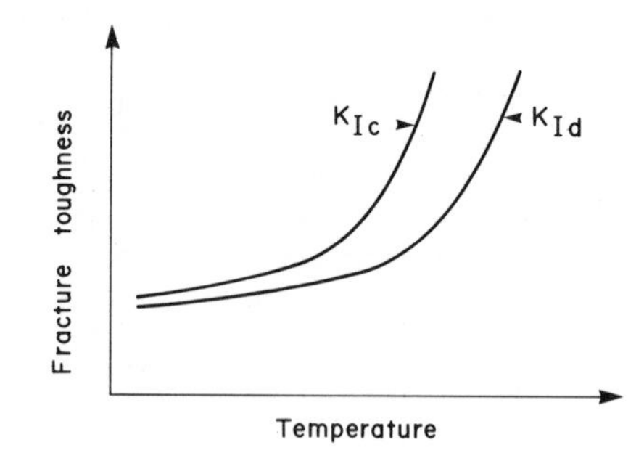

Fig. 9.21 Effect of temperature on fracture toughness

fracture toughness of nuclear reactor steel pressure vessels and other related components can take place. The Charpy V-notch transition temperature is raised by irradiation damage and a plot of K_{Ic} versus T is moved to lower toughness levels and higher temperatures. The extent of embrittlement depends strongly on temperature, and at higher temperatures it may become very small because of rapid annealing out of the damage. The degree of embrittlement of steel is very sensitive to the alloying elements present, and careful selection of steels and control of their quantity is important in such applications.

9.4 FRACTURE MECHANICS AND DESIGN

The application of fracture mechanics principles to the design of a component or structure makes use of the assumption that all materials contain flaws and the knowledge that, from the established fracture toughness of the material from small scale laboratory tests and from consideration of the maximum possible defect size, the stress required to cause this to propagate in a catastrophic manner can be evaluated. In order to do this, detailed stress analysis is necessary in many cases in order to evaluate the loading conditions in critical regions: also non-destructive evaluation (NDE) is generally called for to evaluate flaws which may be present. If no flaws are detected then the maximum flaw size which may be present corresponds to the lower limits of the NDE technique employed.

An alternative procedure to estimate the maximum size of defect present is the proof test, of particular value in testing pressure vessels. In such a test the structure is loaded to a stress greater than that to be employed during normal operation. Upon successful completion of such a test the maximum size of initial flaw, a_i, can be determined from a plot of stress versus flaw size for the particular value of K_{Ic} for the material, as shown in Fig. 9.22. Had there been a flaw greater than a_i, the structure would have failed during the proof test. The allowable crack growth during operation at the working stress level is then (a_c - a_i), and fatigue life, for example, may be estimated where cyclic loading takes place as discussed in Chapter 10.

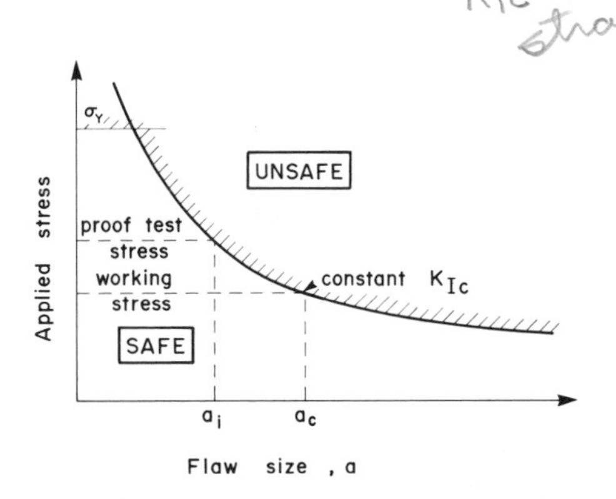

Fig. 9.22 Applied stress versus flaw size showing safe and unsafe regions

In most practical cases the defect which is likely to be present is not a simple through crack in a large evenly stressed plate, but has a more complex shape and loading geometry, requiring evaluation of the stress intensity factor. One common case is that of a through crack formed at the edge of a hole such as a bolt or rivet hole. For the situation shown in Fig. 9.23 and for the crack *not* small compared to the hole, an approximate solution is satisfactory. The

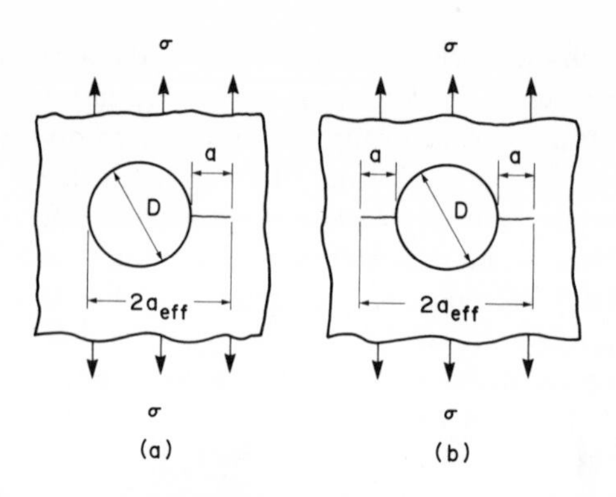

Fig. 9.23 Through cracks emanating from holes

effective crack size is that of the crack plus the diameter of the hole, and the stress intensity factor is given approximately by:

One crack [Fig. 9.23(a)]

$$K_I = \sigma(\pi a_{eff})^{1/2} = \sigma(\pi a)^{1/2}[(D/2a) + 1/2]^{1/2} \quad (9.15a)$$

Two cracks [Fig. 9.23(b)]

$$K_I = \sigma(\pi a_{eff})^{1/2} = \sigma(\pi a)^{1/2}[(D/2a) + 1]^{1/2} \quad (9.15b)$$

For the case where the crack is small compared to the hole, a more accurate solution is required, this being given in the appropriate reference texts [*20, 21*].

A common situation in pressure vessels is the occurrence of a surface flaw of semi-elliptical shape and lying in the axial direction as shown in Fig. 9.24. The stress intensity factor is given by [*22*]

$$K_I = 1.12\,\sigma(\pi a/Q)^{1/2} \qquad \text{(external surface)} \quad (9.16a)$$

or by

$$K_I = 1.12\,(\sigma + p)(\pi a/Q)^{1/2} \qquad \text{(internal surface)} \quad (9.16b)$$

where $Q = \Phi_o^2$, and Φ_o is a complete elliptical integral of the second kind given by

$$\Phi_o = \int_0^{\pi/2} \{1 - [(c^2 - a^2)/c^2]\sin^2\theta\}^{1/2} d\theta$$

The factor 1.12 in Eq. (9.16) is a free surface correction factor [*23*], and p is the internal pressure. Q is termed the flaw shape parameter and for a semicircular crack ($a = c$) Eq. (9.16a) reduces to:

$$K_I = 1.12\,(\sigma/\sqrt{\pi})2\sqrt{a} \quad (9.17)$$

this being the maximum value of the stress intensity factor at the end of the minor axis of the ellipse.

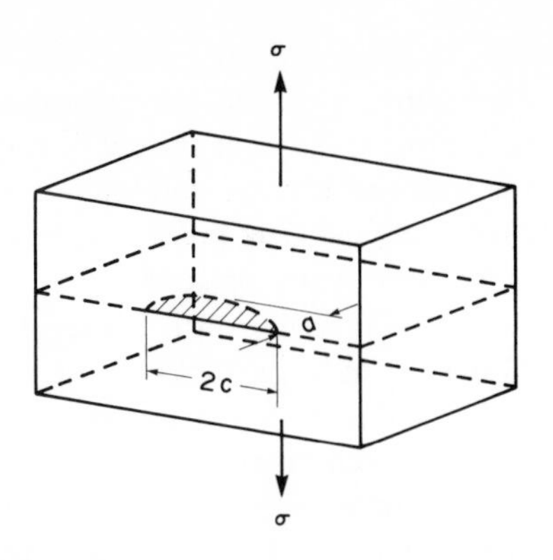

Fig. 9.24 Semi-elliptical surface flaw

The relationship between the flaw shape parameter and the flaw aspect ratio, $a/2c$, is shown in Fig. 9.25 for different ratios of σ/σ_Y. This latter term is important because of the dependence of the stress intensity factor at the crack tip on plastic deformation.

Hence, from a knowledge of σ_Y, K_{Ic}, the possible maximum flaw size and its aspect ratio, the safe working stress can be evaluated on an iterative basis.

9.4.1 The Leak-Before-Break Criterion

A part-through crack in a pressure vessel at the inner surface may grow by stress corrosion or fatigue. If it reaches the outer surface before it achieves the

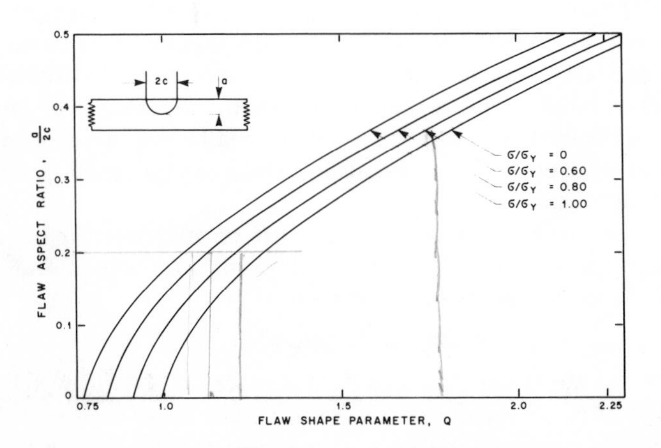

Fig. 9.25 Relation between flaw shape parameter and flaw aspect ratio for surface cracks

critical size for fracture, leakage will take place and this can generally be detected. If, on the other hand, it reaches the critical size first, catastrophic failure will occur without prior leakage. It should be noted that, after a crack passes right through the wall and leakage commences, the through thickness crack may grow to reach a critical size: generally, however, leakage will be detected prior to this condition being reached.

Clearly, the leak-before-break criterion is a useful and desirable one to maintain if possible. However, in large thick-walled pressure vessels it is not normally possible to achieve this design criterion.

A simplified treatment has been proposed [*24*] in which it is assumed that the crack has a semicircular shape when it reaches the outside surface. For a plate thickness t, the crack has a length $a = t$ and its aspect ratio $a/2a = 1/2$ at breakthough. It is considered that arrest will occur if the stress at breakthrough equals the yield stress of the material, under which condition a plane stress state is assumed to exist. Arrest will occur if

$$K_{Ic} \geqslant \sigma_Y [\pi(t + r_Y)]^{1/2}$$

where the equivalent crack length is $(t + r_Y)$ with r_Y given approximately by Eq. (8.56). Hence, for arrest to take place,

$$(K_{Ic})^2 \geqslant 1.5\sigma_Y^2 \pi t \tag{9.18}$$

More rigorous analysis of the criterion is given by Rolfe and Barsom [*25*] and by Broek [*26*], but the general principles have been outlined here.

9.4.2 Choice of Material

Generally, the higher the yield strength for a particular type of material (e.g., quenched and tempered steels, maraging steels, aluminum alloys), the lower is its plane strain fracture toughness. In using high strength materials where these are required to meet the design requirements based on stress analysis and what may be termed the applied mechanics approach to design, care must be taken to ensure that the maximum tolerable defect size is within the limits of detection available. Certainly, small cracks are more difficult to detect that large ones, so that a material which can tolerate larger cracks can be advantageous.

An appropriate basis for comparison between materials is to consider their operating stress level to be the same fraction of yield stress. Thus, for each material the critical crack size is given by

$$a_c = (K_{Ic}/\alpha\sigma_Y)^2/\pi$$

where α is the ratio of operating to yield stress (σ/σ_Y). The critical defect sizes for a series of materials stressed to the same fraction of their respective yield stresses are proportional to $(K_{Ic}/\sigma_Y)^2$, these quantities being an indication of the relative sizes of their plastic zones at fracture.

An interesting comparison of three materials, steel, an aluminum alloy, and a titanium alloy, is given by May [27] from data supplied by E.T. Wessel. The design considerations include minimum weight, and the material properties are shown in Table 9.2. Consideration of a semi-elliptical flaw with $a/2c = 0.2$ and tensile loading with $\sigma = \sigma_Y/2$ gives rise to critical flaw depths, a, of 7.39 mm, 4.42 mm and 10.82 mm, for the steel, aluminum and titanium, respectively. Thus, the titanium alloy is the best choice.

On consideration of equivalent flaws with $a = 3.81$ mm and $a/2c = 0.2$ as before, the fracture stresses can be evaluated. Again using $\sigma = \sigma_Y/2$ as the operating stress level, the safety factors for each alloy (= fracture stress/design stress) come out to be 1.15, 1.01 and 1.60, respectively. Again the titanium is the best alloy in this example, but the very small factors of safety for the first two, when dealing with a defect of a practical size, should be noted.

To conclude this discussion it should again be noted that the fracture toughness can be reduced greatly under corrosive conditions and consideration of this should be given at the design stage.

Table 9.2 Properties of Three Alloys Examined

Alloy	Density (Mg/m^3)	σ_Y (MPa)	σ_Y/Density ($10^3\ m^2/s^2$)	K_{Ic} ($MPa\sqrt{m}$)
Steel	7.86	1724	219	110
Aluminum	2.71	586	216	33
Titanium	4.51	965	214	88

9.4.3 Concluding Remarks

It must be emphasized that the fracture mechanics approach to design does not replace the more traditional design procedures involving stress analysis, strength of materials formulae and the synthesis of an engineering structure: rather, it complements it. In the fracture mechanics approach, materials selection and design stress determination are based on the assumption that local discontinuities are always present, or that they may grow from regions of high local stress under the influence of cyclic loading or SCC.

An area into which defects are readily introduced and from which they may grow is a weld. In addition to the applied stresses, the effects of residual stresses may be particularly great in such areas and must be taken into account in the fracture mechanics analysis and in estimating fatigue crack growth.

It is not only in aircraft, rocket motor casings and other sophisticated structures that fracture mechanics is a valid and necessary design approach today. It is used and required in pipelines, ship structures, large holding tanks and automotive components. For further discussion of practical applications, the reader should consult the many specialized texts which are now available [*25*, *26*, *28-30*].

REFERENCES

1. HAHN, G. T., and ROSENFIELD, A. R., *Acta Met.*, **13**, 293 (1965).
2. ASTM Standard E 399-78, *Standard Test Method for Plane-Strain Fracture Toughness of Metallic Materials*, ASTM, Philadelphia (1978).
3. BROWN, W. F., and SRAWLEY, J. E., *Plane Strain Crack Toughness Testing of High Strength Metallic Materials*, STP 410, ASTM, Philadelphia (1966).
4. WESSEL, E. T., *Eng. Fracture Mechanics*, **7**, 77 (1968).
5. *Tentative Recommended Practice for R-Curve Determination*, ASTM, Philadelphia (1976).
6. KRAFFT, J. M., SULLIVAN, A. M., and BOYLE, R. W., in *Proceedings of the Crack Propagation Symposium*, Vol. 1, p. 8, The College of Aeronautics, Cranfield (1961).
7. WELLS, A. A., in *Proceedings of the Crack Propagation Symposium*, Vol. 1, p. 210, The College of Aeronautics, Cranfield (1961).
8. DUGDALE, D. S., *J. Mech. Phys. Solids*, **8**, 100 (1960).
9. GOODIER, J. N., and FIELD, F. A., in *Fracture of Solids*, p. 103, (D. C. Drucker and J. J. Gilman, eds.), Interscience, New York, (1963).
10. RICE, J. R., *Third International Congress on Fracture*, Munich, (1973).
11. RICE, J. R., *Trans. ASME, J. Appl. Mech.*, **35**, 379 (1968).
12. RICE, J. R., in *Fracture: An Advanced Treatise*, (H. Liebowitz, ed.), Vol. 2, *Mathematical Fundamentals*, p. 191, Academic Press, New York (1968).
13. BEGLEY, J. A., and LANDES, J. D., in *Fracture Toughness*, STP 514, p. 1, ASTM, Philadelphia (1972).
14. LANDES, J. D., and BEGLEY, J. A., in *Fracture Toughness*, STP 514, p. 24, ASTM, Philadelphia (1972).
15. BROWN, B. F., and BEACHEM, C. D., *Corrosion Science*, **5**, 745 (1965).
16. BROWN, B. F., *Materials Research and Standards*, **6**, 129 (1966).
17. NOVAK, S. R., and ROLFE, S. T., *J. of Materials*, **4**, 701 (1969).
18. CAMBELL, J. E., BERRY, W. E., and FEDDERSEN, C. E., *Damage Tolerant Design Handbook*, MCIC-HB-01 (Sept. 1973).
19. SHOEMAKER, A. K., and ROLFE, S. T., *Trans. ASME, J. Basic Eng.*, **91**, 512 (1969).
20. ROOKE, D. P., and CARTWRIGHT, D. J., *Compendium of Stress Intensity Factors*, HMSO, London (1976).
21. TADA, H., PARIS, P. C., and IRWIN, G. R., *The Stress Analysis of Cracks Handbook*, Del Research Corp., Hellertown, Pa. (1973).
22. IRWIN, G. R., *Trans. ASME, J. Appl. Mech.*, **29**, 651 (1962).
23. PARIS, P. C., and SIH, G. C., in *Fracture Toughness Testing and its Applications*, STP 381, p. 30, ASTM, Philadelphia (1965).
24. "The Slow Growth and Rapid Propagation of Cracks", 2nd Report of a Special ASTM Committee, *Materials Research and Standards*, **1**, 389 (1961).

25. ROLFE, S. T., and BARSOM, J. M., *Fracture and Fatigue Control in Structures*, Prentice-Hall, Englewood Cliffs, N.J (1977).
26. BROEK, D., *Elementary Engineering Fracture Mechanics*, Noordhoff, Leyden (1974).
27. MAY, M. J., *Fracture Toughness*, Chapter 7, Publn. 121, Iron and Steel Inst., London (1968).
28. BURKE, J. J., and WEISS, V., (eds.), *Application of Fracture Mechanics to Design*, Plenum Press, New York (1979).
29. STANLEY, P., (ed.), *Fracture Mechanics in Engineering Practice*, Applied Science Publishers, London (1977).
30. LIEBOWITZ, H. (ed.), *Fracture: An Advanced Treatise*, Vol. 5, *Fracture Design of Structures*, Academic Press, New York (1969).

10

FATIGUE AND FATIGUE DESIGN

10.1 INTRODUCTION

Fatigue is by far the commonest mechanism by which engineering components fail and it may be defined as failure under cyclic or repeated stress, the maximum value of which is lower than the static stress required to cause fracture in the material. It occurs in metals and non-metals alike, and is generally characterized by local crack propagation, the component often showing no obvious signs of failure before the final fracture takes place.

Failure under repeated loading has been known for many years, being reported as long ago as 1838 by Albert [*1*], a number of other studies being made in the next few years, culminating in the classic studies of Wöhler [*2*]. The historical perspective has been reviewed by a number of authors, e.g., [*3-5*]. While most fatigue failures occur in relatively common components such as gear teeth, crankshafts and so on, major and catastrophic failures causing extensive loss of life, such as have occurred in aircraft in recent years, have caused greatly increased public interest to be displayed in the subject. Although today we have a good knowledge of the mechanisms of fatigue damage and fracture, failures still occur. Sometimes these may be attributed to material or manufacturing defects, while in other cases they relate to the statistical nature of the fatigue phenomenon which makes it virtually impossible to prevent all *possibility* of fatigue failure of a component or structure undergoing fluctuating loading. Hence the need for regular and planned non-destructive evaluation (NDE) of critical components and structures exposed to such loading becomes apparent, and should be considered at the design stage.

In this chapter the presentation of fatigue data, including the effects of mean stress, is considered first, followed by a discussion of fatigue properties as they relate to other mechanical properties and microstructural conditions. Some aspects of the mechanisms of fatigue damage are then considered, followed by discussion of the factors governing crack propagation rate. The effect of notches on fatigue is examined, followed by the subject of low-cycle, high-strain fatigue. Then the problem of corrosion fatigue is reviewed, and the chapter concludes with an examination of life prediction under service conditions. Fatigue under multiaxial loading was discussed previously in Chapter 2 when dealing with failure criteria under multiaxial conditions, and elevated temperature fatigue is considered in Chapter 11 when dealing with design for high temperature.

In the space available it is impossible to give a detailed coverage of all the aspects of fatigue mentioned, but it is hoped that the main points are at least touched upon. The appropriate references noted should be examined for more details, or the more general reference texts on fatigue consulted [*6-9*].

10.2 DEFINITIONS AND DATA PRESENTATION

10.2.1 Fatigue Life Data

Fatigue damage is by its nature statistical, and consequently data for fatigue "life" (or number of loading cycles to failure) at a given stress level are not "hard", but subject to variation. Thus, predicted lives can be given for a specified confidence level only: normally data are plotted on a basis of 50 percent probability of failure (probability, $P = 0.5$) in a certain number of cycles at the prescribed stress level.

The applied stress may be split into two components, alternating (σ_a) and mean (σ_m), as indicated in Fig. 10.1. Data obtained under constant cyclic load are normally (and traditionally) plotted on an S-N diagram, where S is the amplitude of the alternating stress (σ_a) and N (or N_f) is the number of cycles to failure (plotted on a logarithmic scale) as shown in Fig. 10.2, the curve being drawn for $P = 0.5$. Figure 10.2 also illustrates the two distinct types of S-N curve and of fatigue behavior observed in engineering materials. In carbon steels and certain other alloys (e.g., titanium and some Al-Mg alloys) a definite *fatigue limit* may result, with the "knee" occurring at between 10^6 and 10^7 cycles, while most metals and nonmetallics do not possess a fatigue limit and it is necessary to define an *endurance limit* for a specified number of cycles, e.g., at 10^7 cycles in Fig. 10.2(b). The reasons why some materials, only, have a definite fatigue limit have been the subject of many studies and it has been determined that this occurs only in alloys in which strain-aging takes

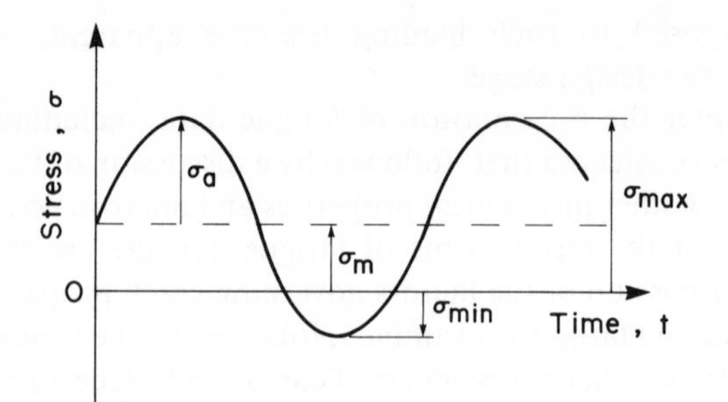

Fig. 10.1 Definition of stress levels in fatigue

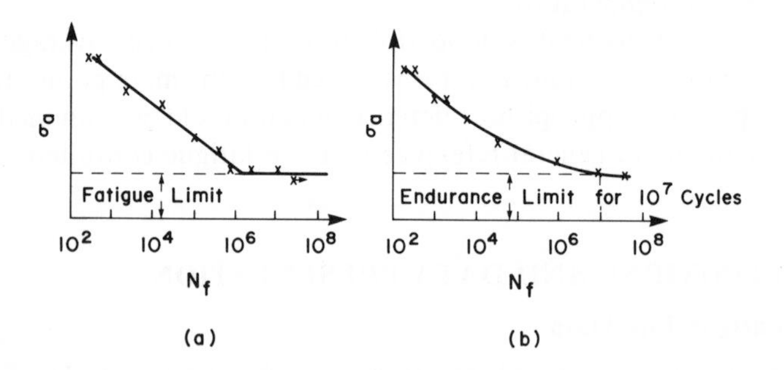

Fig. 10.2 ***S-N*** **curves showing (a) fatigue limit and (b) endurance limit for 10^7 cycles where no fatigue limit exists**

place [*10*]. It should be noted, however, that no fatigue limit is observed in steels and other alloys normally exhibiting one when they are exposed to fatigue loading in a corrosive environment (see Section 10.8).

Strictly speaking, *S-N* plots should not be drawn, but rather *S-N-P* plots as shown in Fig. 10.3. The number of tests required to determine such plots in a rigorous manner is very large, and simplifications must be made: however, the statistical nature of fatigue and correct inclusion of probability factors is of major importance when aircraft structures and components, at least, are designed, and in risk analysis. In dealing with aircraft components a probability of failure of 0.001 may be adopted, whereas probabilities in automotive components may correspond to between 0.001 and 0.05 [*11*]. From determination of histograms for many materials, it has been determined that the frequency distribution of the logarithm of fatigue life has an approximately Normal or Gaussian distribution curve within the limited life portion of the *S-N* curve, and may be represented by

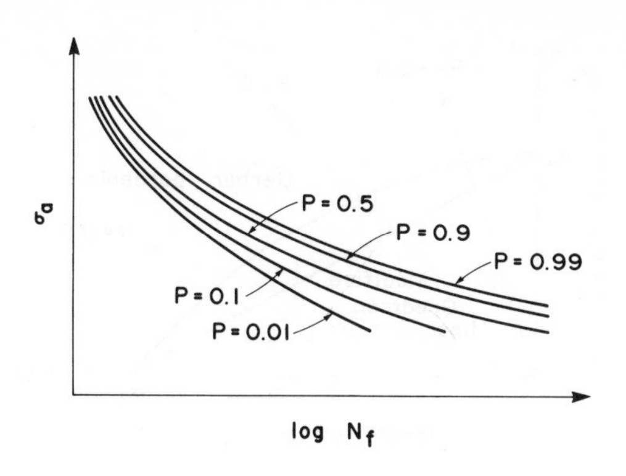

Fig. 10.3 ***S-N-P* plot**

$$f(\log N) = [1/\sigma(2\pi)^{1/2}] \exp[-(\log N - \psi)^2/2\sigma^2] \quad (10.1)$$

where σ is the log standard deviation and ψ is the log mean value of N. The error introduced by drawing the *S-N* curve for $P = 0.5$ through the median values of endurance at each stress level (in the absence of sufficient data for a complete statistical analysis) is small, and such a procedure was recommended by Weibull [*12*].

A distribution function having a better fit than the log Normal one has also been proposed by Weibull [*12, 13*], in the form

$$f(N) = [b/(N_a - N_o)]\,[(N - N_o)/(N_a - N_o)]^{b-1} \exp\{-[(N - N_o)/(N_a - N_o)]^b\} \quad (10.2)$$

where n, N_a and N_o are constants derived from the experimental data. Having three rather than two unknowns, the Weibull function has greater flexibility than the log Normal, but for most practical purposes insufficient test data are available for such refinements in analysis to be attempted.

10.2.2 Effect of Mean Stress

For design purposes it is important to know the effect of mean stress on fatigue life and, rather than have to prepare *S-N* curves for a series of mean stress levels, attempts have been made to develop appropriate mathematical expressions for this so that mean stress can be allowed for in making use of the *S-N* curve for fully reversed loading.

The two relationships which are generally used are based on proposals of Goodman [*14*] and Gerber [*15*], the former's original proposal being modified to consider the experimentally determined fatigue limit under fully reversed loading rather than taking it to be a specified fraction of the UTS. The modified Goodman and Gerber plots are shown in Fig. 10.4, the first having the assumption that the effect of mean stress is a linear one between $\sigma_m = 0$ and the UTS, the second considering the line to be parabolic. The criterion may be

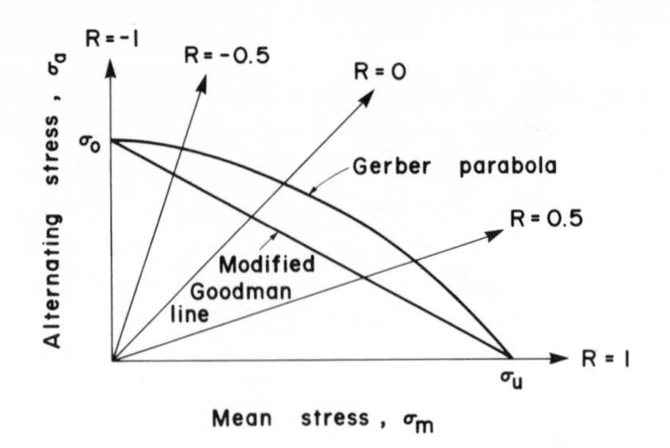

Fig. 10.4 Effect of mean stress on fatigue strength

expressed in the form

$$\sigma_a = \sigma_o \,[1 - (\sigma_m/\sigma_u)^n] \tag{10.3}$$

where σ_o is the fatigue limit or the stress amplitude for specified endurance at $\sigma_m = 0$, σ_u is the UTS, and n has the value 1 for the modified Goodman relationship and 2 for the Gerber.

Loading is often specified in terms of the maximum and minimum values of the cyclic stress, and the ratio of minimum to maximum stress is termed the *stress ratio*, R. Lines of constant R radiate from the origin in Fig. 10.4, several being indicated.

Experimental data generally appear to lie between the modified Goodman and the Gerber lines and the former is usually considered to be a more appropriate and safer criterion to use. If it is considered important that neither yielding nor fatigue failure takes place, σ_u can be replaced in the modified Goodman diagram by σ_Y, the yield stress. The line joining σ_o to σ_Y is referred to as the Soderberg line.

Compressive values of mean stress cause an increase in fatigue strength, and under fully compressive loading fatigue cracks cannot grow. However, in such loading conditions sub-surface cracks may develop, with material subsequently spalling off. This relates to the fact that the stress distribution is not normally uniaxial compression but is complex, with localized compressive loading and bending effects being present. The spalling off of soft bearing metal from steel-backed shell bearings and the spalling of metal from heavily loaded gear teeth are examples of such situations.

10.3 EFFECT OF MATERIAL VARIABLES

The fatigue limit, or the strength at long endurances (10^7 or 10^8 cycles) where there is no fatigue limit, increases with increase in tensile strength of a particular metal or alloy.

In the case of steels the relationship is approximately linear with the *fatigue ratio* (fatigue limit/UTS) of value ~0.5 up to a tensile strength of *circa* 1200 MPa. At levels of tensile strength greater than this the fatigue limit tends to flatten off, or at least increase more slowly, as illustrated in Fig. 10.5 [*16*]. The reasons for this are related to the fact that it becomes relatively easier to initiate a crack at very high static strength levels as decreases in ductility and fracture toughness usually accompany the increase in strength. Cracks tend to form at surface or sub-surface inclusions rather than from surface slip as with lower strength levels. Fatigue properties in high strength steels can, however, be improved considerably by thermo-mechanically processing the material [*17, 18*], these improvements sometimes being much greater than the improvement in static properties. Ausforming is particularly effective, the structure being a very fine dispersion of carbides in a martensitic matrix, while the ductility and fracture toughness are good and combined with high strength. Figure 10.6 illustrates the improvement in fatigue properties from an ausforming treatment, while Fig. 10.7 shows that the maraging steels have similar good fatigue properties.

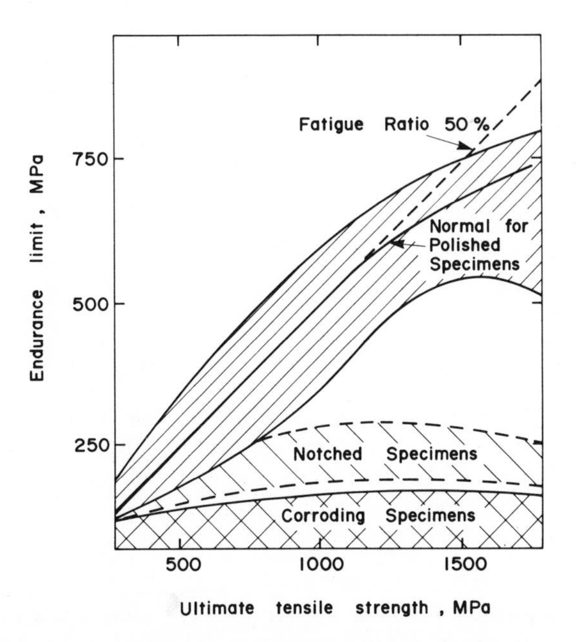

Fig. 10.5 Relation between endurance limit and ultimate tensile strength for steels. After Bullens [*16*]

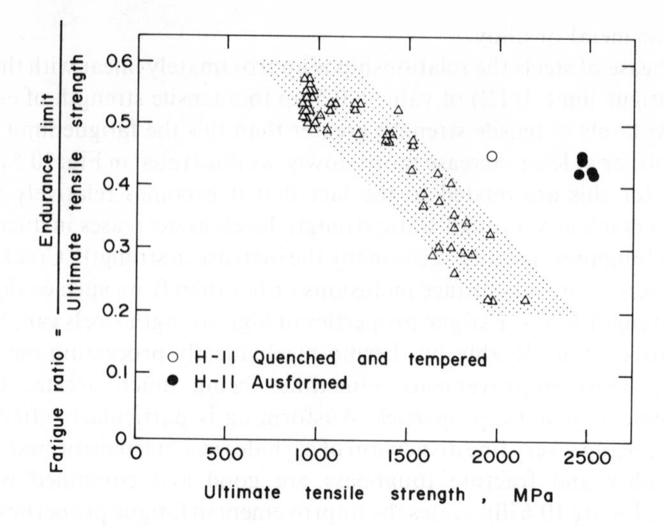

Fig. 10.6 Fatigue ratio versus ultimate tensile strength for conventional heat treated steels, together with results for H-11, showing improvement caused by ausforming. After Borik *et al.* [*17*]

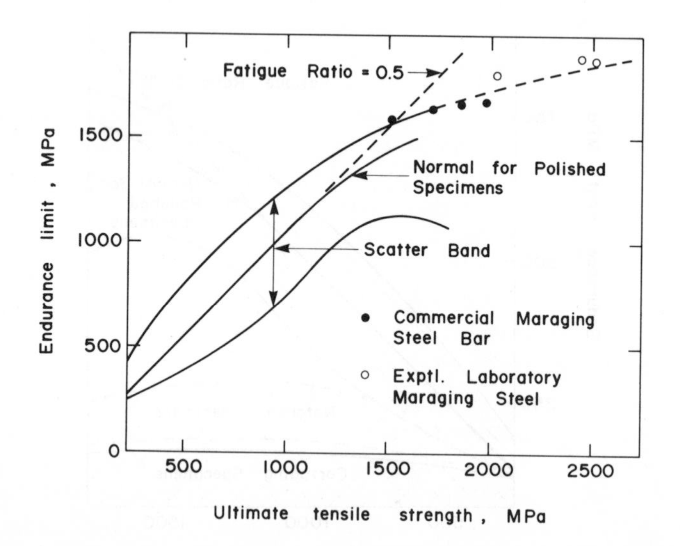

Fig. 10.7 Endurance limit for carbon and low alloy steels together with results for maraging steel (10^7 cycle stress), showing continued rise in endurance limit with tensile strength for specially processed material. After Ref. [*19*]

Cleanliness is also of major importance in high strength steels in ensuring good fatigue properties, and this can best be achieved by vacuum remelting or electroslag refining such steels. Minimizing the number and size of inclusions suppresses the initiation of cracks at them.

The interrelation between strength, ductility and fatigue life is illustrated well by Manson's empirical universal slopes equation [*20*], which has the form

$$\Delta\epsilon = 3.5\,(\sigma_u/E)N_f^{-0.12} + D^{0.6}\,N_f^{-0.6} \tag{10.4}$$

where $\Delta\epsilon$ is the total strain range, σ_u is the UTS, E is the elastic modulus, N_f is the number of cycles to failure, and D is a measure of ductility = $-\ln(1 - RA)$, where RA is the reduction in area at fracture. For large cyclic strains ductility becomes the governing factor, while tensile strength becomes more important for small strain amplitudes and long lives. The general validity of Eq. (10.4) has been demonstrated for many materials, and the optimum choice of heat treatment for a steel for greatest fatigue strength has been demonstrated by Manson [*21*] using this relation.

Tomkins *et al.* [*22*] have developed equations which allow prediction of endurance as a function of strain range in terms of a material's stress-strain properties, and their approach is useful in that it also suggests the parameters which must be changed to improve the fatigue properties of a particular metal.

Further discussion of the relative roles of strength and ductility is given when low-cycle, high-strain fatigue is examined in Section 10.7.

10.4 FATIGUE MECHANISMS

As noted in Section 8.2.5, we must distinguish between the processes of fatigue crack initiation, propagation and final rupture. After a fatigue crack has formed, it may advance under a fluctuating or cyclic stress of a sufficient amplitude, the operative mechanisms having been discussed in Section 8.2.5, while quantitative aspects are examined in Section 10.5. Here, we are concerned primarily with the mechanisms of damage which lead to the formation of a macrocrack. The number of cycles for this initiation process may be either large or small as a fraction of the total life, depending on the level of stressing and the geometry of the component or specimen considered: in particular, if a sharp notch is present the initiation stage may be very short. Generally, in a smooth specimen, the initiation stage may range from less than 10 percent at high stresses to greater than 90% at low stress levels.

When a metal is cyclically loaded between fixed values of stress or strain it will either harden or soften depending on the material's initial condition and the amplitude of the stress or strain. Figure 10.8 illustrates the cyclic stress-strain curves (a) under stress control and (b) under strain control, these curves being drawn for a cyclically strain hardening material. Under strain control, and with cyclic strain hardening, the amplitude of the stress increases rapidly

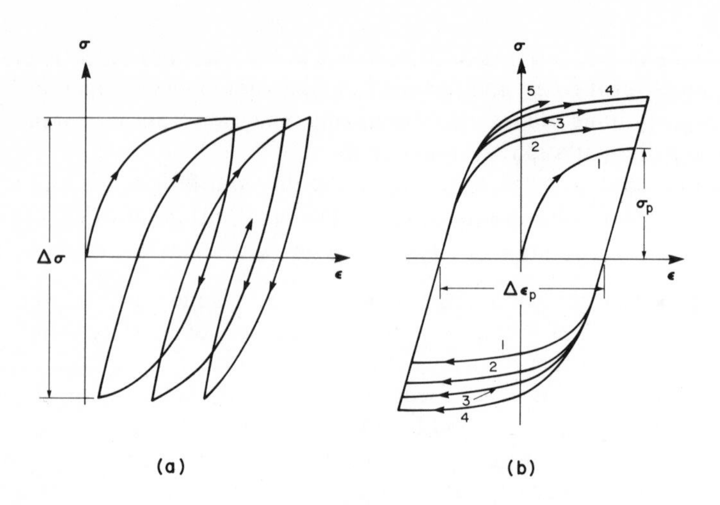

Fig. 10.8 Cyclic stress-strain curves: (a) under stress control; (b) under strain control

during the first few cycles and then settles down to a steady or saturation value, σ_s, as shown in Fig. 10.9(a). Initially cold-worked metals cyclically soften, shaking down to a steady value as shown in Fig. 10.9(b). The value of σ_s reached for a given value of $\Delta\epsilon_p$ is not necessarily the same for the same material in the annealed and cold-worked conditions, frequently being higher for the cold-worked material, although this is not the case for wavy slip mode materials (e.g., Cu, Al and Fe) which have unique values of σ_s for a given $\Delta\epsilon_p$ [*23*].

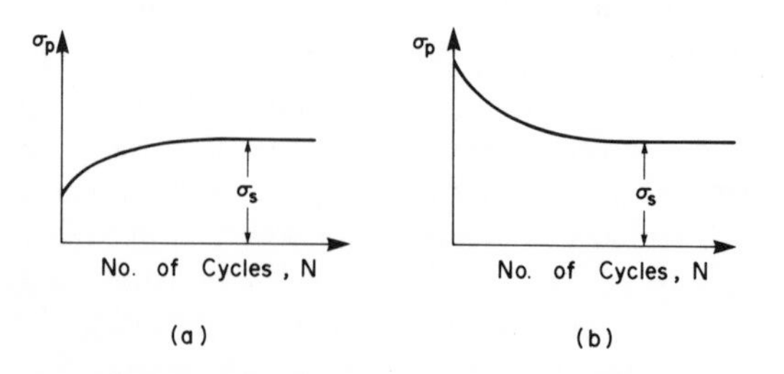

Fig. 10.9 Variation of stress amplitude in strain control testing: (a) annealed material; (b) cold-worked material

From a series of strain-controlled tests made under a range of values of $\Delta\epsilon_p$, we can derive the *cyclic stress-strain curve* for the material, as shown in Fig. 10.10(a). In Fig. 10.10(b) the cyclic and monotonic stress-strain curves for cold-worked and annealed material are compared. As noted above, the two cyclic curves will superimpose for wavy slip metals.

The dislocation structures generated during the fatigue hardening of metals undergoing cyclic straining have been studied by many workers, and an excellent review has been prepared by Grosskreutz [*24*]. During the initial rapid hardening stage, bundles of dislocations are produced, separated by largely dislocation-free regions. With continued cycling the dislocation density within the bundles increases and the spacing of the bundles decreases. At high strain amplitudes a three-dimensional dislocation cell structure is produced, having a spacing characteristic of the strain amplitude.

During the saturation stage, slip bands of inhomogeneous plastic deformation are produced, at least in FCC metals and in specific cases of BCC metals [*25*], provided the cyclic strain amplitude is sufficient. These bands have a different dislocation structure from the matrix and are softer [*26*], thus deformation concentrates in them. Such slip bands were observed on the polished and etched surfaces of fatigue specimens since early in this century [*27*], and it was found that, after electropolishing the surface until slip bands were removed, there were some which re-formed and did not remove easily on subsequent repolishing [*28*]. These were termed *persistent slip bands* (PSB's), and slip band cracks are seen to form at them.

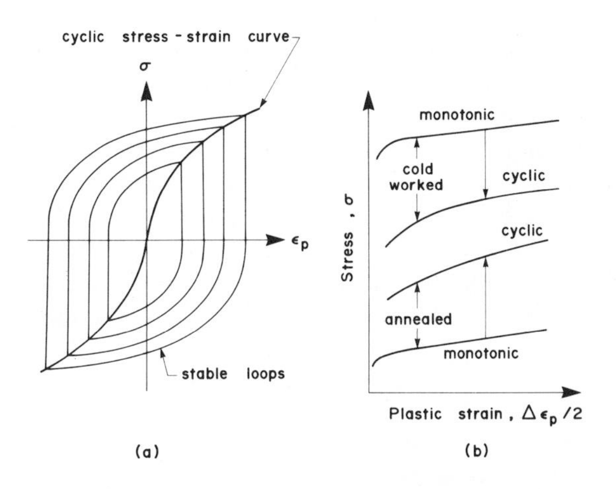

Fig. 10.10 (a) Cyclic stress-strain curve. (b) Monotonic and cyclic stress-strain curves for annealed and cold-worked material

When cyclic strain softening takes place in a cold-worked FCC metal such as pure copper, it has been found that the initially somewhat ragged dislocation cell structure is altered to a cleaner, more sharply defined cell structure as found in the saturated stage of cyclic work hardening. Many precipitation hardened alloys are also subject to cyclic softening and this appears to be caused by the precipitates in the PSB's being cut repeatedly and reverting into solution [*29, 30*]. This explains why the fatigue or endurance ratio in such alloys often tends to be relatively low.

Intensive slip at PSB's can give rise to the formation of the intrusions and extrusions illustrated in Fig. 8.18, from which microcracks may grow, but cracks tend to nucleate wherever inhomogeneous slip is concentrated, including interfaces with second phase particles, grain boundaries or twins, as well as by slip band cracking in PSB's. A very real problem, however, is the definition of a crack, and for practical purposes we can set some arbitrary lower limit of length at, say 0.1 mm, and record the number of cycles to initiate a crack of this size. However, the ratio of cycles to initiate a crack, N_o, to the total number of cycles to failure, N_f, varies widely and tends to zero at high stress levels.

In examining the physical picture of the formation of a macrocrack, we may note that, in general, the surface microcrack grows inwards along crystallographic planes on which the resolved cyclic shear stresses are large. It grows by to-and-fro slip until it reaches a sufficient length that, when the element ahead of its tip fractures, it does not "heal" under compressive stress but turns and grows in a direction normal to the cyclic tensile stress field with its growth rate being dependent on the extent of its cyclic opening. The crack has then changed over from *Stage I* to *Stage II growth* (see Fig. 8.19) using the nomenclature of Forsyth [*31*], or is now propagating in Mode I in fracture mechanics terminology. Stage II growth is much more rapid than Stage I and the life from this point to fracture can be predicted from a knowledge of crack growth relationships for the material: much more difficult is prediction of the cycles to initiate a macrocrack. It should be noted here that a microcrack initiated at a low stress level may not reach its critical depth for change over to a Stage II macrocrack and, in these circumstances, it will not propagate. Also, if the critical size is smaller than the grain size, the crack may be retarded by grain boundaries and grain size can have a considerable effect on fatigue limit [*32*].

10.5 FATIGUE CRACK PROPAGATION

After a macrocrack has been initiated it propagates by an incremental amount during each tensile application of load, provided this is greater than some threshold value. For a plain or smooth specimen the cyclic stress level initiating the crack is sufficient to ensure its propagation. The plasticity required is generated in front of the crack tip by the crack opening, and depends on the difference between the plastic zone sizes at maximum and minimum load. As the plastic zone is submerged in an elastic stress field defined by the stress intensity factor, K, the growth rate can be related to ΔK, the range of stress intensity factor. It has been demonstrated experimentally that in many materials, when subjected to a completely tensile cyclic loading, the crack growth rate is given by [33]:

$$da/dN = C(\Delta K)^m \tag{10.5}$$

where a is crack length, N is number of cycles, and C and m are material parameters. Eq. (10.5) is frequently termed the Paris law [34].

However, it is now clear that the shape of the log-log relation between da/dN and ΔK is not linear, but sigmoidal. Below some threshold value of $\Delta K = \Delta K_T$, no crack growth occurs, while at high values of ΔK, as K_{max} in the cycle tends to K_c for fracture, the growth rate tends to infinity. The general appearance of a fatigue crack growth rate curve is shown in Fig. 10.11, and Eq. (10.5) applies over the midrange. ΔK_T is analogous to the fatigue limit.

Over the region where Eq. (10.5) is valid, the effect of mean stress on crack growth rate is found to be small. However its effect becomes significant at both high and low values of ΔK. With increasing R the value of ΔK_T is decreased, while at high values of ΔK crack growth rates increase as R is raised. Forman *et al.* [35] proposed the relation

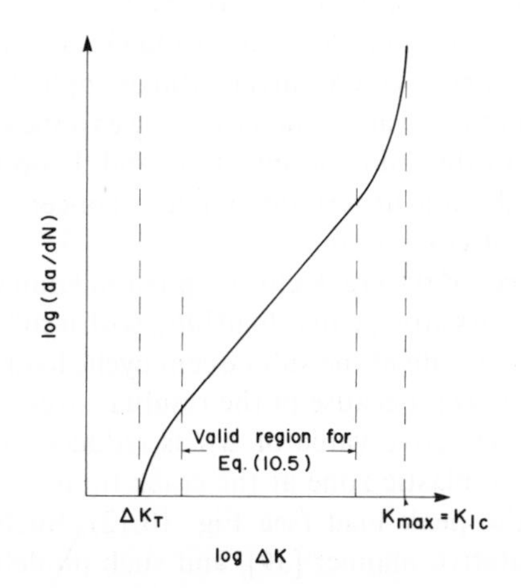

Fig. 10.11 Crack growth rate as a function of ΔK

$$da/dN = C\Delta K^m/[(1 - R)K_c - \Delta K] \tag{10.6a}$$

which can be rearranged in the form

$$da/dN = C\Delta K^{m-1}K_{max}/(K_c - K_{max}) \tag{10.6b}$$

In Eqs. (10.6) da/dN becomes infinite when $K_{max} = K_c$.

Other expressions showing the effect of R have been proposed, but the differences between them are generally small and each may fit satisfactorily for a limited set of data.

For $R < 0$, i.e., when compression occurs during a stress cycle, Eqs. (10.6a) and (10.6b) lose their meaning as no crack growth can occur when a crack is closed and the stress is compressive. Thus, for $R < 0$, it might appear that da/dN should be a function of K_{max}, only, and that ΔK in Eq. (10.5) should be replaced by K_{max}. However, because of residual stresses generated at the crack tip, a crack does not close at the same time as the applied stress becomes compressive: it may have closed prior to this [*36*]. Thus, the crack growth rate is likely to depend on the *effective* range of the stress intensity factor, $\Delta K_{eff} = (K_{max} - K_{op})$, where K_{op} is the value of stress intensity factor when the crack opens fully, and on K_{max}.

Considerable study has been made of the *crack closure phenomenon* since Elber's original papers on the subject [*36, 37*], and it is believed that crack growth rate should in all cases be correlated using either

$$da/dN = C' (\Delta K_{eff})^m \tag{10.7a}$$

or

$$da/dN = C' (U\Delta K)^m \tag{10.7b}$$

where the effective stress range factor, U, equals $(\sigma_{max} - \sigma_{op})/(\sigma_{max} - \sigma_{min})$, σ_{max}, σ_{min} and σ_{op} being the maximum, minimum and crack opening stresses in the cycle. The value of U depends on R, on the ratio of applied stress range to yield stress, $\Delta\sigma/\sigma_Y$, and on the cyclic strain hardening exponent, n, of the material [*38, 39*]. U has been evaluated experimentally and theoretically by a number of authors, but considerable uncertainty remains concerning the detailed and quantitative aspects of crack closure.

One important effect of the crack closure phenomenon is the delay in crack growth rate which occurs after an overload [*40*]: with a sufficient overload and depending on the magnitude of the subsequent cyclic loading, complete crack arrest may take place [*41*]. Because of the residual stress field produced by a tensile overload, the effective value of ΔK is reduced on subsequent stress cycling until the cyclic plastic zone at the crack tip moves out of the plastic zone produced by the peak load (see Fig. 10.12). Such effects have been modeled in a quantitative manner [*39*], and such modeling is of particular importance in predicting crack growth (and life) under spectrum or variable amplitude loading.

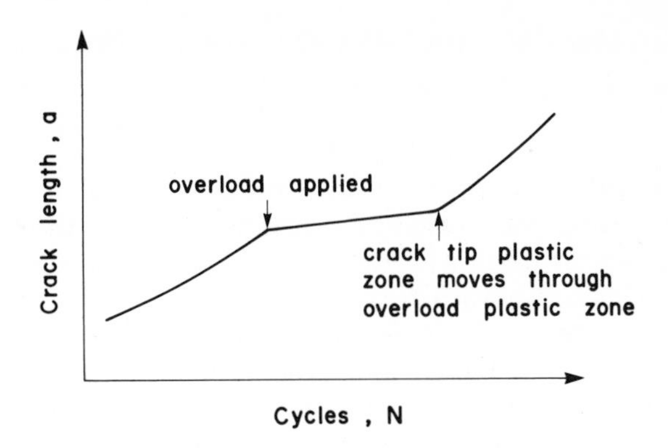

Fig. 10.12 Crack retardation from application of an overload

From a knowledge of the graphical relationship between da/dN and ΔK or of the material constants in Eqs. (10.5), (10.6) or (10.7), the cycles to cause a crack to grow from an initial length a_i to its critical length a_c at fracture (when $K_{max} = K_c$) can be predicted using numerical integration techniques. Thus,

$$N_f = \int_{a_i}^{a_c} (1/C\Delta K^m)da \tag{10.8}$$

For constant amplitude loading at $R = 0$, $\Delta K = K_{max} = \sigma_{max}(\alpha\pi a)^{1/2}$, where α is a geometrical factor, and for $\alpha = 1$ and assuming a power law to hold throughout,

$$N_f = 2(a_c^{(2-m)/2} - a_i^{(2-m)/2})/[(2 - m)C(\sigma_{max}\sqrt{\pi})^m] \tag{10.9}$$

10.5.1 Other Factors Affecting Fatigue Crack Propagation

Fatigue crack propagation rate is affected by many factors relating to service conditions. These include material thickness, frequency of loading, temperature and environment, although environment is generally by far the most important factor.

Fatigue crack growth rate generally increases with increase in specimen thickness. For thin sheets, a crack which starts in a tensile mode, perpendicular to the sheet surface, transforms to a slant mode crack after the plastic zone grows with crack growth so that plane stress conditions obtain. With increase in sheet thickness, the crack length before transition to shear mode increases and the life increases somewhat. In thicker sections where plane strain conditions obtain across most of the section, the edges still develop shear lips and the crack is delayed at the edge relative to the center: thus the crack has a curved front. With increase in thickness the contribution of the restraining slant mode edges decreases, and the crack growth rate may increase somewhat. Also, crack closure occurs primarily in the slant mode regions, while the center of the crack may remain open [*42, 43*], so that the effect of crack closure in reducing the value of ΔK (apparent) is likely to decrease with increasing thickness. These effects are generally small when compared with the uncertainty arising from material variability.

The effect of frequency by itself is generally small over the range 0.25-100 Hz, with some increase in crack growth rate being possible at lower frequencies. Generally, when frequency effects appear to be present, they relate to effects of environment, which become more severe at low frequencies when attack by or interaction of the atmosphere with the fresh surface exposed at the crack tip under tensile loading may occur.

The effect of low temperature is generally to reduce crack propagation rate [*44, 45*], although it may increase below the ductile-brittle transition temperature in steel [*44*]. High temperatures reduce fatigue strength and are dealt with separately in Chapter 11 when considering design for elevated temperature conditions.

Environmental effects on crack growth rate can be very large, with relative humidity (RH) being the major factor for tests of some alloys, such as high strength steels, in moist air [*46, 47*], the crack growth rate increasing markedly with increase in RH. The increase in crack growth rate in the presence of environments such as moist air or liquid metals can be explained by chemisorbed atoms at the crack tip which facilitate the nucleation of dislocations [*48*]. Alternatively, the operative mechanism may be the generation of atomic hydrogen at the fresh crack tip surface in moist atmospheres, and its subsequent diffusion into the metal ahead of the crack causing embrittlement.

The wider subject of corrosion fatigue is discussed in Section 10.8.

10.6 FATIGUE IN NOTCHED MEMBERS

Fatigue cracks always initiate at a structural discontinuity. Hence, there has been continued interest in the initiation and growth of fatigue cracks from notches in structural members. Such discontinuities reduce the fatigue resistance of a component. In the past it was considered appropriate to base the design of a notched member on the magnitude of the maximum stress at the notch root, based on an elastic stress concentration factor, k_t, the bulk of the material being at a stress below its yield stress. However, it has been found that the fatigue strength of a component does not decrease by as large a factor as k_t from that of a smooth specimen, because the conditions at the notch root are not elastic, the estimated high local stress being reduced by plastic flow. The reduction in fatigue strength at N cycles is defined by a factor k_f, *the fatigue strength reduction factor*, which represents the effective stress concentration factor.

Various relations between k_f and k_t have been developed, including Peterson's relation [49]:

$$k_f = 1 + (k_t - 1)/[1 + (\alpha/\rho)^{1/2}] \tag{10.10}$$

where α is a material constant determined experimentally, and ρ is the notch radius.

In order to provide a basis of comparison between materials, the *fatigue notch sensitivity factor*, q, was introduced, defined by

$$q = (k_f - 1)/(k_t - 1) \tag{10.11}$$

When $q = 0$, $k_f = 1$, and there is no notch effect. When $q = 1$, $k_f = k_t$, and the notch has its maximum possible effect. However, empirical approaches such as this are unsatisfactory in that size effects are ignored, large components failing at a lesser number of cycles than predicted on the basis of small specimen laboratory data.

10.6.1 The Neuber Rule Approach

An alternative approach has been developed by Topper *et al.* [50], based on the ideas of Neuber [51], who postulated that during plastic deformation the geometric mean of the stress and strain concentration factors, k_σ, and k_ϵ respectively, was constant and equal to k_t, i.e.,

$$k_t = (k_\sigma k_\epsilon)^{1/2} \tag{10.12}$$

To apply Neuber's rule to fatigue, k_t is replaced by k_f and k_σ and k_ϵ are written in terms of stress and strain ranges. Hence

$$k_f = [(\Delta\sigma/\Delta S)(\Delta\epsilon/\Delta e)]^{1/2} \tag{10.13}$$

or

$$k_f (\Delta S \Delta e E)^{1/2} = (\Delta\sigma\Delta\epsilon E)^{1/2} \tag{10.14}$$

where ΔS and Δe are the nominal stress and strain ranges in the member and $\Delta\sigma$ and $\Delta\epsilon$ are the local stress and strain ranges at the notch root. For elastic conditions in the member,

$$k_f \Delta S = (\Delta\sigma\Delta\epsilon E)^{1/2} \tag{10.15}$$

To predict the fatigue behavior of notched members, Topper *et al.* [50] plotted $(\Delta\sigma\Delta\epsilon E)^{1/2}$ for smooth specimens versus reversals to failure, $2N_f$, and then substituted $k_f (\Delta S \Delta e E)^{1/2}$ on the ordinate to cover notched members, their lives being estimated for given values of ΔS and Δe. In order to draw up the plot in the high strain range where there is plastic deformation in the member, the cyclic stress-strain curve is used (see Fig. 10.10), however this is unnecessary for the long life region of the *S-N* curve.

The Neuber approach has been shown to predict the notch-fatigue behavior of materials ranging from 2024 aluminum to high strength steel with a reasonable degree of success [50, 52]. However, it does consider the crack propagation stage to be negligible and ignores the effect of residual stresses. Thus it is more applicable to studies of crack initiation only. Otherwise size effects are ignored as with the approaches involving a fatigue notch sensitivity factor.

Crack initiation prediction using the Neuber approach involves consideration of the cyclic deformation of the material in the region of the crack tip. This is simulated by means of a smooth specimen [*53*] as indicated in Fig. 10.13. The notched component is subject to cyclic stress but the plastic region associated with the notch is under cyclic strain control derived from the elastic deformation of the bulk material. Hence the crack initiation region is under strain control as is the test specimen simulating this area. A plot of strain range, $\Delta\epsilon$, versus number of cycles to initiate a crack, or to develop a crack of some small specified length, can be drawn up as shown in Fig. 10.14(a). Now from Neuber's relation [Eq. (10.12)] for a notched specimen,

$$k_t(\Delta S \Delta e E)^{1/2} = (\Delta\sigma\Delta\epsilon E)^{1/2} \qquad (10.16)$$

or for the case when the nominal stress levels are elastic, as is normally the case,

$$k_t\Delta S = (\Delta\sigma\Delta\epsilon E)^{1/2} \qquad (10.17)$$

From the cyclic stress-strain curve for the material (see Fig. 10.10) a plot of $(\Delta\sigma\Delta\epsilon E)^{1/2}$ versus $\Delta\epsilon E$ can be derived as indicated in Fig. 10.14(b). The value of $\Delta\epsilon E$ corresponding to a specified value of $k_t\Delta S$ $(=\Delta\sigma\Delta\epsilon E)^{1/2}$ can be read off and the appropriate value of $\Delta\epsilon$ inserted in Fig. 10.14(a), allowing the cycles to initiate a crack to be read off.

10.6.2 The Approach of Smith and Miller [*56*]

A number of observations relating to notched specimen fatigue behavior have caused problems in applying any of the above predictive methods. Specifically, it has been observed that cracks may initiate at sharp notches and then cease to propagate after some distance [*54, 55*]. Also, for notches of constant k_t but varying depth, k_f increases as the specimen size increases. Smith and Miller [*56*] have developed an analysis which serves to explain these effects and to provide a framework for prediction of whether or not cracks will initiate or propagate, or both. The following discussion is based on their analysis.

Consider a fatigue crack at a notch as shown in Fig. 10.15. It may be initiated in the first instance because of the local cyclic plasticity produced by the high stress field at the notch. As the crack grows it will move out of the influence of this local stress field into the much lower stress field of the bulk material. The crack will continue to grow only if it generates sufficient crack tip plasticity, the limiting condition being that this must be greater than that corresponding to the threshold stress intensity factor range, ΔK_T.

When the crack is within the notch stress field, the stress intensity factor has

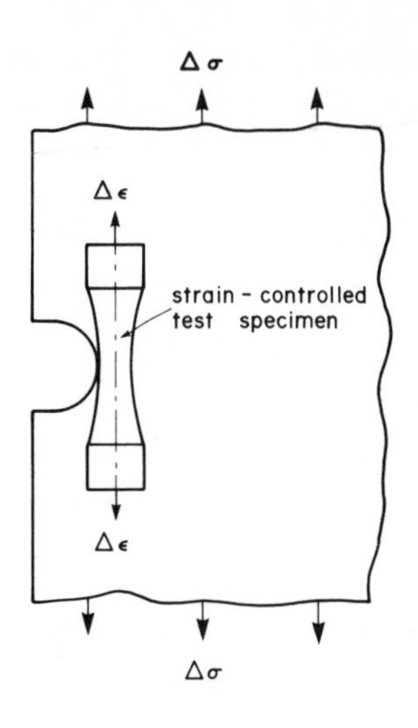

Fig. 10.13 Strain-controlled smooth specimen simulation of notch root behavior

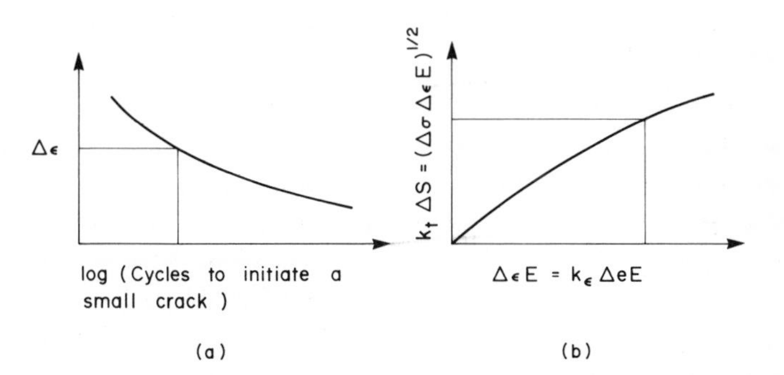

Fig. 10.14 (a) Cycles to initiate a small crack in a smooth specimen as a function of applied strain range. (b) Plot derived from cyclic stress-strain curve

the form

$$K = A\ k_t\ \sigma(\pi l)^{1/2} \tag{10.18}$$

where A is a geometrical constant, σ is the nominal stress, and l is the crack length from the free surface. At very short crack lengths, $K \to 0$, and no growth would be expected for $\Delta K < \Delta K_T$. Hence, we can consider the initiation and early growth of a fatigue crack to be controlled by the plastic strain field at the notch rather than K. The strain field decreases with distance from the notch

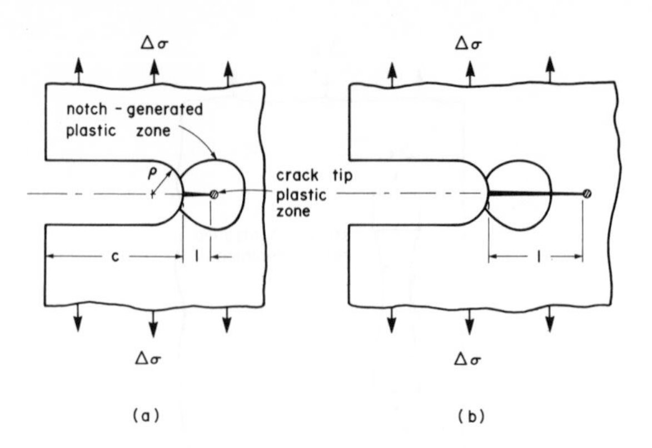

Fig. 10.15 Crack growth from a notch. (a) Crack within notch-generated plastic zone: growth controlled by cyclic plasticity. (b) Crack beyond notch-generated plastic zone: growth controlled by crack tip plasticity

and the crack propagation rate can be expected to fall as the crack grows, as indicated in Fig. 10.16, unless ΔK_T is exceeded by crack-generated plasticity. If this is the case, the crack will grow and accelerate with increasing ΔK, K being defined by

$$K = \sigma\,[\alpha\pi(c + l)]^{1/2} \tag{10.19}$$

where α is a geometrical factor and c is the notch depth, considering a surface notch as shown in Fig. 10.15. At the point when control of crack growth is taken over by crack tip plasticity, and considering zero-to-tension loading, together with the fact that, for small crack lengths and sharp notches, $l \ll c$, the range of stress intensity factor is given approximately by

$$\Delta K = 1.12\,\sigma(\pi c)^{1/2} = \Delta K_T \tag{10.20}$$

the factor 1.12 being introduced as the crack is considered at a free surface. Thus, the minimum nominal stress required to cause continued crack propagation is

$$\sigma = 0.5\Delta K_{TC}^{-1/2} \tag{10.21}$$

The stress required to initiate the crack can be taken approximately as σ_o/k_t, where σ_o is the fatigue limit, however Smith and Miller [57] have derived a criterion based on the situation when the growth rates of cracks of lengths L and l in unnotched and notched specimens, respectively, are identical under the same bulk stress, σ. The crack in the notched specimens is considered to have an *equivalent length* $L = l + e$. Equating ΔK for both cases (da/dN being identical), the notch contribution, e, to equivalent crack length for the region within the notch stress field ($l < 0.13(c\rho)^{1/2}$, where ρ is the notch radius) was determined to be

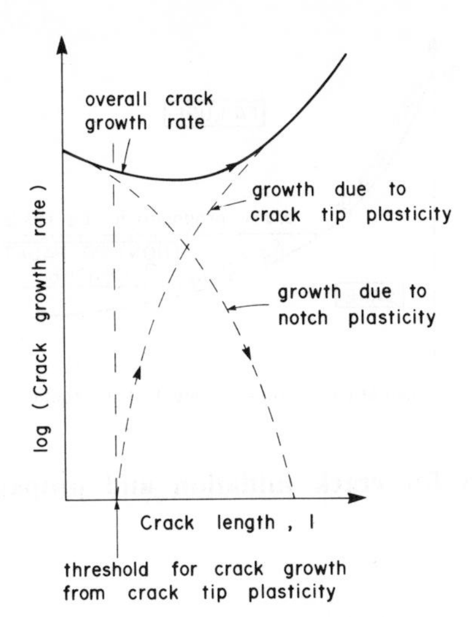

Fig. 10.16 Crack growth rate from a notch

$$e = 7.69l(c/\rho)^{1/2}$$

and the stress intensity factor for the equivalent crack is

$$K = [1 + 7.69(c/\rho)^{1/2}]^{1/2}\ \sigma\ (\pi l)^{1/2} \tag{10.22}$$

Let

$$K = k_{\text{fatigue}}\ \sigma(\pi l)^{1/2}$$

where k_{fatigue} is a fatigue concentration factor comparable to k_t in

$$K = k_t\ \sigma(\pi l)^{1/2}$$

and defined by

$$k_{\text{fatigue}} = [1 + 7.69(c/\rho)^{1/2}]^{1/2} \tag{10.23}$$

The condition for crack initiation can be rewritten as

$$\sigma \geqslant \sigma_o / k_{\text{fatigue}} \tag{10.24}$$

Figure 10.17 shows the form of the criteria for crack initiation and crack propagation on the basis of Smith and Miller's approach. The conditions for non-propagating cracks are clear, and comparison with experimental observations [*54, 58*] has demonstrated the general applicability of the approach. In addition, the proposal that cracks initiating at a notch initially slow down in growth rate before accelerating again after they leave the notch field (providing the bulk stress is sufficiently high), has been confirmed experimentally [*59, 60*]. This approach appears to form a good basis for design, but caution must be exercised in cases where the stress is not completely reversed as both σ_o and ΔK_T are functions of R.

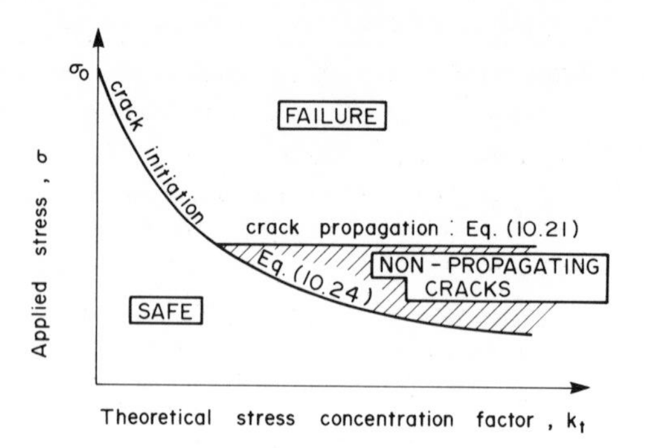

Fig. 10.17 Criteria for crack initiation and propagation to failure in notched material

Life prediction for notched components stressed above the limit for non-propagating cracks may be made approximately be determining the number of cycles to initiate a small crack of length approximately $0.13(c\rho)^{1/2}$, and then by integrating the Paris equation [Eq. (10.5)] with an appropriate expression for ΔK for a long crack from this initial length to a critical length corresponding to K_c.

10.7 LOW-CYCLE, HIGH-STRAIN FATIGUE

There is increased concern today about fatigue failures which occur under large cyclic strains taking place at low frequency and which may lead to failure in from a few hundred to not more than 50 000 cycles. Such conditions may be met with in pressure tubes and vessels, forging presses and dies, gun barrels and brake drums, for example. Also of importance are situations where design is for strictly limited life as in rocket motor hardware.

Data concerning low-cycle, high-strain fatigue normally are obtained by cycling test specimens under controlled strain limits as illustrated in Fig. 10.8(b). From such tests, a plot of $\Delta\epsilon$, or more appropriately, $\Delta\epsilon_p$, the plastic strain range, versus cycles to failure is drawn up. Under conditions of constant plastic strain range, the number of cycles to failure is found to be related to the strain range by the expression

$$\Delta\epsilon_p/2 = \epsilon_f' (2N_f)^c \tag{10.25}$$

where ϵ_f' is the *fatigue ductility coefficient*, i.e., the true plastic strain at one reversal ($2N_f = 1$), and c, the *fatigue ductility exponent*, is the slope of the $\log\Delta\epsilon_p$ versus log(life) plot. Eq. (10.25) is termed the Coffin-Manson relationship [*61, 62*] and is based on uniaxial test specimens cyclically strained in the plastic regime, and on a recognition that the plastic component of the total strain is the important factor in fatigue. It has been shown that Eq. (10.25) applies to a wide range of materials having great differences in hardness and ductility (see, e.g., [*63*]).

Both ϵ_f' and c are *fatigue ductility properties* of the material, and a reasonable approximation of ϵ_f' is obtained by equating it to ϵ_f, the true monotonic fracture ductility [*64*]. The value of c has been found to depend on material conditions and ductility, increasing to some extent as ductility decreases. It generally has a value in the range - 0.5 to - 0.7 [*65*], and Morrow [*65*] has related it to the *cyclic strain hardening exponent*, n', in the equation for the cyclic stress-strain curve (Fig. 10.10), which can be written as

$$\sigma_a = K'(\Delta\epsilon_p/2)^{n'} \tag{10.26}$$

where K' is the *cyclic strength coefficient*. The relationship proposed between c and n' is

$$c = -1/(1 + 5n') \tag{10.27}$$

The reason for the applicability of the Coffin-Manson relationship lies in the repetitive nature of the fatigue process. At high strain ranges fatigue cracks initiate very early in the life and most of the latter is spent in crack propagation. The total number of cycles to failure is

$$N_f = \int_{a_o}^{a_c} (dN/da)\, da + N_o$$

and for $N_f >> N_o$,

$$N_f \simeq \int_{a_o}^{a_c} (dN/da)\, da \tag{10.28}$$

By substitution of a suitable expression for da/dN is terms of material properties and the plastic strain range, and integration of Eq. (10.28), an equation of similar form to Eq. (10.25) is found. For example, using the analysis of crack growth rate made by Tomkins [*66*], we obtain

$$\Delta\epsilon_p(N_f)^{1/1+2n'} = \text{constant}$$

or, from comparison with Eq. (10.25),

$$c = -1(1 + 2n') \tag{10.29}$$

For $n' \simeq 0.15$, as in many wavy slip materials, c is overestimated from Eq. (10.29) when compared with measured values, but the mechanistic basis of the Coffin-Manson relation is clear. Generally, the empirical value for c given by Eq. (10.27) is to be preferred in the absence of directly measured values.

A linear relationship between log(stress amplitude) and log(fatigue life) in the high cycle range was proposed by Basquin [*67*], and may be written as

$$\sigma_a = \sigma_f' (2/N_f)^b = E\Delta\epsilon_e/2 \tag{10.30}$$

where σ_f' is the *fatigue strength coefficient*, i.e., the stress intercept at one reversal ($2N_f = 1$), and b, the *fatigue strength exponent*, is the slope of the $\log\sigma_a$ versus log(life) plot. The terms σ_f' and b are *fatigue strength properties* of the material. The value of σ_f' may be approximated by σ_f, the true monotonic fracture strength, corrected for necking, while b may be found experimentally, or from the cyclic strain hardening exponent, n', through the relationship derived by Morrow [*65*],

$$b = -n'/(1 + 5n') \tag{10.31}$$

From comparison of Eqs. (10.27) and (10.31), we see that

$$n' = b/c \tag{10.32}$$

Manson and Hirschberg [*68*] showed that the resistance of a metal to total strain cycling was given by the sum of its elastic and plastic strain resistance, or, from Eqs. (10.25) and (10.30),

$$\Delta\epsilon/2 = (\sigma_f'/E)(2N_f)^b + \epsilon_f'(2N_f)^c \tag{10.33}$$

This has been shown to apply well to data for many alloys, and Eq. (10.33), together with Eqs. (10.25) and (10.30), is shown schematically in Fig. 10.18. The *transition fatigue life*, $2N_t$, defines the point at which strength and ductility properties contribute equally to fatigue resistance. For high strength steels $2N_t$ may be less than 10 reversals, while for lower strength steels it may exceed 10^5 reversals, this being shown in Fig. 10.19 [*64*]. Note that in Fig. 10.18 a horizontal line has been drawn to intersect the $\Delta\epsilon$-$2N_f$ curve at $2N_f = 2\times10^6$ cycles: such a horizontal line, intersecting at this point or at $2N_f \simeq 10^7$ cycles, is appropriate where the material has a definite fatigue limit.

It may be seen that the strain amplitude-life curve can be *estimated* from a limited number of material properties. Specifically, we require ϵ_f, n', σ_f and E, the only cyclic property being n'. This forms a useful method of estimating low cycle fatigue resistance of a material.

Manson [*20*] rewrote the Coffin-Manson equation as

$$\Delta\epsilon_p = (N_f/D)^{-0.6} \tag{10.34}$$

where D is the ductility, defined by

$$D = \ln[1/(1 - RA)]$$

where RA is the reduction in area from a tensile test. Taking the relationship between elastic strain range and cycles to failure as [*20*] [cf. Eq. (10.30)],

$$\Delta\epsilon_p = 3.5\,(\sigma_u/E)\,N_f^{-0.12} \tag{10.35}$$

where σ_u is the UTS, total strain range versus cycles to failure is obtained by adding Eqs. (10.34) and (10.35), resulting in

$$\Delta\epsilon = 3.5\,(\sigma_u/E)N_f^{-0.12} + D^{0.6}\,N_f^{-0.6}$$

which is Eq. (10.4), Manson's universal slopes equation.

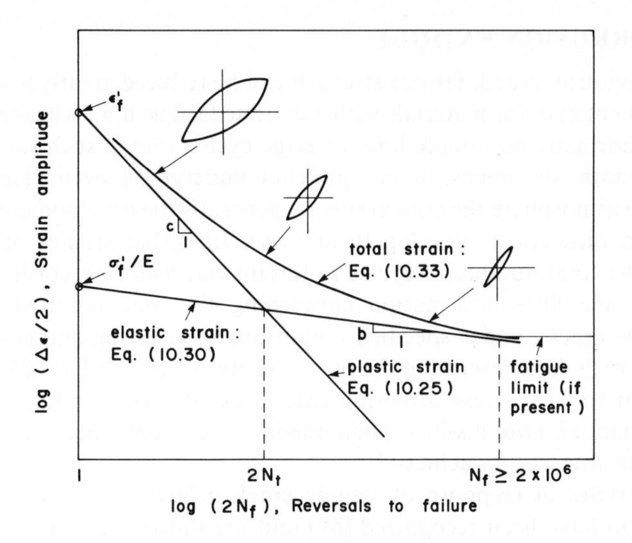

Fig. 10.18 Relation between cyclic strain amplitude and number of reversals to failure

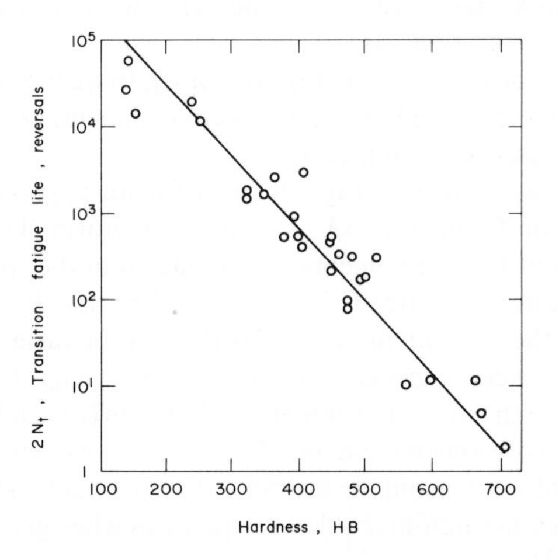

Fig. 10.19 Transition life as a function of hardness for steel. After Landgraf [*64*]

10.8 CORROSION FATIGUE

As previously noted, fatigue strength can be reduced greatly in a corrosive environment and, for materials with a definite fatigue limit when tested in air, there is normally no fatigue limit at large cycle times in such surroundings.

In smooth specimens or components undergoing cyclic loading in a corrosive atmosphere the critical event is generally the development of surface corrosion pits: cracks develop from these. In earlier studies of corrosion fatigue the total number of cycles to failure was usually recorded, but such data provide little information concerning the time required to initiate detectable cracks under specified conditions of stressing and environment, together with data describing the relation between ΔK and da/dN. Also, it is important to treat stress corrosion cracking and corrosion fatigue together rather than as two distinct phenomena, since completely static loading conditions are seldom achieved.

Three types of response of fatigue crack growth rate to environmental interaction have been recognized [*69*] and are illustrated in Fig. 10.20: they may conveniently be discussed with relation to K_{Iscc}. In Fig. 10.20(a) behavior corresponding to the aluminum-water system is shown, this resulting from the interaction of fatigue and environmental attack. Figure 10.20(b) illustrates the behavior observed in the hydrogen-steel system where environmental crack growth is related directly to the superposition of steady-state crack growth on fatigue crack growth, there being no interaction effects. In Fig. 10.20(c) the behavior of most alloy-environment systems is illustrated, with behavior above K_{Iscc} approximating to the second type, while below K_{Iscc} interactive effects occur as in the first case.

With the first type of behavior [Fig. 10.20(a)], frequency effects are present to a small extent in saturated and aqueous environments [*70*] and may be large in partially saturated environments.

For behavior as shown in Fig. 10.20(b) for an aggressive environment, da/dN depends on frequency, ΔK, K_{max} and waveform. These effects can be taken into account using an algebraic summation of the "pure" fatigue crack growth and that due to static SCC during the time that $K > K_{Iscc}$ [*71*]. Figure 10.21 illustrates the method: in Fig. 10.21(a) the cyclic variation of K is shown, while (b) shows the crack growth rate in SCC above K_{Iscc}. The contribution to fatigue crack growth from environment is shown in (c), while (d) indicates the integrated effects of environment and K_{max} on fatigue crack growth rate. This procedure provides a reasonable estimate in many cases, although the cyclic contribution becomes minimal at low frequencies when growth is primarily as in pure SCC [*72*].

The fact that environmentally induced effects can cause an increase in crack growth rate at values of $K < K_{Iscc}$ is thought to relate to the plastic strain occurring at the crack tip during tensile loading. Additional clean surface is

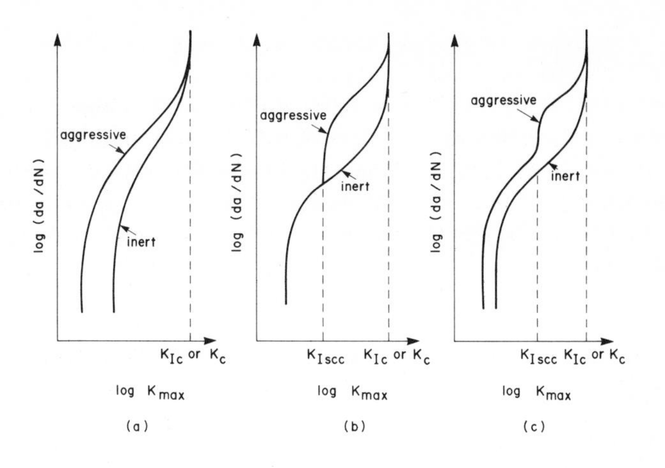

Fig. 10.20 Effect of environment on fatigue crack growth rate

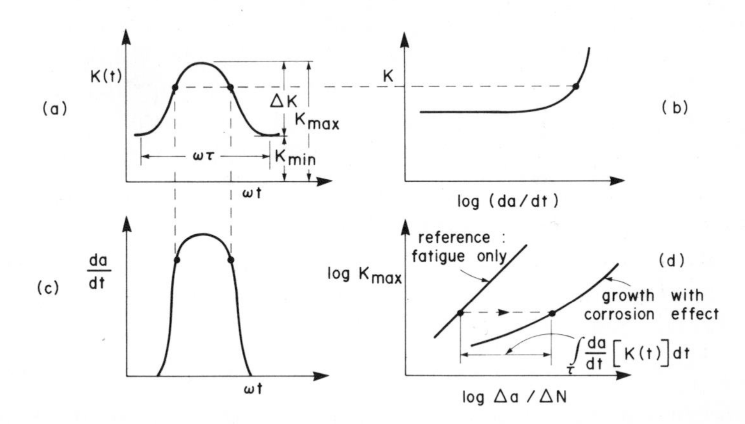

Fig. 10.21 Wei-Landes superposition model for environmentally enhanced fatigue crack growth

generated upon each load cycle and can be reacted upon by the environment. In static loading for the same K value, the plastic deformation taking place originally will be balanced by strain hardening and no additional surface will be exposed.

From the point of view of the materials engineer specifying materials for service under conditions of corrosion fatigue it is important to conduct tests on the specific material/environment combination of interest. When data must be extrapolated to provide predictions at longer times it is important to note that the short time tests should be made at higher stress, *not* in a more aggressive atmosphere. This is because material/environment couples are very specific and small variations in the composition of either can lead to large changes in life. In an ideal case the design should involve a stress below the threshold stress for crack initiation and below that for propagation from the largest flaw liable to be present. However, such ideal conditions can seldom be achieved in practice.

10.9 LIFE PREDICTION UNDER SERVICE CONDITIONS

10.9.1 Cumulative Damage

In practice, most structures and components are subjected to cyclic loadings of varying amplitude, rather than the cyclic stress of constant amplitude used in most conventional fatigue testing machines. For complete safety a component should be so designed that at no time will there be a cyclic stress greater than the fatigue or endurance limit of the material, but this leads to a ridiculously conservative design in many cases, particularly where high cyclic loads occur only occasionally. However, an approach is needed to design components for finite life under conditions of varying stress amplitude, and to assess the *cumulative damage* produced by the loading.

A linear cumulative damage rule was proposed by Palmgren [*73*] and Miner [*74*] and, at failure, it may be written as

$$D = D(n_1/N_1) + D(n_2/N_2) + D(n_3/N_3) + \ldots.$$

where D is the damage at fracture, n_1 is the number of cycles at a cyclic stress amplitude σ_1 at which N_1 is the number of cycles to failure, and so on for n_2, etc. Thus, the failure criterion is

$$1 = (n_1/N_1) + (n_2/N_2) + (n_3/N_3) + \ldots. = \Sigma(n_i/N_i) \tag{10.36}$$

Although there is not a good physical picture of the "damage" represented by D, and there are many sets of data which do not give a value of $\Sigma(n/N) = 1$ at failure, the Palmgren-Miner law does at least give a first approximation of the expected life under variable amplitude loading. When the first block of load cycles is at low stress, followed by blocks of higher stress level, Eq. (10.36) tends to be conservative. However, if an initial block of load cycles is at high stress, causing early crack initiation, $\Sigma(n/N)$ may be < 1 at failure. Hence, it is preferable to test components using loading sequences or histories derived from service records. Other test procedures involve blocks of load cycles or random load testing, any of these being readily applied using servo-hydraulic closed loop testing systems.

It is of interest to note that the Palmgren-Miner law can be derived on the basis of summing the number of cycles for crack growth increments for a_i to a_1

to a_2....to a_c at different stress levels and utilizing the Paris equation [Eq. (10.5)] [9]. Thus, when a proof test has been conducted on, or NDE made of a component to determine the maximum size of initial flaw present and K_{Ic} is known, the fatigue life can be predicted for the crack to grow from a_i to a_c in Fig. 9.22, either from Eq. (10.8) in the case of simple loading, or by evaluating

$$a_i + \Sigma\Delta a = a_c$$

with Δa_1 being the incremental crack growth in n_1 cycles at σ_1, etc., and given by

$$\Delta a_1 = \int_{N=0}^{N=n_1} (da/dN)dN$$

da/dN being at σ_1.

10.9.2 Multiaxial Stressing

Fatigue under multiaxial stressing was discussed at some length in Section 2.7 from the point of view of defining a failure criterion. However, it is appropriate to comment here on the more recent approach of Miller [75, 76] which has to some extent a physical basis.

Brown and Miller [75] delineate the possible crack paths by noting that crack initiation and early growth occur on shear planes and relating the three planes of maximum shear strain to the orientation of the free surface at which cracks will initiate. Cracks are considered to be of two types, A and B, and these, together with Stages I and II crack growth in each, are illustrated in Fig. 10.22. It is considered that the governing parameters are the maximum shear strain, $\Delta\gamma = (\Delta\epsilon_1 - \Delta\epsilon_3)/2$, and the tensile strain, $\Delta\epsilon_n$, generated across the plane of maximum shear and which will assist in decohesion. Thus, it resembles the empirical theory of McDiarmid [77] described in Chapter 2.

Miller plots his results on the Γ-plane, on a plot of maximum shear strain versus the normal tensile strain, and the form of the results obtained is shown in Fig. 10.23, the contours being for constant crack growth rate or lifetime.

While this work is still not in a fully developed state, it appears that it should be given serious consideration as effects of anisotropy and out of phase stressing can be included [76].

10.9.3 Fatigue Crack Growth in Pressure Vessels

Fatigue in pressure vessels was also discussed in Section 2.7 in reviewing multiaxial fatigue and it is obviously amenable to study using the approach of Miller outlined above. However, a more common approach is one based on fracture mechanics and which may be used to predict the growth of a defect by fatigue to the critical size for fracture.

In a pressure vessel containing an internal semi-elliptical surface flaw, as shown in Fig. 9.23, the stress intensity factor at the end of the minor axis was given by Eq. (9.16b) as

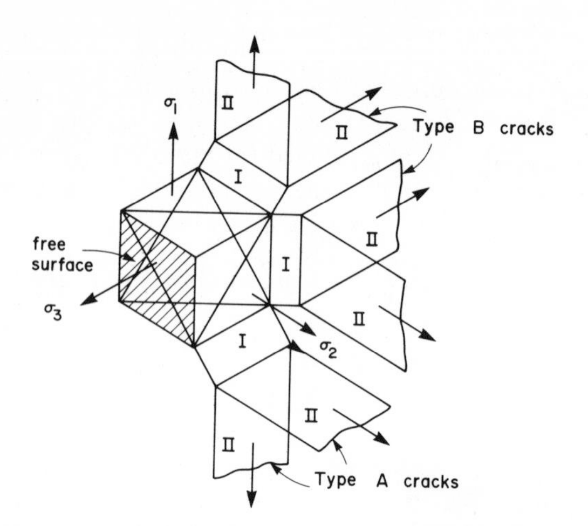

Fig. 10.22 Type A and Type B cracks developed in a three-dimensional stress system. Type B cracks are usually more dangerous. Note that cracks do not normally develop in the plane of the surface

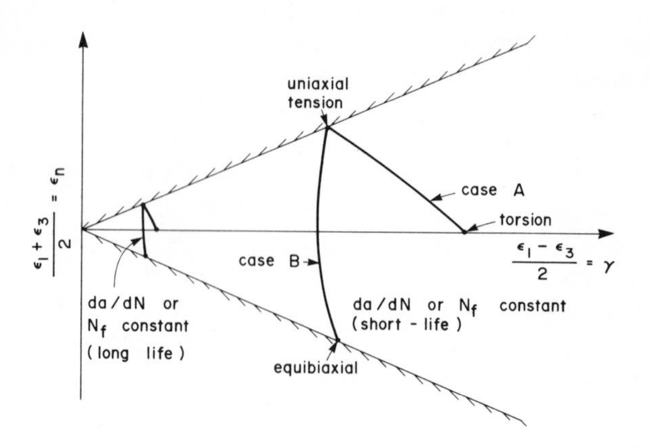

Fig. 10.23 The Γ-plane showing plots of equal lifetime or crack growth rate

$$K_{\mathrm{I}} = 1.12\,(\sigma + p)\,(\pi a/Q)^{1/2}$$

If the wall is thin with respect to the vessel diameter, so that the hoop stress is essentially constant through its thickness, the growth of such a crack to a critical depth can be predicted from a knowledge of the constants in the Paris equation for the material and integration of this equation with substitution for

ΔK. Numerical techniques would be employed appropriately to do this.

However, in the case of a thick walled vessel there is a significant stress gradient through the wall thickness, the maximum hoop stress being at the inside surface. Hence, the stress intensity factor at the end of the major axis of the semi-elliptical flaw, given by

$$K_I = 1.12\,(\sigma + p)\,(\pi a^2/Qc)^{1/2}$$

may be larger than that at the end of the minor axis which is the maximum value for σ not varying with thickness. This will depend on the ratio a/c and on the stress gradient, i.e., the difference between the values of σ in the two equations for K_I. Thus, growth of the crack should be predicted in the directions of the two axes of the ellipse.

Similar complications arise when treating the growth of a fatigue crack from a region of local stress concentration, and numerical solutions using finite element techniques may be required to determine how K varies with crack growth and to evaluate the crack growth rates and consequent shape change of the crack. Alternatively, experimental studies may be justified to determine the actual changes from an initial flaw shape and hence allow determination of stress intensity factors.

Tomkins [*78, 79*] has correlated fatigue in thick cylinders with simple push-pull or torsion data. Assuming that crack initiation occurs early, the crack growth rate was written in terms of the maximum hoop stress, σ_t, and the value of the bore shear strain range, $\Delta\gamma$, as

$$da/dN = B\Delta\gamma(\sigma_t)^2 a \tag{10.37}$$

The effect of pressurized fluid entering the cracks was allowed for by rewriting this as

$$da/dN = B\Delta\gamma(\sigma_t + p)^2 a \tag{10.38}$$

where p is the fluid pressure.

For repeated or fully reversed straining, this can be expressed in terms of shear range components as

$$da/dN = B'\Delta\gamma\Delta\tau^2 a \tag{10.39}$$

Similarly, crack propagation in push-pull or torsion fatigue tests can be expressed in terms of shear range components and integration of such relations gives the endurance life. Experimental data for thick walled cylinders correlated well with the predictions from push-pull and torsion fatigue data [*79*].

REFERENCES

1. ALBERT, W. A. J., *Arch. Miner. Geognosie Berg. Hüttenkunde,* **10**, 215 (1838).
2. WÖHLER, A., *Z. Bauw.*, **8**, 642 (1858); **10**, 583 (1860); **13**, 233 (1863); **16**, 67 (1866); **20**, 74 (1870).

3. Burstall, A. F., *A History of Mechanical Engineering*, Faber and Faber, London (1963).
4. Gough, H. J., *The Fatigue of Metals*, Scott, Greenwood, and Son, London (1924).
5. Mann, J. Y., *Fatigue of Materials*, Melbourne Univ. Press, Melbourne (1967).
6. Kennedy, A. J., *Processes of Creep and Fatigue in Metals*, Oliver and Boyd, Edinburgh (1962).
7. Forrest, P. G., *Fatigue of Metals*, Pergamon, Oxford (1962).
8. Cazaud, R., *Fatigue of Metals*, Chapman and Hall, London (1953).
9. Frost, N. E., Marsh, K. J., and Pook, L. P., *Metal Fatigue*, Clarendon Press, Oxford (1974).
10. Sinclair, G. M., *Proc. ASTM*, **52**, 743 (1952).
11. Lipson, C., and Krafve, A. H., *Proc. 9th Int. Auto. Tech. Congress London*, p. 207, (G. Eley, ed.), London (1962).
12. Weibull, W., *Fatigue Testing and Analysis of Results*, Pergamon, Oxford (1961).
13. Weibull, W., *Trans. ASME, J. App. Mech.*, **73**, 293 (1951).
14. Goodman, J., *Mechanics Applied to Engineering*, Longman, Green, and Company, London (1899).
15. Gerber, W., *Z. bayer Archit. Ing. Ver.*, **6**, 101 (1874).
16. Bullens, D. K., *Steel and Its Heat Treatment*, Vol. 1, 5th Ed., Wiley, New York (1948).
17. Borik, F., Justusson, W. M., and Zackay, V. F., *Trans. ASM*, **56**, 327 (1948).
18. Craik, R. L., May, M. J., and Latham, D. J., *Metals Eng. Quarterly*, **9**, no. 3, 12 (1969).
19. Tetelman, A. S., and McEvily, A. J., Jr., in *Fracture: An Advanced Treatise*, Vol. 6, p. 137, (H. Liebowitz, ed.) Academic Press, New York (1971).
20. Manson, S. S., *Experimental Mechanics*, **5**, 193 (1971).
21. Manson, S. S., *Trans. ASME, J. of Basic Eng.*, **84**, 533 (1962).
22. Tomkins, B., Sumner, G., and Wareing, J., in *Fracture 1969*, p. 712 (P. L. Pratt *et al.* eds.), Chapman and Hall, London (1969).
23. Feltner, C. E., and Laird, C., *Acta Met.*, **15**, 1621 (1967).
24. Grosskreutz, J. C., *Phys. Stat. Sol.*, (b), **47**, 11 (1971); *Phys. Stat. Sol.*, (b), **47**, 359 (1971).
25. Mughrabi, H., Ackermann, F., and Herz, K., in *Fatigue Mechanisms*, p. 69, (J. T. Fong, ed.), STP 675, ASTM, Philadelphia (1979).
26. Helgeland, O., *J. Inst. Metals*, **93**, 570 (1965).
27. Ewing, J. A., and Humphrey, J. C. W., *Phil. Trans.* **A200**, 241 (1903).
28. Thompson, N., Wadsworth, N. J., and Louat, N., *Phil. Mag.*, **1**, 113 (1956).
29. Clark, J. B., and McEvily, A. J., Jr., *Acta Met.*, **12**, 1359 (1964).
30. Stubbington, C. A., *Acta Met.*, **12**, 931 (1964).
31. Forsyth, P. J. E., in *Proceedings of the Crack Propagation Symposium*, p. 76, The College of Aeronautics, Cranfield (1961).
32. Taira, S., Tanaka, K., and Hoshina, M., in *Fatigue Mechanisms*, p. 135 (J. T. Fong, ed.) STP 675, ASTM, Philadelphia (1979).

33. PARIS, P. C., and ERDOGAN, F., *Trans. ASME, J. Basic Eng.*, **85**, 528 (1963).
34. PARIS, P. C., in *Fatigue - An Interdisciplinary Approach, Proc. 10th Sagamore Conf.*, p. 125, Syracuse Univ. Press, Syrucuse, N.Y. (1964).
35. FORMAN, R. G., KEARNEY, V. E., and ENGLE, R. M., *Trans. ASME, J. Basic Eng.*, **89**, 459 (1967).
36. ELBER, W., *Eng. Fracture Mechanics*, **2**, 37 (1970).
37. ELBER, W., in *Damage Tolerance in Aircraft Structures*, p. 230, STP 486, ASTM, Philadelphia (1971).
38. LAL, K. M., GARG, S. B. L., and LE MAY, I., *Trans. ASME, J. Eng. Matls. and Technology*, **102**, 147 (1980).
39. LAL, K. M., and LE MAY, I., *Fatigue of Eng. Materials and Structures* (in press).
40. SCHIJVE, J., and BROEK, D., *Aircraft Eng.*, **34**, 314 (1962).
41. PROBST, E. P., and HILLBERRY, B. M., *J. AIAA*, **12**, 330 (1974).
42. LINDLEY, T. C., and RICHARDS, C. E., *Matls. Sci. and Eng.*, **14**, 281 (1973).
43. SHAW, W. J. D., and LE MAY, I., in *Fracture Mechanics*, p. 247 (C. W. Smith, ed.), STP 677, ASTM, Philadelphia (1979).
44. STEPHENS, R. I., CHUNG, J. H., and GLINKA, G., *Low Temperature Fatigue Behavior of Steels - A Review*, SAE Paper No. 790517 (1979).
45. LE MAY, I., and LUI M.-W., in *The Influence of Environment on Fatigue*, p. 117, I. Mech. E., London (1977).
46. SHAW, W. J. D., and LE MAY, I., in *The Influence of Environment on Fatigue*, p. 93, I. Mech. E., London (1977).
47. LE MAY, I., in *Fatigue Mechanisms*, p. 873 (J. T. Fong, ed.), STP 675, ASTM, Philadelphia (1979).
48. LYNCH, S. P., in *Fatigue Mechanisms*, p. 174, (J. T. Fong, ed.), STP 675, ASTM, Philadelphia (1979).
49. PETERSON, R. E., in *Metal Fatigue*, p. 293 (G. Sines and J. L. Waisman, eds.), McGraw-Hill, New York (1959).
50. TOPPER, T. H., WETZEL, R. M., and MORROW, J., *J. of Materials*, **4**, 200 (1969).
51. NEUBER, H., *Trans. ASME, J. App. Mech.*, **28**, 544 (1961).
52. LE MAY, I., WHELAN, J.M., and SHAW, W.J.D., ASME Paper No. 75-DE-31 (1975).
53. STADNICK S. J., and MORROW, J., in *Testing for Prediction of Material Performance in Structures and Components*, p. 229, STP 515, ASTM, Philadelphia (1972).
54. FROST, N. E., and DUGDALE, D. S., *J. Mech. and Phys. of Solids*, **5**, 182 (1957).
55. FROST, N. E., *J. Mech. Eng. Sci.*, **2**, 109 (1960).
56. SMITH, R. A., and MILLER, K. J., *Int. J. Mech. Sci.*, **20**, 201 (1978).
57. SMITH, R. A., and MILLER, K. J., *Int. J. Mech. Sci.*, **19**, 11 (1977).
58. FROST, N. E., *Proc. I. Mech. E.*, **173**, 811 (1959).
59. BROEK, D., NLR TR 72134U, National Aerospace Laboratory, The Netherlands (1972).
60. EL HADDAD, M. H., SMITH, K. N., and TOPPER, T. H., in *Fracture Mechanics*, p. 274, (C. W. Smith, ed.), STP 677, ASTM, Philadelphia (1979).

61. Coffin, L. F., *Trans. ASME*, **76**, 438 (1959).
62. Manson, S. S., NACA TN 2933 (1954).
63. Coffin, L. F., *Proc. I. Mech. E.*, **188**, 109 (1974).
64. Landgraf, R. W., in *Achievement of High Fatigue Resistance in Metals and Alloys*, p. 3, STP 467, ASTM, Philadelphia (1970).
65. Morrow, J., in *Internal Friction, Damping, and Cyclic Plasticity*, p. 45, STP 378, ASTM, Philadelphia (1965).
66. Tomkins, B., *Phil. Mag.*, **18**, 1041 (1968).
67. Basquin O. H., *Proc. ASTM*,**10**, 625 (1910).
68. Manson, S. S., and Hirschberg, M. H., in *Fatigue - An Interdisciplinary Approach, Proc. 10th Sagamore Conf.*, p. 133, Syracuse Univ. Press, Syracuse, N.Y. (1964).
69. McEvily, A. J., and Wei, R. P., in *Corrosion Fatigue: Chemistry, Mechanics and Microstructure*, NACE-2, p. 381, National Association of Corrosion Engineers, Houston (1972).
70. Wei, R. P., *Trans. ASME, J. Eng. Materials and Technology*, **100**, 113 (1978).
71. Wei, R. P., and Landes, J. D., *Materials Research and Standards*, **9**, 25 (1969).
72. McMahon, C. J., Jr., *Trans. ASME, J. Eng. Materials and Technology*, **95**, 142 (1973).
73. Palmgren, A., *Z. Ver. dt. Ing.*, **68**, 339 (1924).
74. Miner, M. A., *Trans. ASME, J. App. Mech.*, **67**, A-159 (1945).
75. Brown, M. W., and Miller, K. J., *Proc. I. Mech. E.*, **187**, 745 (1973).
76. Miller, K. J., in *Fatigue Testing and Design*, Vol. 1, p. 13.1 (R. G. Bathgate, ed.), Soc. of Environmental Engrs., London (1976).
77. McDiarmid, D. L., in *Proc. of the 2nd Int. Conf. on Pressure Vessel Technology*, Part II, p. 851, ASME, New York (1973).
78. Tomkins, B., *Int. J. on Pressure Vessels and Piping*, **1**, 37 (1973).
79. Tomkins, B., in *Proc. of the 2nd Int. Conf. on Pressure Vessel Technology*, Part II, p. 835, ASME, New York (1973).

11

CREEP AND DESIGN FOR ELEVATED TEMPERATURE

11.1 INTRODUCTION

Creep may be defined as time dependent deformation under an applied load. In general, engineering alloys do not exhibit time dependent deformation under normal service conditions unless used in an application where the temperature is relatively high, as in boiler superheater tubes, turbine blades and disks, nuclear reactor fuel element cladding, and so on. There are many exceptions, however, to this generalization; for example, lead creeps at room temperature and this has caused problems with water pipes in the past. Hence, we must consider temperature as "high" or "low" in relation to the melting point of the material, and it is found that metals exhibit much of same behavior at the same value of homologous temperature (T/T_m).

Figure 11.1 shows a schematic creep curve which may be obtained from either a metal or a non-metal, strain being plotted against time. On loading, an initial strain occurs, this including both elastic and plastic components. This is followed by a stage in which the creep rate decreases with time[1] - the *transient*

---

[1] In some cases primary creep has been found to be inverse in nature, the strain rate increasing to a steady value. However, this is exceptional and depends on the material condition, having been observed in Al alloys [*1*] and austenitic stainless steel welds [*2*].

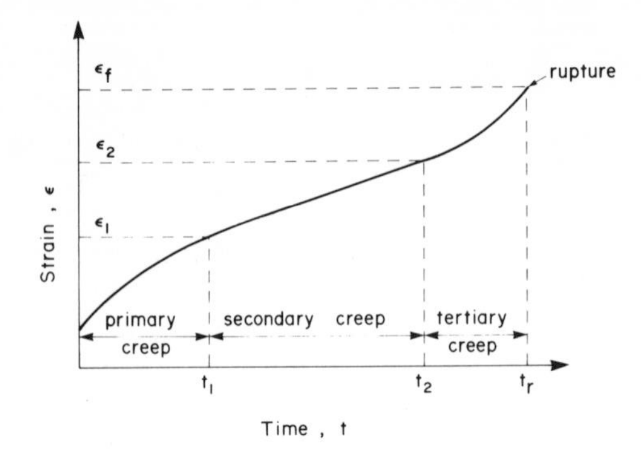

Fig. 11.1 Schematic creep curve

or *primary stage* of creep. Beyond some strain value (ϵ_1 on Fig. 11.1) the creep rate remains constant during the *secondary stage* of creep until at a later time and strain (t_2 and ϵ_2 on Fig. 11.1) it starts to accelerate during *tertiary* creep, fracture finally taking place at the rupture strain and time (ϵ_f and t_r, respectively). Most laboratory creep tests are made under conditions of *constant load*, but this means that the stress in the test specimen is increasing continuously as the cross-sectional area falls off. Hence, a more fundamental test should be made under conditions of *constant stress*, which can be achieved very simply using a cam designed to reduce the load as the specimen extends, and assuming constancy of volume during plastic deformation. When constant stress testing is done, it is found that the strain-time curve may differ substantially from that produced under constant load conditions; in particular, the initiation of tertiary creep is postponed.

If the stress is removed after a component has undergone creep deformation, as shown in Fig. 11.2, some of the primary creep may be recovered, as part of the primary creep is *anelastic* or *recoverable* in nature. This occurs because the internal stress opposing the movement of dislocations increases gradually due to the bowing out of dislocations and to the interactions between dislocations. Hence, when the applied stress is removed, the internal stress acts to move dislocations back and to return (in part, only) the material to its original dimensions.

The presence of any of the stages of creep shown in Fig. 11.1 is a function of stress, temperature, and material structure, and to obtain a full understanding of creep behavior we must establish a relationship between strain and time, including the effects of stress, stress state, temperature, microstructure (including material history as it affects this), composition and environment on this. The matter is a complex one, and it is proposed first to examine the

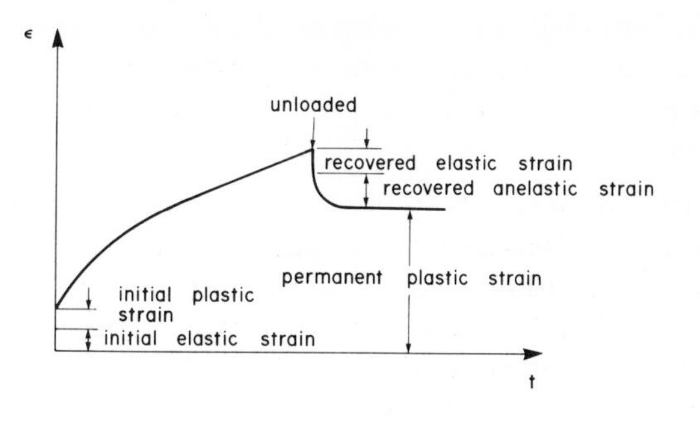

Fig. 11.2 Effects of stress removal during creep

mechanisms of deformation during primary and secondary creep, including consideration of resulting strain-time relationships, before reviewing the observed laws of creep relating stress, strain, time and temperature. This is followed by discussion of tertiary creep and fracture, extrapolation methods for creep-rupture data, multiaxial creep deformation, and creep-fatigue interaction, before moving on to discuss briefly the effects of environment, including different atmospheres as well as radiation.

11.2 MECHANISMS OF CREEP DEFORMATION

The literature concerning the mechanisms of creep deformation is a vast one, and it is impossible to discuss or review other than a few of the more important mechanisms and theories here. For more detailed consideration of the various theories, see recent reviews, e.g. [*3-5*].

The major mechanisms by which creep can occur are as follows:

- *Dislocation glide*, involving dislocations gliding along slip planes, and overcoming barriers by thermal activation;
- *Dislocation creep*, involving the movement of dislocations which overcome barriers, such as other dislocations, by thermally assisted mechanisms involving the movement of vacancies or interstitials to or from them;
- *Diffusion creep*, involving the flow of point defects through a crystal under the influence of an applied stress;
- *Grain boundary sliding*, involving the sliding of grains in a polycrystal past each other;
- *Irradiation creep*, involving the production of *internal stresses* greater than the external stress by means of irradiation with high energy particles, the material then creeping under the influence of this internal stress;

- *Thermal cycling creep*, involving the production of internal stresses greater than the external stress by means of thermal expansion of anisotropic grains in a polycrystal. As in the previous case, creep may occur to relax these stresses.

Creep deformation by dislocation glide occurs at high stress levels relative to those normally considered for creep deformation. The movement of dislocations is impeded by obstacles such as precipitates, grain boundaries, solute atoms and other dislocations, but at temperatures above absolute zero there is a certain probability of a dislocation bypassing an obstacle by thermal activation. The resultant strain rate, $\dot{\epsilon}$, may be written as [*6*]

$$\dot{\epsilon} = \dot{\epsilon}_o \exp\left[-ba_a\,(Gb/L - \sigma)/kT\right] \tag{11.1}$$

on the basis that the creep rate due to dislocation glide depends on temperature in accordance with Boltzmann statistics. In Eq. (11.1), b is the Burgers vector, L the interobstacle spacing, G the shear modulus, a_a the activation area, σ the applied stress, T the absolute temperature, and $\dot{\epsilon} = \dot{\epsilon}_o$ when $\sigma \simeq Gb/L$ (at absolute zero).

In dislocation creep the movement of dislocations takes place by climb or glide and the time taken by the dislocations to overcome obstacles is dependent on the flow of point defects under the action of applied stress to or from jogs along their length.

Diffusion creep is of two basic types, namely Nabarro-Herring [*7*, *8*] or Coble [*9*], in which the stress-directed diffusional flow of vacancies takes place respectively in a transcrystalline manner or along grain boundaries. In order to preserve compatibility at grain boundaries, grain boundary sliding *must* take place in conjunction with diffusional flow [*10*]. Thus, we may consider the operation of several mechanisms occurring at the same time: they may operate either in parallel (i.e., independently) or in series (i.e., sequentially) [*11*].

For i mechanisms in parallel the steady state creep rate is given by

$$\dot{\epsilon} = \sum_i \dot{\epsilon}_i \tag{11.2}$$

where $\dot{\epsilon}_i$ is the creep rate for the ith process, and the fastest mechanism will dominate the creep behavior. Alternatively, for i mechanisms in series,

$$1/\dot{\epsilon} = \sum_i (1/\dot{\epsilon}_i) \tag{11.3}$$

and the slowest mechanism will dominate.

We shall now discuss dislocation creep, diffusion creep and grain boundary sliding in some more detail. Particular attention will be paid to the first, as this covers a major portion of the conditions of practical importance for engineering alloys, and the theory is developed to the extent that it can predict deformation under complex stress-temperature histories with a considerable degree of success. Irradiation creep is discussed briefly in Section 11.8.2.

11.2.1 Dislocation Creep

Most of the theories which have been developed to allow the formulation of constitutive equations for dislocation creep are based on the thesis of Bailey [*12*] and Orowan [*13*], who considered steady state creep to be a consequence of the competing effects of strain hardening and thermal softening being in balance. The creep rate can be considered to depend on the number of dislocations which are able to glide when a specific stress system is applied, i.e., on the density of *mobile dislocations*. These will move under the influence of the *active* or *effective stress*[2], σ_a, acting on them, defined as

$$\sigma_a = \sigma - \sigma_i \tag{11.4}$$

where σ is the applied stress and σ_i is the *internal stress* acting between dislocations and inhibiting their movement. This latter stress is responsible for the promotion of recovery by rearrangement and annihilation of dislocations, and it also controls strain hardening, while glide is controlled by the active stress. Hence, we may set out the Bailey-Orowan equation for steady creep as

$$\dot{\epsilon}_s = r/h = -(\partial\sigma_i/\partial t)_{\sigma,\epsilon,T}/(\partial\sigma_i/\partial\epsilon)_{\sigma,t,T} \tag{11.5}$$

r and h representing the rates of recovery and hardening, respectively.

We require to formulate accurate physical models for the processes of recovery and strain hardening, such that the resulting mathematical relations correctly predict the shape of the creep deformation curve for the transient and steady state conditions, predict stress and temperature dependence correctly, and also provide correct information regarding the effect of stress changes on deformation rate. Many separate theories have been formulated [*14-24*], and most have been reviewed by Gittus [*3*]. Here we shall concentrate attention on one specific model due to Gittus [*3, 24*] which has been used to provide good correlation between predictions and experimental observations. Other models follow the same general principles to a great extent. The Gittus model includes within it the important point of anelastic strain occurring during primary creep, which is of concern in stress cycling or varying stress situations.

During plastic deformation, dislocations may form a complex three-dimensional network, and this is frequently a valid description of the substructure produced during creep [*5*]. The network spacing is a function of dislocation density, and Fig. 11.3 illustrates the effect of application of stress on the network [*25*]. The link P_1P_2 is the first to move, and nodes P_1 and P_2 pass through P_3 and P_4, two positions of the link as it moves being marked at

---

[2] The term, *active stress*, is preferred in this discussion to avoid confusion with the von Mises effective stress (σ_{eff}) used in multiaxial creep (see Section 11.6).

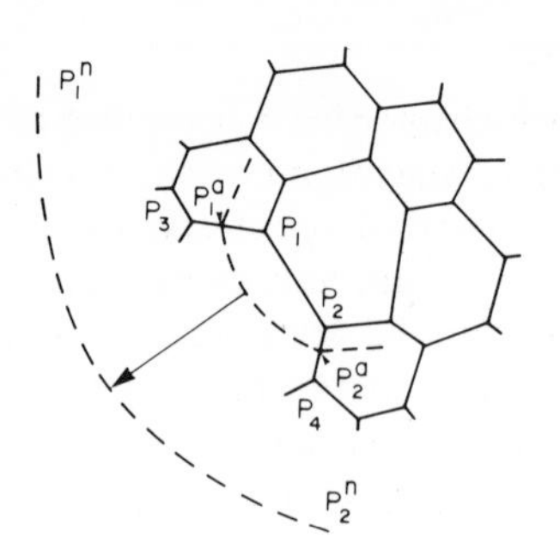

Fig. 11.3 McLean's model of effect of stress on a dislocation network

$P_1^a P_2^a$ and $P_1^n P_2^n$. At the later position the link is held up by interaction with other dislocations in the network, the internal stress due to dislocation interaction between the moving link and other parts of the network having increased to the level of the applied stress.

In steady state creep, we consider the dislocation links to glide with a mean velocity v over an average distance L_s before being held up. The velocity, v, is given by $\dot{\epsilon}_s/b$, where b is the Burgers vector and ρ the mobile dislocation density. The frequency with which the mobile dislocations are halted is $v/L_s = x\dot{\epsilon}_s/b\sqrt{\rho}$, where x is the factor relating the network spacing to L_s, the latter being taken to be several times the network spacing[*19*]. Recent observations of dislocation densities and creep rates indicate that the relation $v/L_s \simeq 10\dot{\epsilon}_s/b\sqrt{\rho_0}$, holds, where ρ_0 is the asymptote towards which the dislocation density tends as $\sigma \to 0$ [*3*]. The instantaneous rate of immobilization when there are ρ mobile dislocations is $i\rho$ where

$$i = \lambda\dot{\epsilon}_s \tag{11.6}$$

and λ is a materials parameter given by

$$\lambda \simeq 10/b\sqrt{\rho_0} \tag{11.7}$$

If the supply of dislocations can continue even without network recovery, the density of the mobile dislocations will tend asymptotically to ρ_∞. Thus, the rate of change of mobile dislocation density, $\dot{\rho}_A$, due to immobilization of dislocations at internal stress peaks and generation at sources is

$$\dot{\rho}_A = \lambda\dot{\epsilon}_s(\rho_\infty - \rho) \tag{11.8}$$

Recovery processes will take place to reduce the density of dislocations in the network where it is a maximum. Friedel ([*26*], p. 239) discussed the coarsening of the Frank network in the absence of applied stress where the movement of dislocations under the influence of σ_i is controlled by vacancy flow. He showed that the rate of coarsening, dr_s/dt, is given by

$$dr_s/dt \simeq D_v G b^3 c_j / r_s k T \tag{11.9}$$

where r_s is the dislocation spacing, D_v is the self-diffusion coefficient, c_j is the jog concentration and k is Boltzmann's constant. Now $r_s = \rho^{-1/2}$, thus

$$dr_s/d\rho = -(\rho^{-3/2})/2 \tag{11.10}$$

Substituting in Eq. (11.9), we obtain the rate of *coarsening* in terms of dislocation density, $-d\rho/dt$, as

$$-d\rho/dt \simeq (2D_v G b^3 c_j/kT)\rho_o^2 \tag{11.11}$$

in which ρ_o is substituted for ρ, as the stress $\sigma \rightarrow o$. Let ρ_i be the immobile dislocation density ($\rho_i < \rho_o$ for low stress) then

$$\dot{\rho}_i \simeq \gamma\rho_i \simeq \gamma(\rho_o - \rho) \tag{11.12}$$

where

$$\gamma = -(2D_v G b^3 c_j/kT)\rho_o \tag{11.13}$$

Whenever an immobilized link is eliminated by recovery, another one can glide. Hence the total rate of change of density of mobile dislocations is $\dot{\rho}_A + \dot{\rho}_i$, or [from Eqs. (11.8) and (11.12)],

$$\lambda\dot{\epsilon}_s\,(\rho_\infty - \rho) + \gamma\,(\rho_o - \rho) \tag{11.14}$$

which may be written as

$$\dot{\bar{\rho}} = \lambda\dot{\epsilon}_s\,(\bar{\rho}_\infty - \bar{\rho}) + \gamma\,(1 - \bar{\rho}) \tag{11.14}$$

where

$$\bar{\rho} = \rho/\rho_o,\ \bar{\rho}_\infty = \rho_\infty/\rho_o \tag{11.15}$$

Now consider the effect of removing or reversing the applied stress. Those links which have been immobilized will move back under the influence of internal stress until they are immobilized once more by other internal stress peaks. The number of links which move on reversing stress will be the number of immobilized links, $\rho_i = (\rho_o - \rho)$, and [from Eq. (11.6)] the rate of their further immobilization will be $i(\rho_o - \rho)$. If the stress is reversed a second time, they will be remobilized once more; thus we see that the density of dislocations which are mobile under tensile stress is increased at a rate of $i(\rho_o - \rho)$ during compression. Simultaneously, recovery and consequent network coarsening will continue, and there will be an increase in the number of mobile dislocations due to this, as given by Eq. (11.12). Hence, the total rate of change of density of tensile-mobile dislocations during compression can be expressed as

$$\dot{\bar{\rho}} = [i(\rho_o - \rho) + \gamma(\rho_o - \rho)]/\rho_o$$

or

$$\dot{\bar{\rho}}\,\mu\,(\lambda\dot{\epsilon}_{sc} + \gamma)\,(1 - \bar{\rho}) \tag{11.16}$$

where $\dot{\epsilon}_{sc}$ is the steady state compressive creep rate under the compressive stress applied. Similarly, we can obtain $\dot{\bar{\rho}}_c$, the rate of change of density of dislocations which are mobile under compressive stress.

When a dislocation is immobilized it is acted upon by internal stress caused by the bowing and elastic displacement of neighboring dislocations in the network. This bowing and elastic displacement can be recovered by the removal of the immobilized dislocation segment either by stress reversal or by recovery. This recoverable strain is time-dependent and must be distinguished from the Hookean elastic strain in the lattice: hence it is *anelastic* strain, taking place *after* stress removal or reversal. Its magnitude can be approximated as $br_s/2$ per dislocation, where r_s is the average spacing between adjacent dislocations, and its sign depends on whether the dislocations were immobilized under the action of tensile or compressive stress. Thus the recoverable or anelastic strain is

$$\xi = br_s\,[(\rho_o - \rho) - (\rho_o - \rho_c)]/2$$

or

$$\xi = b\sqrt{\rho_o}\,(\bar{\rho}_c - \bar{\rho})/2 \tag{11.17}$$

where ρ_c is the mobile dislocation density in compression, and substituting $r_s = \rho_o^{-1/2}$.

We can now write expressions for the total creep rate ($\dot{\epsilon}_t$) and total creep strain (ϵ_t) (excluding the Hookean elastic component). The creep strain rate is the sum of (a) that due to ρ mobile dislocations moving at a velocity $v\,(\dot{\epsilon})$, and (b) that due to the recoil of immobilized dislocations ($\dot{\xi}$), or

$$\dot{\epsilon}_t = \dot{\epsilon} + \dot{\xi} \tag{11.18}$$

Creep strain is similarly

$$\epsilon_t = \epsilon + \xi \tag{11.19}$$

We obtain $\dot{\xi}$ by differentiating Eq. (11.17) and substituting for $\dot{\bar{\rho}}$ from Eqs. (11.14) and (11.15), $\dot{\bar{\rho}}_c$ being found from similar relations for compressive creep. In defining $\dot{\epsilon}$, we assume that the mobile dislocations are sufficiently distant from internal stress peaks that the stress exerted on them by these peaks may be neglected. This is reasonable, as the strain controlled by movements within these peaks (ξ) is estimated separately [*3*]. The glide velocity, v, of dislocations is a function of stress [*3, 15*]; thus for the particular conditions of temperature and stress considered, we may write

$$\dot{\epsilon}/\bar{\rho} = \dot{\epsilon}_s/\bar{\rho}_L \tag{11.20}$$

where $\bar{\rho}_L$ and $\dot{\epsilon}_s$ are respectively the mobile dislocation density and creep rate, as $t \to \infty$.

Consider the situation when a specimen is subjected to a sequence of m time periods in each of which the temperature is T_m and the stress σ_m. $\dot{\epsilon}_{tsm}$ is the steady state tensile plastic strain rate during period m of duration t_m, and $\dot{\epsilon}_{csm}$ is the compressive strain rate during period m of duration C_m. Also, let

$$\left.\begin{array}{l} \dot{\epsilon}_{tsm} \text{ imply } |\dot{\epsilon}_{tsm}| \\ \dot{\epsilon}_{csm} \text{ imply } |\dot{\epsilon}_{csm}| \\ \\ \dot{\epsilon}_{csm} \text{ and } C_m = 0, \text{ if } \sigma_m \geqslant 0 \\ \dot{\epsilon}_{tsm} \text{ and } t_m = 0, \text{ if } \sigma_m \leqslant 0 \end{array}\right\} \quad (11.21)$$

and let

$$\mu_{tm} = \lambda \dot{\epsilon}_{tsm} + \gamma_m \quad (11.22)$$

where γ_m is the value of γ from Eq. (11.13) for the mth period. The limiting value of the mobile dislocation density as $t_m \rightarrow \infty$ is (from Eq. (11.14) with $\dot{\bar{\rho}} = 0$)

$$\bar{\rho}_{tLm} = (\lambda \dot{\epsilon}_{tsm} \bar{\rho}_\infty + \gamma_m)/\mu_{tm} \quad (11.23)$$

By integrating Eq. (11.14) for the mth period of duration t_m, Gittus obtains (see [*21, 22*])

$$\bar{\rho}_{ti(m+1)} = (\bar{\rho}_{tim} - \bar{\rho}_{tLm}) \exp(-\mu_{tm} t_m) + \bar{\rho}_{tLm} \quad (11.24)$$

where $\bar{\rho}_{ti(m+1)}$ is the tensile mobile dislocation density (the subscript t being added to denote its tensile nature) at the end of period m and beginning of period $(m + 1)$.

Be integrating Eq. (11.20) while Eq. (11.24) is used to provide the values of mobile dislocation density from $t_m = 0$ to $t_m = t_m$, the tensile creep strain during the mth period is obtained as

$$\dot{\epsilon}_{tm} = \dot{\epsilon}_{tsm} \{[(\bar{\rho}_{tim} - \bar{\rho}_{ti(m+1)})/\mu_{tm} \bar{\rho}_{tLm}] + t_m\} \quad (11.25)$$

The effect of compressive straining on increasing the tensile mobility of immobilized dislocations is obtained by integrating Eq. (11.16) to give

$$\bar{\rho}_{ti(m+1)} = (\bar{\rho}_{tim} - 1)\exp(-\mu_{cm} t_m) + 1 \quad (11.26)$$

where $\mu_{cm} = \lambda \dot{\epsilon}_{csm} + \gamma_m$, analogous to Eq. (11.22).

For compressive creep deformation, and the effects on this of tensile strain, similar equations to Eqs. (11.24), (11.25) and (11.26) can be written [*3*].

The creep strain is then obtained from Eq. (11.19) with ϵ defined by successive applications of Eq. (11.25) and the equivalent expressions for ϵ_{cm} and ξ obtained from Eq. (11.17).

Gittus [*3, 24, 27*] has applied the foregoing analysis and predictive method to cover primary and secondary creep under constant stress; the effects of stress reduction, removal and reversal (where the Bauschinger effect is found); and to creep under multiaxial stress. It appears that the approach gives predictions which correspond closely with experimental observations, although the model is sensitive to the correct choice of the various constants (λ, γ, etc.) obtained by fitting the creep equations to data from creep tests. In the case of a stress reduction during primary creep, the model correctly predicts that the creep rate will decrease when the reduction is less than some critical vaue, while creep will be reversed (reduction in length occurring) when the reduction is greater than this critical amount. With stress reduced by the critical amount, so that the applied stress and internal stress momentarily balance, the creep rate will fall to zero immediately after the stress reduction.

Transient creep under constant stress is treated as follows: the plastic strain is obtained by substituting Eq. (11.24) in Eq. (11.25) to give

$$\epsilon = (\dot{\epsilon}_s/\mu)\,[(1 - \bar{\rho}_L)/\bar{\rho}_L]\,[1 - \exp(-\mu t)] + \dot{\epsilon}_s t \tag{11.27}$$

Similarly, the anelastic strain is obtained by substituting for $\bar{\rho}$ in Eq. (11.17) with $\bar{\rho}_c = 1$, as

$$\xi = l_a\,(1 - \bar{\rho}_L)\,[1 - \exp(-\mu t)] \tag{11.28}$$

where $l_a = b\sqrt{\rho_o}/2$. Hence, from Eq. (11.19)

$$\epsilon_t = (1 - \bar{\rho}_L)\,[(\dot{\epsilon}_s/\mu\bar{\rho}_L) + l_a]\,[1 - \exp(-\mu t)] + \dot{\epsilon}_s t$$

or

$$\epsilon_t = \epsilon_p\,[1 - \exp(-\mu t)] + \dot{\epsilon}_s t \tag{11.29}$$

where ϵ_p represents the total primary creep strain when secondary creep commences, defined by

$$\epsilon_p = (1 - \bar{\rho}_L)\,[(\dot{\epsilon}_s/\mu\bar{\rho}_L) + l_a] \tag{11.30}$$

Eq. (11.29) is well-supported by experimental evidence, and an empirical equation having this form was proposed by McVetty [*28*] in 1934, although it is more commonly referred to as the Garofalo equation [*4*]. Equations of similar form can be derived from fundamental approaches other than that considered here: for example, from Li's dislocation model for mobility creep [*29*], or from Morrison's approach which concerns the time of flight of dislocations and their waiting time for remobilization [*30, 31*]. Further discussion of strain-time relations is given in Section 11.3.

In examining the modeling of steady state creep, we shall consider the process of viscous glide of dislocations rather than climb rate as being the controlling factor. We consider there to be a friction stress which the dislocation must overcome to move through the lattice, this drag being caused by solute atoms or second phase particles, and of value F. The stress required to move the dislocation is

$$\sigma - F = Gb/2\pi r_s \tag{11.31}$$

and the steady state creep rate is given by

$$\dot{\epsilon}_s \simeq bv/r_s^2 \tag{11.32}$$

where v is the velocity of the mobile dislocations.

The change in dislocation mesh size is given by

$$dr_s = (\partial r_s/\partial t)\,dt + (\partial r_s/\partial \epsilon)d\epsilon \qquad (11.33)$$

and for $\dot{\epsilon} = \dot{\epsilon}_s$, $dr_s/dt = 0$. Thus

$$\dot{\epsilon}_s = -(\partial r_s/\partial t)/(\partial r_s/\partial \epsilon) \qquad (11.34)$$

From the strain given by,

$$\epsilon = \rho b r_s, \qquad (11.35)$$

$$\partial r_s/\partial \epsilon = -r_s^2/2b \qquad (11.36)$$

Hence, inserting Eqs. (11.9) and (11.36) in Eq. (11.34) we obtain

$$\dot{\epsilon}_s = (2D_V G b^4 c_j/kT)/r_s^3 \qquad (11.37)$$

From Eq. (11.31)

$$1/r_s^3 = (8\pi^3/G^3b^3)\,(\sigma - F)^3 \qquad (11.38)$$

and inserting this in Eq. (11.37), we have

$$\dot{\epsilon}_s = (16\pi^3 D_v b c_j/G^2kT)\,(\sigma - F)^3 \qquad (11.39)$$

The value of F will be given by ([*3*], p. 88)

$$F = Av/b \qquad (11.40)$$

where A is a temperature dependent constant whose value depends on the particular mechanism controlling the dislocation motion. Eq. (11.39) may be rearranged to

$$\dot{\epsilon}_s = (16\pi^3 c_j D_v G b/kT)/[(\sigma - F)/G]^3 \qquad (11.41)$$

and in the absence of a friction stress,

$$\dot{\epsilon}_s = (16\pi^3 c_j D_v G b/kT)\,(\sigma/G)^3 \qquad (11.42)$$

Other dislocation theories for steady state creep involving the climb of edge dislocations or the movement of jogged screw dislocations give rise to equations of a similar form to Eq. (11.42), with stress exponents ranging from 3 to 6. Thus, we may write the general equation for steady state dislocation creep as

$$\dot{\epsilon}_s = B\,(D_v G b/kT)\,(\sigma/G)^n \qquad (11.43)$$

where B and n are material constants. Since the diffusion coefficient, D_v, can be described by

$$D_v = D_o \exp(-Q/RT)$$

where Q is the activation energy for self-diffusion, we can rewrite Eq. (11.43) as

$$\dot{\epsilon}_s = B'\sigma^n \exp(-Q/RT) \qquad (11.44)$$

where B' is a constant which may be structure dependent.

In practice, values of n of from 4 to 7 have been reported for many engineering alloys, rising to as high as 40 for some precipitation or dispersion hardened alloys, accompanied by activation energies greatly exceeding those for self-diffusion.

Although complex models have been postulated to justify high values of n and of apparent activation energy for creep, it has been demonstrated that introduction of the concept of a friction stress or back stress, as used in developing Eq. (11.39), allows the exponent to be reduced to the order of that in the theoretical treatments (3 to 4) with concomitant decrease of the apparent activation energy to approximate that for self-diffusion [*32-35*]. There is, however, a lack of agreement on the methods for measuring back stress or friction stress, and relatively few reliable measurements have been made. Thus, considerable development is to be expected in this area.

11.2.2 Diffusion Creep

In transcrystalline diffusion creep of the Nabarro-Herring type [*7, 8*], vacancies are considered to move as shown in Fig. 11.4 from tensile-stressed to compression boundaries. In the absence of applied stress, the probability of a vacancy being present is given by Eq. (4.22) as

$$c_{eq} = n/N \simeq \exp(-E_f/kT)$$

With stress, σ, present on the boundary, the formation of a vacancy is aided as the work to form it will be reduced by $\Omega\sigma$ where Ω is the volume of a vacancy, and the probability of its lying at a grain boundary will be

$$(n/N)_\sigma \simeq \exp(-E_f/kT)\exp(\Omega\sigma/kT) = c_{eq}\exp(\Omega\sigma/kT) \qquad (11.45)$$

Consider a single crystal stressed as in Fig. 11.4. The vacancy concentrations on faces AB and AD will be respectively raised and lowered to $c_{eq}\exp(\Omega\sigma/kT)$ and $c_{eq}\exp(-\Omega\sigma/kT)$. There will be a concentration gradient

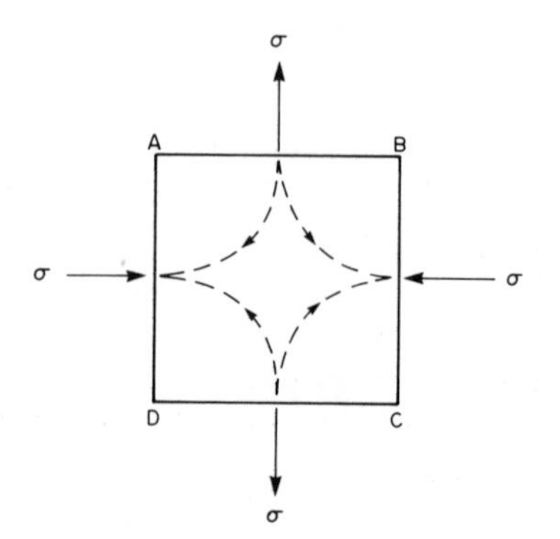

Fig. 11.4 Schematic illustration of Nabarro-Herring creep with vacancy flow paths indicated

which, over a path of mean length L between AB and AD is given approximately by

$$(1/L\Omega)\ c_{eq}\sinh(\Omega\sigma/kT)$$

Assuming concentrations to be constant at the boundaries, the net flux of vacancies per unit area, J_v, is

$$J_v = -D\partial c/\partial x \simeq (2Dc_{eq}/L\Omega)\ (\Omega\sigma/kT) \tag{11.46}$$

where D is the vacancy diffusion coefficient, and taking sinh as equal to its argument.

The face AB of the crystal will move by $J_v\Omega$ in unit time. Hence, taking $\Omega \simeq b^3$, noting that L is directly related to d, the grain size, substituting D_v, the self-diffusion coefficient, = Dc_{eq}, we may define the vacancy creep rate as

$$\dot{\epsilon} = A_1\ (D_vGb/kT)\ (b/d)^2\ (\sigma/G) \tag{11.47}$$

where A_1 is a constant.

A detailed analysis of diffusion creep in polycrystals has been made by Raj and Ashby [*36*], considering simultaneous grain boundary sliding. They examined both Nabarro-Herring and Coble [*9*] creep, in the second of which grain boundary diffusion dominates. Setting the creep equation in the form

$$\sigma = \eta\dot{\epsilon} \tag{11.48}$$

in accordance with Eq. (11.47), and considering η as a Newtonian viscosity term, Raj and Ashby obtained values as follows:

$$\begin{aligned} &\text{Nabarro-Herring creep:} \quad && \eta_v = (1/42)\ (d^2kT/D_v\Omega) \\ &\text{Coble creep:} \quad && \eta_b = (1/132)\ (d^3kT/\delta D_b\Omega) \end{aligned} \tag{11.49}$$

where D_B is the grain boundary diffusion coefficient and δ is the effective width of the grain boundary. Hence, Coble creep can be represented by an equation similar to Eq. (11.47) as

$$\dot{\epsilon} = A_2\ (D_BGb/kT)\ (b/d)^3\ (\sigma/G) \tag{11.50}$$

Diffusion creep becomes the controlling process at high temperatures and relatively low stresses, Coble creep being dominant for small grains and Nabarro-Herring for larger grain sizes.

A third form of diffusion creep can occur at high temperature under low stress where sinks and sources for vacancies operate *within* grains. The mechanism is termed Harper-Dorn diffusion creep [*37*], and can occur where the grain size is large, with subgrain boundaries and climbing dislocations acting as the vacancy sources and sinks. The diffusion path (L) becomes the spacing between dislocations, and the Nabarro-Herring equation [Eq. (11.47)] is rewritten as

$$\dot{\epsilon} = A_3\ (D_vGb/kT)\ (\sigma/G) \tag{11.51}$$

In all three types of diffusion creep, represented by Eqs. (11.47), (11.50) and (11.51), the direct dependence of $\dot{\epsilon}$ on σ should be noted.

11.2.3 Grain Boundary Sliding

Ignoring, for the moment, the need for intragranular deformation to provide accommodation when grains in polycrystalline material slide over one another, and considering sliding only, as can occur in a bicrystal, we can derive an expression for grain boundary sliding.

The process can be regarded as Newtonian viscous in nature, and the relevant diffusion coefficient will be that for the grain boundaries, D_B. The rate of sliding will be proportional to the number of boundaries present, i.e., to the inverse of grain size. Thus the rate of grain boundary sliding may be written as

$$\dot{\epsilon}_{gbs} = A_4 (D_B G b / kT)(b/d)(\sigma/G) \tag{11.52}$$

where A_4 is a constant.

Detailed consideration of the process of grain boundary sliding in conjunction with the diffusional flow of atoms, as occurs in superplastic deformation, has been given by Ashby and Verrall [*38*]. However, in the more common situation existing during creep it appears that accommodation will take place by means of dislocation climb. The rate of accommodation may be written from Eq. (11.42) as

$$\dot{\epsilon}_{acc} = A_5 (D_v G b / kT)(b/d)(\sigma/G)^n \tag{11.53}$$

and they may be considered as sequential processes, applying Eq. (11.3).

While grain boundary sliding plays a secondary role during steady state creep under most conditions in engineering alloys, it plays an important role in leading to fracture in tertiary creep, where accommodation does not take place fully and intercrystalline cracks develop. Thus, for conditions where grain bounday sliding plays an important part as a sequential process during secondary creep, early development of intercrystalline cracks and premature failure may result. This matter has not been studied to any great extent, and deserves greater attention.

11.2.4 Deformation Maps

Deformation maps provide a concise means of displaying the *dominant* deformation mechanisms as a function of grain size, temperature and stress. All mechanisms will operate to a greater or lesser extent, but generally one outweighs the other, and creep deformation can be predicted satisfactorily using the constitutive equation appropriate to it alone.

Early work on the subject was done by Weertman [*39*] but much of the development and application of deformation mapping is due to Ashby [*40*], while an alternate method of presentation has been developed by Mohamed and Langdon [*41*]. Here, we shall discuss briefly the formulation of such maps, giving representative examples and noting some applications. A detailed discussion of Ashby-type maps, together with their extension to irradiation creep has been given by Gittus [3].

The different deformation mechanisms which may apply have been discussed in the preceding sections, and the governing constitutive equations will be summarized here, as follows:

Defectless flow, involving glide on slip planes without the aid of dislocations — a limiting case. Creep rate is

$$\dot{\epsilon} = \infty; \quad \sigma/G \geqslant \sigma_{max}/G$$
$$\dot{\epsilon} = 0; \quad \sigma/G < \sigma_{max}/G \qquad (11.54)$$

where σ_{max} is the tensile stress corresponding to the theoretical stress for slip [Eq. (4.4)].

Dislocation glide, involving dislocations overcoming obstacles by thermal activation. This will occur at stresses greater than some cut-off or limiting value, σ_o, and for FCC and some HCP metals can be represented by Eq. (11.1) as

$$\dot{\epsilon} = \dot{\epsilon}_o \exp\left[-ba_a (Gb/L - \sigma)/kT\right]; \sigma \geqslant \sigma_o$$
$$\dot{\epsilon} = 0; \quad \sigma < \sigma_o \qquad (11.55)$$

For BCC metals the process is more complex and will not be considered further here.

Dislocation creep, involving the climb and glide of dislocations by means of stress-assisted vacancy movement. The creep rate is given by Eq. (11.43) as

$$\dot{\epsilon} = B\,(D_v Gb/kT)\,(\sigma/G)^n$$

Diffusion creep. From Eqs. (11.47), (11.50) and (11.51), we have

Nabarro-Herring: $\dot{\epsilon} = A_1\,(D_v Gb/kT)\,(b/d)^2\,(\sigma/G)$

Coble: $\dot{\epsilon} = A_2\,(D_B Gb/kT)\,(b/d)^3\,(\sigma/G)$

Harper-Dorn: $\dot{\epsilon} = A_3\,(D_v Gb/kT)\,(\sigma/G)$

Grain boundary sliding. Eqs. (11.52) and (11.53) are substituted in Eq. (11.3) to evaluate its operation as a sequential process, giving

$$\dot{\epsilon} = \dot{\epsilon}_{gbs}\,\dot{\epsilon}_{acc}/(\dot{\epsilon}_{gbs} - \dot{\epsilon}_{acc}) \qquad (11.56)$$

The dominant mechanism is found by solving the various equations for specified conditions of grain size, stress and temperature, using known values of the constants. In an Ashby-type map, as shown in Fig. 11.5, information for a specific grain size is plotted with coordinate axes of normalized stress, (σ/G), and homologous temperature, and the boundaries between the various regions indicate the conditions under which two mechanisms give equal values

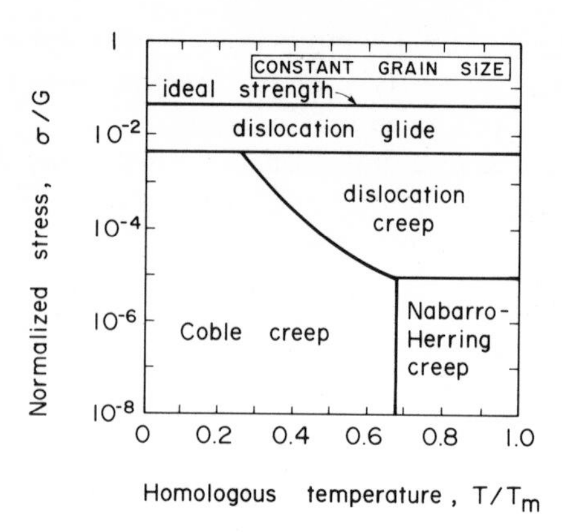

Fig. 11.5 Ashby-type deformation mechanism map

of secondary creep rate. The boundary between elastic and creep regimes is normally based on a creep rate of 10^{-8}/s.

In Mohamed and Langdon's procedure [*41*], the axes are normalized grain size, (d/b), and normalized stress, and the plot is for a specified temperature as shown in Fig. 11.6. The inclusion of grain boundary sliding, considered as a sequential process, is shown in Fig. 11.7 [*11*]. In the latter two figures, two lines of constant strain rate are shown, 10^{-9}/s and 10^{-10}/s, the first representing approximately the lowest practicable laboratory creep rate, while the second represents the lower limiting rate for many practical and design purposes. It may be seen that the diagrams of Mohamed and Langdon are somewhat simpler to compute and plot than are those of the Ashby type as all lines are straight, except where grain boundary sliding is included.

It is a simple matter to identify the controlling mechanism of deformation for a material under specific conditions using the appropriate deformation map, provided the constitutive equations (and the constants within them) have been determined accurately. A knowledge of the dominant deformation mechanism is useful in choosing a strengthening process or mechanism which might be applied to improve the material's performance.

Another area of application lies in the use of extrapolation procedures, as a particular extrapolation method should be used preferably over a range of conditions where the same mechanism is dominant (see Section 11.5). For example, using accelerated creep tests and the Sherby-Dorn procedure for extrapolation to service conditions [*42*] involves testing at the required stress and temperature (σ_1 and T_1, respectively) to produce a partial creep curve in a time of $\sim 10^3$h, and subsequently testing another specimen at σ_1 but at higher temperature, T_2, chosen so that rupture occurs in a time of $\geqslant 10^3$h. Writing

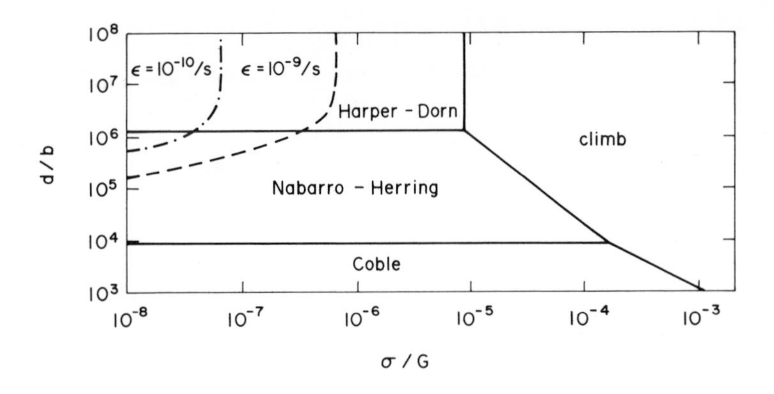

Fig. 11.6 Modified deformation mechanism map for pure aluminum at $0.5T_m$ [11]

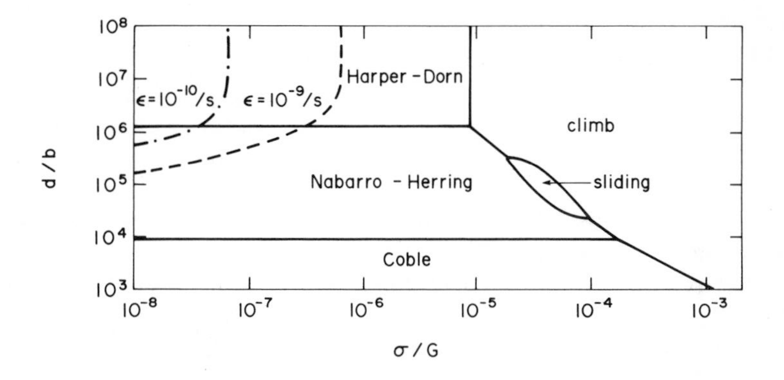

Fig. 11.7 Grain boundary sliding, treated as a sequential process and included in the deformation mechanism map for pure aluminum at $0.5T_m$ [11]

$$t_1\exp(-Q_c/RT_1) = t_2\exp(-Q_c/RT_2) \tag{11.57}$$

where t_1 and t_2 are the times to reach a specified strain at T_1 and T_2, respectively, Q_c may be found. Subsequently, a plot can be drawn of strain versus $\theta = t\exp(-Q_c/RT)$, the temperature-corrected time, and strain can be estimated at T_1 for long times. The procedure is valid only if the controlling mechanisms are the same in both tests, and deformation maps provide a simple means of checking this, and a rational basis for the choice of test conditions.

11.3 LAWS OF CREEP DEFORMATION

In discussing the constitutive equations governing creep behavior in the preceding section, we were concerned primarily with the controlling mechanisms and the development of relations based on these. While this is pedagogically the correct approach, it does not satisfy the needs of the designer for simple, if empirical, equations which can be used to predict creep deformation or undertake stress analysis in a structure or component undergoing creep, and we shall now consider some such empirical relations. These should, however, be correlated with physical mechanisms in an attempt to ensure that predictive information derived from them can be treated with confidence, and one more detailed phenomenological model which does depend directly on mechanisms will be reviewed.

Owing to the complexity of the processes involved, it appears unlikely that we shall be able to avoid empiricism and depend entirely on physically-based relations for a considerable time to come. Not only must the equations be reasonably simple to understand, but they must also be amenable to manipulation to cover effects of changes in load and temperature and the resulting strain transients.

11.3.1 Time, Temperature and Stress Functions

For tensile creep deformation under constant stress, the general equation can be written as

$$\epsilon = f(t, T, \sigma) \tag{11.58}$$

and this can be simplified by separating the functions for time, temperature and stress dependence, to give

$$\epsilon = f_1(t) f_2(T) f_3(\sigma) \tag{11.59}$$

Separation in this manner has been found to be generally satisfactory for evaluation of creep strain in many components.

In his early quantitative studies of creep, Andrade [*43, 44*] showed that deformation could be represented by

$$l = l_o(1 + \beta t^{1/3}) \exp(kt) \tag{11.60}$$

for a number of alloys, where l is length at time t, l_o is the original gage length, and β and k are constants depending on stress and temperature. Many other time functions have been proposed, and they are reviewed by Kennedy [*45*]. Representative ones are as follows:

$\epsilon = a(1 + \beta t^{1/3}) \exp(kt)$	Andrade [*43, 44*]	(11.61)
$\epsilon = a + bt^n$	Bailey [*46*]	(11.62)
$\epsilon = a + bt + c \exp(-dt)$	McVetty [*28*]	(11.63)
$\epsilon = at^m + bt^n + ct^p + \ldots.$	Graham [*47*]	(11.64)
$\epsilon = a \log t + bt^n + ct$	Wyatt [*48*]	(11.65)

where $a, b, c, d, m, n, p, \beta$ and k are constants. Obviously, all such relations cannot be correct, although they have been observed to apply over specific ranges of temperature and stress for particular alloys.

Where secondary creep occurs, generally at temperatures greater than $\sim 0.4T_m$, it is necessary that a linear term be included in the deformation-time equation adopted, if it is not already present. For many metals and alloys, a particularly useful relation is given by the equation of Garofalo [*4*] which is developed from McVetty's earlier work [*28*]. In this, the total strain, ϵ_t, is written as the sum of initial strain on loading, ϵ_0, transient and steady state terms, in the form

$$\epsilon_t = \epsilon_0 + \epsilon_p [1 - \exp(-mt)] + \dot{\epsilon}_s t \tag{11.66}$$

where ϵ_p is the limiting transient creep strain and m is a material constant. This equation can also be derived from various dislocation theories of creep [see Section 11.2.1; and cf. Eq. (11.29)].

The effect of temperature on creep rate is complex. At low temperatures, $T < 0.3T_m$, logarithmic creep of the form, $\epsilon = \alpha \log t + C$, where α and C are constants, takes place. At higher temperatures diffusion processes become more important and recovery mechanisms operate more easily, and exponential functions involving temperature and an activation energy (generally close to that for self-diffusion) have particular significance, as discussed in Section 11.2.1. Dorn [*49*] suggested that temperature dependence should be expressed in the form, $\exp(-Q_c/RT)$, Q_c being the activation energy for the creep process. Because of the interaction between time and temperature in rate processes, Dorn proposed the use of the parameter $t\exp(-Q_c/kT)$, and showed that creep data from tests made at different temperatures could be plotted as a single line on the basis of ϵ versus $t\exp(-Q_c/kT)$. Thus, a convenient, though approximate, relation for creep strain at high temperature where diffusion processes are controlling, and at specified stress, takes the form

$$\epsilon = a[t\exp(-Q_c/kT)]^m \tag{11.67}$$

Many expressions have been proposed for the stress dependence of creep strain. Generally, they are expressed in terms of creep rate, and the following are representative:

$$\dot{\epsilon} = B\sigma^n \quad \text{Norton [50]} \tag{11.68}$$

$$\dot{\epsilon} = C\sinh(b\sigma) \quad \text{McVetty [51]} \tag{11.69}$$

$$\dot{\epsilon} = D\exp[(b\sigma) - 1] \quad \text{Soderberg [52]} \tag{11.70}$$

In many cases, Norton's law [Eq. (11.68)] provides a satisfactory fit to experimental data, and it is the simplest equation to use. While most data have been correlated for steady state creep, the relations can also be applied to transient creep rates. More detailed discussion of stress dependence is given by

Garofalo [4].

Hence, a general expression for creep deformation may be written as

$$\epsilon = A'\sigma^n [t\exp(-Q_c/kT)]^m \tag{11.71}$$

where A' is a material constant dependent on temperature. For constant temperature, Eq. (11.71) may be expressed as

$$\epsilon = A\sigma^n t^m \tag{11.72}$$

The creep rate is

$$\dot{\epsilon} = mA\sigma^n t^{(m-1)} \tag{11.73}$$

when Eq. (11.72) is differentiated directly with respect to time. Alternatively, taking the mth root of both sides of Eq. (11.72) and differentiating, we obtain

$$\dot{\epsilon} = m\mathrm{A}^{1/m}\sigma^{n/m}\epsilon^{(m-1)/m} \tag{11.74}$$

Eqs. (11.73) and (11.74) respectively represent *time hardening* and *strain hardening* behavior. The former assumes creep rate to be a function of stress and time only, while the latter assumes it to be a function of stress and the current strain. Neither makes allowance for the path by which the current state was reached. For constant stress both give identical results, but in cases of varying stress the strain hardening equation gives markedly better results. Unfortunately, however, it is more difficult to handle analytically, thus the time hardening relation is still used in practice on many occasions, although there is less excuse for this with widespread use of computers in stress and strain analysis.

A matter which has been ignored so far in the discussion of empirical creep equations is anelastic strain which may be recovered on unloading or reducing the stress level. Inclusion of consideration of such effects can be important in situations where stress and temperature change during service operation, and a brief discussion will now be given of *creep recovery*, as such effects are generally termed, and inclusion of it in quasi-empirical creep equations.

11.3.2 Inclusion of Creep Recovery or Anelastic Deformation

Creep recovery on unloading was illustrated in Fig. 11.2. As it is an anelastic effect, it can be fitted by a function of the type

$$\xi_R = \sum_i A_i [1 - \exp(-t/\lambda_i)] \tag{11.75}$$

where ξ_R is the recovered anelastic strain, A is a constant, t is time and λ_i represent constants from the relaxation spectrum (see Chapter 5). Thus, creep recovery can be modelled using spring-dashpot models with a wide range of relaxation times.

The extent to which transient creep can be recovered on unloading depends on stress, temperature and time: at low stresses, essentially all transient creep may be recovered; at high stresses virtually none may be, with the transition depending on temperature and the extent of prior deformation. Detailed discussion of what they termed "anelastic creep" is given by Lubahn and Felgar [53], while Goldhoff [54] has discussed its engineering signficance, particularly relating to heat resistant steels.

Several attempts have been made to include anelastic effects in creep strain-time relations. For example, Pao and Marin [55] summed the strains resulting from a Maxwell element, and a (reversible) Voigt or Kelvin element. However, this considers all the transient component of creep strain to be recovered on unloading, which is not consistent with observations.

Recently, Miller [56] has developed a model which is phenomenological in nature but based on physical mechanisms. The model is general, covering hot working, cyclic loading, relaxation etc., and while further developments along the same general lines may be expected, it is considered appropriate to review the model and relations briefly here.

The basic premises are as follows:

- The nonelastic strain rate is given by

$$\dot{\epsilon} = f\,[(\sigma - R)/D] \tag{11.76}$$

where σ is applied stress, R is a rest stress related to kinematic or directional hardening, and D is a characteristic drag stress (the isotropic hardening variable).

- The steady state creep rate, $\dot{\epsilon}_s$, is determined by

$$\dot{\epsilon}_s = B'\,[\sinh(A\sigma)]^n \tag{11.77}$$

where B' depends on temperature and A does not.

- Temperature dependence is determined by the factor θ', defined by

$$B\theta' = B' \tag{11.78}$$

where B is a constant, and θ' is given by

$$\theta' = \exp(-Q/0.6kT_m)\,[\ln(0.6T_m/T) + 1], \quad T \leqslant 0.6\,T_m \tag{11.79a}$$

and

$$\theta' = \exp\,(-Q/kT), \quad T \geqslant 0.6\,T_m \tag{11.79b}$$

where Q is the activation energy for plastic flow at high temperature.

From consideration of warm working and steady state creep deformation, the following governing equations are derived:

$$\dot{\epsilon} = B\theta'\,[\sinh\,(|\sigma - R|/D)^{1.5}]^n\,\mathrm{sgn}\,(\sigma - R) \tag{11.80a}$$

$$\dot{R} = H_1\dot{\epsilon} - H_1B\theta'\,[\sinh\,(A_1|R|)^n\,\mathrm{sgn}\,(R) \tag{11.80b}$$

$$\dot{D} = H_2\,|\dot{\epsilon}|\,[C_2 + |R| - (A_2/A_1)D^3] - H_2C_2B\theta'\,[\sinh\,(A_2D^3)]^n \tag{11.80c}$$

where A_1, A_2, C_2, H_1 and H_2 are all material constants.

Miller [56] discusses in some detail the determination of the various constants from standard test data, and demonstrates the applicability of the model to Type 304 stainless steel. Certainly, the model does appear to cover a very wide range of conditions and the fact that it does take anelastic effects into account should facilitate its applicability to situations where stress changes take place. It is also consistent with mechanistic models of dislocation creep such as that of Gittus discussed in Section 11.2.1.

11.3.3 Stress Relaxation

When total strain is maintained constant under conditions of constant elevated temperature, as in the case of a bolt in a bolted joint, the stress will decrease with time. The phenomenon is known as *stress relaxation*, and involves the conversion of initial elastic strain into identical plastic strain as the stress decreases. At first sight it appears that prediction of stress relaxation could be made on the basis of creep data, but this is complicated because of the anelastic component of strain. Thus, stress decrease in a specimen held at constant total strain results from both plastic and anelastic strains.

Prestrain to a higher stress than the bolting stress can be used to counteract and delay the normal decrease of stress by relaxation. In such a situation, the load *decrement* causes anelastic strain to *decrease* with time, causing an increase in the elastic strain and an increase in stress with time. Thus, after prestraining and subsequent loading to fixed total strain, stress can increase at first if the anelastic decrement predominates, followed by stress decrease at a later time provided the residual stress is sufficiently large to cause significant creep strain. This has been demonstrated experimentally [*54, 57*].

During stress relaxation, the strain at any time is constant, and the sum of the elastic and creep components of strain (ϵ_e and ϵ_c, respectively) will equal the initial elastic strain, ϵ_{eo}, neglecting anelastic effects. Thus, $\epsilon_{eo} = \epsilon_e + \epsilon_c$. Differentiating,

$$(1/E)\, d\sigma/dt + \dot{\epsilon}_c = 0 \qquad (11.81)$$

Rigorous analysis of stress relaxation phenomena is not simple, but a first approximation can be obtained by relating creep rate to stress alone as in Eq. (11.68). Substituting for $\dot{\epsilon}_c$ and integrating, the time for the stress to relax from σ_o to σ is given by

$$t = [1/BE\,(n-1)]\,[(1/\sigma^{n-1}) - (1/\sigma_o^{\,n-1})] \qquad (11.82)$$

Serious overestimation of the stress remaining at shorter relaxation times is given by Eq. (11.82) because of the consideration of steady state creep only, and better predictions can be made by considering time hardening or strain hardening empirical relations for the particular material of interest, i.e., using relations such as Eqs. (11.73) or (11.74) for $\dot{\epsilon}_c$ and integrating.

It should be noted that stress relaxation data and relations can be applied to the relief of residual stresses in components which have been welded, cast or forged.

11.4 TERTIARY CREEP AND FRACTURE

11.4.1 Tertiary Creep

Tertiary creep is associated either with local reduction of cross-sectional area through necking or, more generally, with the nucleation and growth of grain boundary voids. However, such voids, which may be of either the *w* or *r*-type as discussed in Section 8.1, have also been observed to have initiated during secondary creep [*58*], although their major growth occurs during the tertiary stage. Hence, we may consider tertiary creep as involving the processes of fracture which lead, eventually, to final failure.

The extent to which grain boundary voids occur is dependent on stress and temperature. In general, they form under conditions of lower stress and higher temperature, with *r*-type voids occurring at the higher end of the temperature scale. Thus, longer lives at high temperatures have a tendency to lead to intercrystalline fracture, with transcrystalline fracture being observed under conditions of higher stress and lower temperature. In the past this led to the concept of an "equicohesive temperature", below which fracture was intragranular and above which it was intercrystalline [*59*]: whilst the concept is a gross simplification, it does illustrate the broad differences between high and lower temperature behavior. Figure 11.8 shows the idea of an equicohesive temperature, its value being increased with increase in strain rate: this may be considered as a forerunner of the fracture mechanism map for creep discussed in Section 11.4.2.

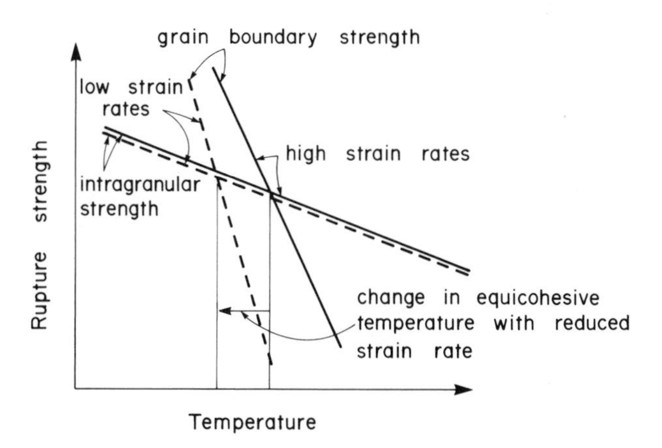

Fig. 11.8 Concept of the equicohesive temperature showing effect of strain rate on fracture mode

11.4.2 Fracture Mechanisms

If we examine the creep fracture processes in more detail, we find that they may be classified as follows: transgranular fracture; intergranular fracture; and high temperature rupture. These will be discussed briefly, in turn, following the treatment of Miller and Langdon [*60*].

Transgranular creep fracture is similar to low temperature ductile fracture, involving the nucleation of holes at inclusions or second phase particles in the matrix, with subsequent growth until coalescence and fracture result. For all deformation occurring under steady state conditions, the time to fracture is

$$t_r = \epsilon_f / \dot{\epsilon}_s \tag{11.83}$$

The total elongation at fracture, ϵ_f, is composed of nucleation and growth strains, ϵ_n and ϵ_g, respectively. Ashby [*61*] showed that

$$\epsilon_g = (1/1.8)\,[n/(n-1)]\,\ln\,(0.7 f_v^{-1/2} - 1) \tag{11.84}$$

where f_v is the volume fraction of inclusions and n is the strain rate exponent. Hence,

$$t_r = \{\epsilon_n + (1/1.8)\,[n/(n-1)]\,\ln\,(0.7 f_v^{-1/2} - 1)\}\,(\dot{\epsilon}_s)^{-1} \tag{11.85}$$

assuming ϵ_n is constant and ignoring the possible effect of external necking.

In intergranular creep failure, we must consider both *w* and *r*-type voids as noted above. Considering the former first, Fig. 11.9 shows a wedge crack under tensile stress with grain boundary sliding occurring. The rate of change of wedge height with time is given approximately by

$$dh/dt \simeq \dot{\epsilon}_{gbs} d \tag{11.86}$$

Crack instability occurs when

$$h\sigma \geqslant 2\gamma \tag{11.87}$$

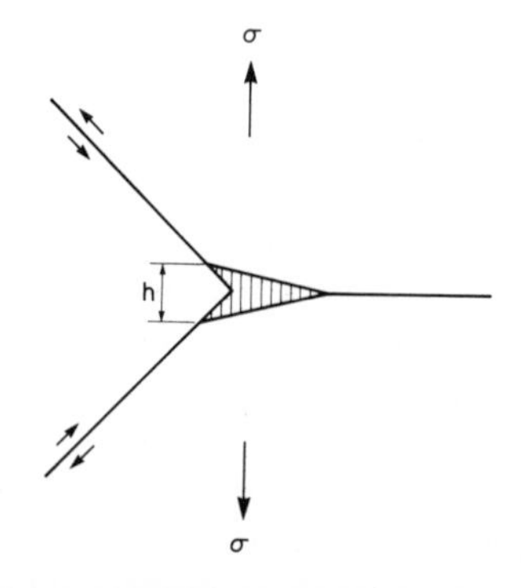

Fig. 11.9 Wedge crack growing under tensile stress

where γ is the effective surface energy of fracture. Thus,

$$t_r = (2\gamma/\sigma d)\,(\dot{\epsilon}_{gbs})^{-1}$$

or

$$t_r = (2\gamma/Gd\zeta)\,(\sigma/G)^{-1}(\dot{\epsilon}_s)^{-1} \tag{11.88}$$

where $\zeta = \dot{\epsilon}_{gbs}/\dot{\epsilon}_s$, the contribution of grain boundary sliding to total strain.

The nucleation and growth of *r*-type cavities on grain boundaries have been the subject of considerable recent study. However, although not entirely understood, it is thought that nucleation arises from stress concentrations produced during grain boundary sliding. Making the assumption that this is a continuous process during creep deformation, growth may be considered either from diffusional flow of vacancies or by power law dislocation creep in the matrix.

Vacancies can be considered to move along the grain boundary into the cavity, which is equivalent to the grains being pushed apart by means of the return flow of atoms. For the number of cavities per unit grain boundary area constant, the corrected model of Raj and Ashby [*62, 63*] predicts the time to fracture as

$$t_r = 6 \times 10^{-3} kTl^3/\delta D_B \sigma \Omega \tag{11.89}$$

where l is the average void spacing on the boundary, and the other terms are as defined in Section 11.2.2. Taking $\Omega = b^3$,

$$t_r = (6 \times 10^{-3} l^3/\delta D_B b^2)\,(kT/Gb)\,(\sigma/G)^{-1} \tag{11.90}$$

Where power law creep controls the growth of holes, Raj [*64, 65*] developed an analysis leading to

$$t_r = [(1 - A_o)/1.48]\,(\dot{\epsilon}_s)^{-1} \tag{11.91}$$

where A_o is the initial area fraction of cavities on the boundary. If the pores form at second phase particles, A_o is equivalent to the area fraction of these on the boundary, or

$$A_o = (\pi/4)\,(p_o/l_o)$$

where p_o and l_o are respectively average particle diameter and spacing on the boundary.

Creep rate in Eqs. (11.85), (11.88) or (11.91) is given by Eq. (11.43) at high temperatures but, where diffusion is along dislocation cores at low temperatures, this is modified to some extent [*60*].

The concept of a fracture map, displaying conditions where a particular fracture mode existed, was introduced by Wray [*66*] and developed by Ashby and Raj [*67*]. Making the simplifying assumption that the fracture mechanisms are independent, t_r can be determined for the various fracture processes and the operative one is taken to be that leading to the lowest value of t_r. Thus, Fig. 11.10 shows a fracture map for Type 316 stainless steel of 150

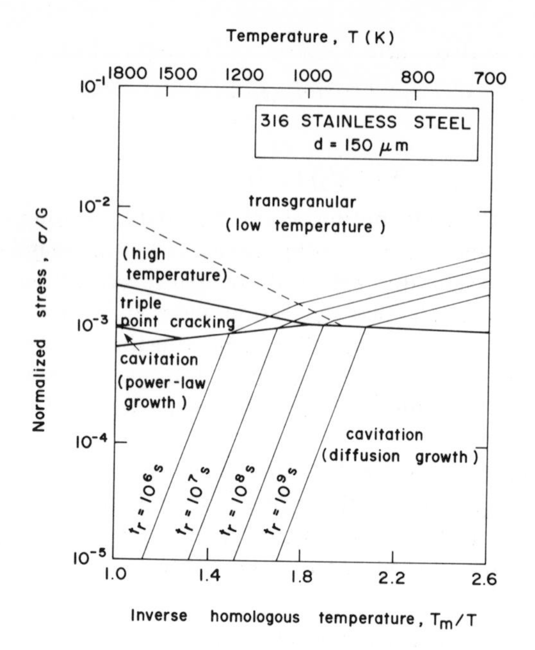

Fig. 11.10 Fracture map for Type 316 stainless steel (After Miller and Langdon [*60*])

μm grain size plotted on the basis of inverse homologous temperature, rather than homologous temperature, as this simplifies its construction [*60*]. Lines of constant time to rupture are superimposed, and the dotted line shows the distinction between high and low temperature power law creep.

As with deformation mechanism maps, fracture maps indicate possible dangers in extrapolation procedures which involve crossing field boundaries, and suggest the limits which may be imposed on extrapolation parameters.

11.5 EXTRAPOLATION OF CREEP AND CREEP-RUPTURE DATA

11.5.1 Introduction

Components must be designed, in many cases, for service lives which are much greater than those for which experimental creep data are available; hence there is a need to be able to predict the stress and temperature requirements for long lives on the basis of much shorter time data. For example, a plant operating at high temperature may have a design life of 30-40 years (~260 000 - 350 000 h), while creep data may not be available for the particular alloys of interest beyond lives of, perhaps, 30 000 h. Thus, procedures for extrapolation of creep data and, in particular, of creep-rupture data to longer times have been studied extensively over the past three decades. Such procedures are also of great value in interpolating creep data and in minimizing the effects of the inevitable scatter in experimental data on predictions of appropriate stress and temperature conditions for the required life.

A great many time-temperature parametric methods have been developed and there is, as yet, no universally accepted "best" or standardized procedure, although a considerable attempt has been made to develop such procedures in the past few years by a Task Force of the ASTM/ASME/MPC Joint Committee on the Effects of Temperature on the Properties of Metals [*68*]. In the present discussion it is proposed to review briefly the more commonly used parametric methods, while for a more detailed review covering other methods which have been proposed or developed, the interested reader is referred to the work of Conway [*69*], and Manson [*70*].

11.5.2 The Larson-Miller (LM) Parameter

This parameter originated in the tempering studies of Hollomon and Jaffe [*71*], who attempted to correlate hardness as a function of tempering temperature and time. They first assumed hardness, H, to be a function of the parameter, $t\exp(-Q/RT)$, i.e.,

$$H = f_1\,[t\exp(-Q/RT)] \tag{11.92}$$

Q being considered constant. However, it was found that

$$Q = f_2(H) \tag{11.93}$$

and

$$t\exp(-Q/RT) = t_o = \text{constant} \tag{11.94}$$

From Eqs. (11.92), (11.93) and (11.94),

$$H = f[T(C + \log t)] \tag{11.95}$$

where C is a constant, and Larson and Miller [*72*] postulated that the creep behavior of metals had some similarity to tempering phenomena. Thus, they suggested that

$$T\,(C + \log t_r) = \text{constant} \tag{11.96}$$

where t_r is the time to rupture.

If we plot creep-rupture data for a series of stress levels, such a plot will produce a series of straight lines with common intercept, $-C$, provided the data conform to the LM parameter expressed by Eq. (11.96). This is shown in Fig. 11.11.(*a*).

Extrapolation involves plotting experimental data as shown in Fig. 11.11(a), determining the value of C, and then plotting a master curve on the coordinate axes log (σ) versus $T(C + \log t_r)$ as shown in Fig. 11.12(a). From this a set of isothermal curves may be plotted as shown in Fig. 11.13a, or else the

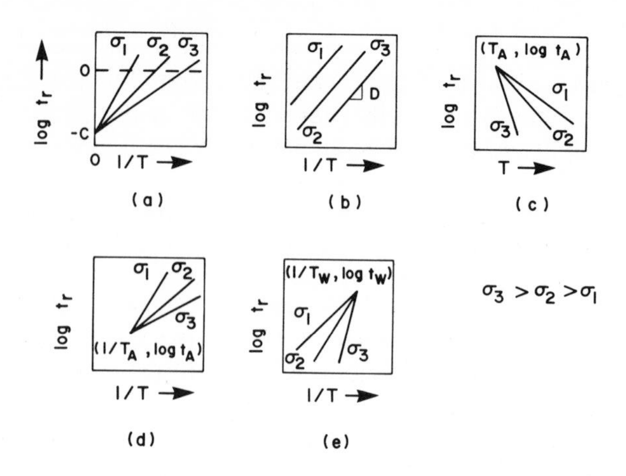

Fig. 11.11 Schematic time-temperature parametric plots for creep-rupture data: (a) Larson and Miller; (b) Orr-Sherby-Dorn; (c) Manson and Haferd; (d) Goldhoff-Sherby; (e) White-Le May

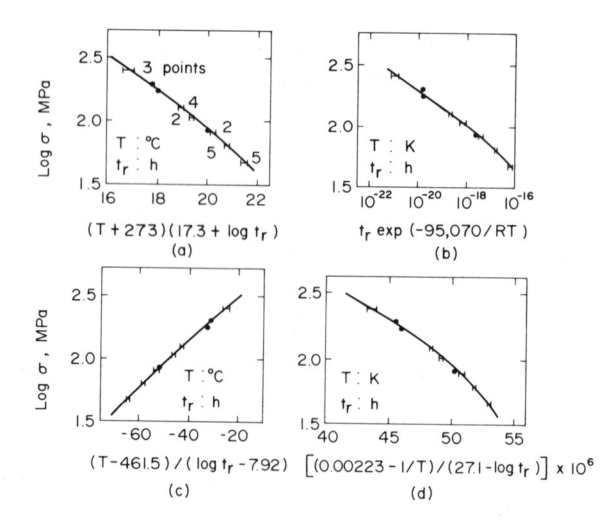

Fig. 11.12 Master curves for Type 316 weldment creep data [76]; (a) Larson and Miller; (b) Orr-Sherby-Dorn; (c) Manson and Haferd; (d) White and Le May

master curve may be used directly, substituting the required time-to-rupture, temperature and value of C to read off the corresponding stress to cause rupture.

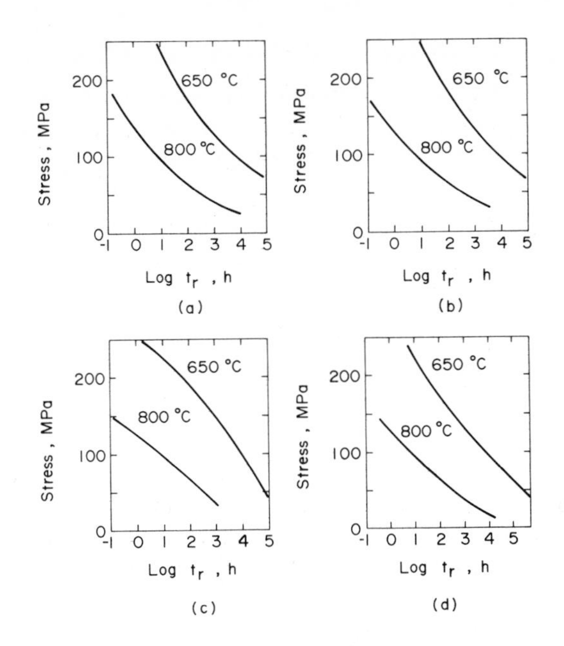

Fig. 11.13 650° C and 800° C isotherms derived from Fig. 11.12: (a) LM plot, (b) OSD plot; (c) MH plot; (d) WL plot

11.5.3 The Orr-Sherby-Dorn (OSD) Parameter

During the course of studies of the creep of a range of pure metals and alloys, Dorn and his co-workers [*73*] observed that creep strain could be correlated by the parameter

$$\theta = t\exp(-Q_c/RT) \tag{11.97}$$

where Q_c is the activation energy for creep in the particular alloy. They also observed that

$$\epsilon = f(\theta, \sigma) \tag{11.98}$$

where ϵ is the instantaneous strain, being a function of θ and stress. It was observed that Q_c remained constant above 0.45 T_m.

If Eq. (11.98) is valid up to fracture, the strain at rupture depends on applied stress, for Q_c constant. Thus

$$\epsilon_r = \theta(\sigma)$$

or

$$\theta_r = F(\sigma) = t_r \exp(-Q_c/RT) \tag{11.99}$$

from Eq. (11.97). For a given stress and two rupture points, we have

$$Q_c = [R \ln (t_{r1}/t_{r2})]/(1/T_1 - 1/T_2) \tag{11.100}$$

After Q_c has been determined, the parameter θ_r can be obtained, and a master curve on a plot of log σ versus θ can be developed.

Figure 11.11(b) shows the form of the OSD plot for rupture data at a series of stress levels, the parallel lines of slope D allowing Q_c to be evaluated, while an OSD master curve is shown in Fig. 11.12(b), with the resulting isothermals in Fig. 11.13(b).

11.5.4 The Manson-Haferd (MH) Parameter

Because deviations were noted in linearity of plots of t_r versus $1/T$ used in the LM parametric method, Manson and Haferd [*74*] proposed the empirical parameter

$$\phi = (T - T_A)/(\log t_r - \log t_A) \tag{11.101}$$

where ϕ is the reciprocal of the slope of a given constant stress line, and T_A and $\log t_A$ are material constants defining the point of intersection of the isostress lines, as shown in Fig. 11.11(c). Such a plot appeared to give a best approximation to straight lines, in confirmation with earlier ideas of Bailey [*46*].

A master curve for the MH parameter is shown in Fig. 11.12(c), and resulting isothermals in Fig. 11.13(c).

11.5.5 The Goldhoff-Sherby (GS) Parameter

The GS parametric method was proposed initially by Manson [*70*] and applied by Goldhoff and Hahn [*75*]. It is similar to the method of Larson and Miller, except that, instead of restricting the point of intersection of families of isostress lines to the ordinate axis [see Fig. 11.11(a)], the intersection point is allowed to float as shown in Fig. 11.11(d). The parameter has the form

$$P = (\log t - \log t_A)/(1/T - 1/T_A) \tag{11.102}$$

where $(1/T_A, \log t_A)$ defines the point of intersection. Application of the method is similar to that described for the other parametric methods.

11.5.6 The White-Le May (WL) Parameter

In examining the creep-rupture behavior of austenitic stainless steel weldments and parent metal it was found that the isostress lines plotted on coordinate axes of $\log t_r$ versus $1/T$ neither converged to the point (0, $-C$) as required by the LM method, nor were they parallel as required by OSD, but they *diverged*, having increasing slope with increasing stress [*76*]. This led White and Le May [*76*] to propose the alternative parameter

$$P' = (1/T_w - 1/T)/(\log t_w - \log t_r) \tag{11.103}$$

where $(1/T_w, \log t_w)$ defines the point of intersection of the isostress lines as shown in Fig. 11.11(e).

The approach is essentially a modification of the Goldhoff-Sherby parametric method, and a WL master curve and resulting isothermal data plots are shown in Figs. 11.12(d) and 11.13(d), respectively.

11.5.7 The Minimum Commitment Method (MCM)

There are considerable dangers inherent in assuming that a particular parametric method should be applied to a specific data set, as marked differences can be produced in the values of stress appropriate to long-life conditions by using different parameters. Thus, the concept of the Minimum-Commitment Method (MCM) was developed [*77*], which avoids forcing the pattern of the data and allows the best-fitting parameter to be used for the data being considered [*78*]. The MCM parameter has the form:

$$\log t_r + AP(T)\log t_r + P(T) = G(\log \sigma) \qquad (11.104)$$

where A is a constant relating to structural stability and P and G are functions of temperature and stress, respectively. Eq. (11.104) contains all of the time-temperature parameters previously discussed [*79*], and the value of A will reflect the applicable parameter, being negative for MH or WL parameters, positive in the case of LM and GS, and zero for the OSD parameter. The appropriate value of A is determined either by selecting trial values for it and solving the MCM equations so that a plot of A versus standard deviation between input data points and predicted values can be drawn to determine A for minimum error [*78*], or alternatively, a graphical procedure ("focal-point convergence") may be used [*80*].

11.5.8 Parametric Methods for Creep Data

Although much effort has been expended on applying time-temperature parametric methods to rupture data during the past two decades, they can also be applied to creep strain and strain rate data. In fact, in the early period of their formulation, data for creep rate and creep strain in specified time were utilized [*72, 73*] in parametric analysis, and there is evidence that increased interest in the use of parameters for correlating and extrapolating such data is likely in future.

Substitution of secondary creep rate for rupture time may be based on the Monkman and Grant [*81*] relationship which is found to apply for many metals and alloys, namely:

$$\dot{\epsilon}_s t_r = \text{Constant} \qquad (11.105)$$

Hence $1/\dot{\epsilon}_s$ may be substituted for t_r in each parametric method, and it has been pointed out that such a procedure is less empirical than might be apparent at first sight [*82*]. Figure 11.14 shows the WL parameter applied to data for $\dot{\epsilon}_s$ for a Type 316 stainless steel, while Fig. 11.15 shows isotherms derived from this analysis with superimposed experimental data points [*83*].

As noted earlier, it appears important to avoid extrapolation across

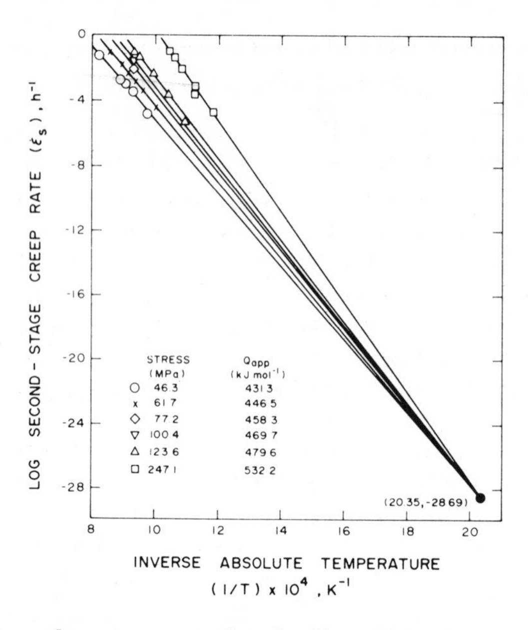

Fig. 11.14 Secondary creep rate data for Type 316 stainless steel plotted for correlation using WL parameter [*83*]

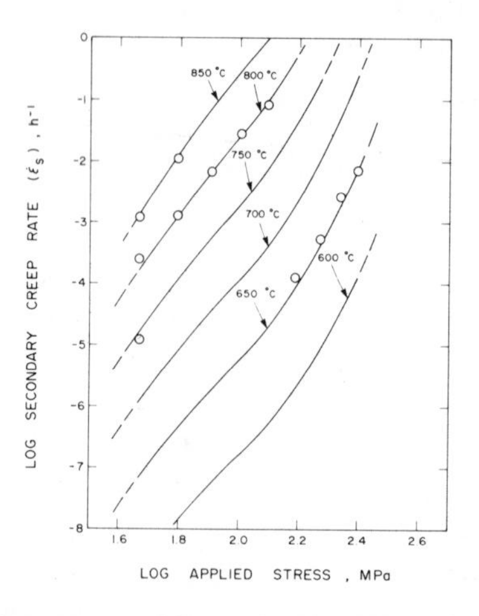

Fig. 11.15 Isotherms for creep rate generated from Fig. 11.14 with superimposed experimental data points

boundaries into regions where different creep mechanisms are dominant, and this matter requires much more attention than it has been given in the past.

11.6 CREEP AND MULTIAXIAL STRESS

11.6.1 Creep Deformation Under Multiaxial Stress

In practical engineering situations components are generally subject to multiaxial, or at least biaxial stressing; it is unusual for them to be loaded in uniaxial tension. The creep data which are available for the purposes of design will have been obtained, in the majority of cases, from uniaxial constant stress tests (or, in practice, constant load tests which will not differ appreciably in terms of strain rate measurements, except at large strains). The problem is then to predict multiaxial creep deformations arising from complex stress on the basis of uniaxial data. A related problem is the prediction of the redistribution of stress in a component due to creep deformation. In the analysis of static equilibrium structures, the method of superposition can be used to combine the effects of the individual components of stress on deformation, but this is not possible under creep conditions where the relations between stress and strain rate are generally non-linear. Hence, the effect of the components of stress must be considered for them in combination. Thus, the analysis is far from simple in all but the most elementary cases.

Many analyses of multiaxial creep deformation have been made, and the early studies of Bailey [*46*], Odqvist [*84*], and Soderberg [*85*] among others, may be noted. However, much of our knowledge of multiaxial creep stems from the work of the late Dr. A. E. Johnson and his associates at the National Engineering Laboratory (NEL), Scotland, and Johnson's careful experimental studies and achievements have been summarized and reviewed in a Memorial Volume [*86*]. His experiments on materials subjected to multiaxial stress have provided the justification for methods of prediction of multiaxial deformation based solely on the more normally available uniaxial test data.

In considering deformation under multiaxial stress, there are essentially two basic questions which must be resolved. The first concerns the equivalent effect of complex stress to produce a specified magnitude of strain, while the second relates to the distribution of deformation between the three principal directions. We shall examine these in turn.

When discussing plasticity in Chapter 2 the concept of a yield criterion was introduced, and we can introduce the analogous concepts of effective stress and effective strain for creep deformation [*87*]. For isotropic materials it is usual to assume that the effective stress has the same form as the von Mises stress for yielding and this is given by Eq. (1.60). While other definitions and criteria for σ_{eff} are possible, the von Mises-type relation appears to be a reasonable one, based on the available experimental data for multiaxial creep.

The second question is again resolved by analogy with plasticity theory, on the assumption of constancy of volume and the coincidence of the principal axes of stress and strain. Thus, the strain increments may be written in terms of the deviatoric components of stress in the form

$$d\epsilon_{ij} = (3/2)\sigma'_{ij} d\lambda \tag{11.106}$$

[cf. Eqs. (1.59)] and, equivalent to plasticity theory, the effective strain increment, $d\epsilon_{eff}$, is given by [see Eq. (1.61)]

$$d\epsilon_{eff} = (\sqrt{2}/3)\,[(d\epsilon_1 - d\epsilon_2)^2 + (d\epsilon_2 - d\epsilon_3)^2 + (d\epsilon_3 - d\epsilon_1)^2]^{1/2} \tag{11.107}$$

For steady state creep under complex stress, we may write [*88*]

$$\dot{\epsilon}_{eff} = f(\sigma_{eff}) \tag{11.108}$$

and the steady state component creep rates are [cf. Eqs. (1.71)]

$$\dot{\epsilon}_{ij} = (3/2)\,(\dot{\epsilon}_{eff}/\sigma_{eff})\,\sigma'_{ij} = (\partial\sigma_{eff}/\partial\sigma_{ij})\dot{\epsilon}_{eff} \tag{11.109}$$

If the stress dependence of creep rate can be written in the form of a power function, Eq. (11.68) can be rewritten as

$$\dot{\epsilon}_{eff} = B\sigma^n_{eff} \tag{11.110}$$

and the component creep rates are

$$\dot{\epsilon}_{ij} = (3/2)\,B\,(\sigma_{eff})^{n-1}\,\sigma'_{ij} \tag{11.111}$$

The strain occurring during multiaxial steady state creep conditions can be evaluated from Eq. (11.111) with a knowledge of the stress state and of the constants B and n, which may be obtained from uniaxial tensile data.

For creep under transient conditions, we require an appropriate equation for the creep deformation. From the work of Johnson *et al.* [*89*], it has been shown that for several metals and over a range of temperatures multiaxial primary creep can be represented by the equation

$$\dot{\epsilon}_{ij} = F(J_2)\,\sigma'_{ij}\,\phi\,(t) \tag{11.112}$$

where J_2 is the second invarient of the stress deviation tensor = $\Sigma\,(\sigma_1 - \sigma_2)^2/6$. On the basis of the von Mises equivalent stress and equivalent effective strain,

$$\dot{\epsilon}_{eff} = (2/\sqrt{3})\,F\,(J_2)\,(\sqrt{J_2})\,\phi\,(t) \tag{11.113}$$

or on the basis of octahedral shear strain ($\gamma_{oct} = \sqrt{2}\epsilon_{eff}$), the normal strain on the octahedral planes being zero, we may write

$$\gamma_{oct} = 2\,\sqrt{2/3}\,F\,(J_2)\,(\sqrt{J_2})\,\phi\,(t) \tag{11.114}$$

Eq. (11.112) representing "time-hardening" creep, may be rewritten as

$$\dot{\epsilon}_{ij} = f_1\,[F\,(J_2)\,\sigma'_{ij}]/f_2\,(\dot{\epsilon}_{ij}) \tag{11.115}$$

which is the "strain-hardening" form of the creep equation. For conditions of constant stress, both give identical results.

A reasonable equation to represent primary creep in uniaxial tension was given by Eq. (11.72), and the corresponding time-hardening and strain-hardening equations for creep rate, expressed on the basis of effective stress and strain, are

Time hardening: $\dot{\epsilon}_{eff} = mA(\sigma_{eff})^n t^{m-1}$ (11.116)

Strain hardening: $\dot{\epsilon}_{eff} = mA^{1/m}(\sigma_{eff})^{n/m}(\epsilon_{eff})^{1-1/m}$ (11.117)

Hence, the component creep rates can be written as

Time hardening: $\dot{\epsilon}_{ij} = (3/2)m\,A(\sigma_{eff})^{n-1}\sigma'_{ij}\,t^{m-1}$ (11.118)

Strain hardening: $\dot{\epsilon}_{ij} = (3/2)m\,A^{1/m}(\sigma_{eff})^{(n/m)-1}(\epsilon_{eff})^{1-1/m}\,\sigma'_{ij}\,t^{m-1}$ (11.119)

Solutions for the principal strain rates may be obtained more easily from Eq. (11.118), but they are less accurate when stresses change during creep, because of the inherent limitations of the time hardening approach. Eq. (11.118) has the same form as that for steady state creep [Eq. (11.111)] and its use leads to the same values of stress as are obtained from Eq. (11.111) when these are to be determined. It is also useful in determining time-dependent creep rates from steady state solutions, at least as a first approximation. Use of Eq. (11.119) requires numerical solution because of the accumulation of plastic strain.

Creep relaxation under multiaxial conditons of load has been studied by Johnson *et al.* [*90*], and experimental studies have also been reported by Henderson and Snedden [*91*]. The latter authors demonstrated that creep relaxation under multiaxial conditions can be predicted satisfactorily (at least for one material — copper) by means of tensile relaxation tests, and that it can also be predicted to a reasonable degree of accuracy from tensile primary creep data and the strain hardening equation. Results were plotted on the basis of octahedral shear stress as a function of time, and some of their data are shown in Fig. 11.16. It appears probable that the method can be used generally, although tests should certainly be made using a variety of materials.

Many analytical solutions to multiaxial creep problems appear in the literature, although there are limited experimental data available. Detailed consideration of creep in components having thick sections has been given by Johnson and his co-workers, including the cases of thick tubes [*92*], spheres [*93*], pressure vessels [*94, 95*], and solid bars under combined loads [*96, 97*], and these analyses allowed prediction of the changes in stress distribution with time. Subsequently, Smith [*98*] published a general treatment based on a finite difference solution of the equilibrium equations, applying the method to cylinders, spheres and thin disks. This method is amenable to computer solution.

An important concept in predicting the creep of components from limited tensile creep data is that of a *reference stress*. It is found that the deformation of a component can be simulated by a single test at a unique stress level, this being termed the reference stress. Its value is insensitive to the exact stress dependence of creep, hence a prior knowledge of $\dot{\epsilon}$ as a function of σ is not required, and component and reference stress creep test curves are found to be

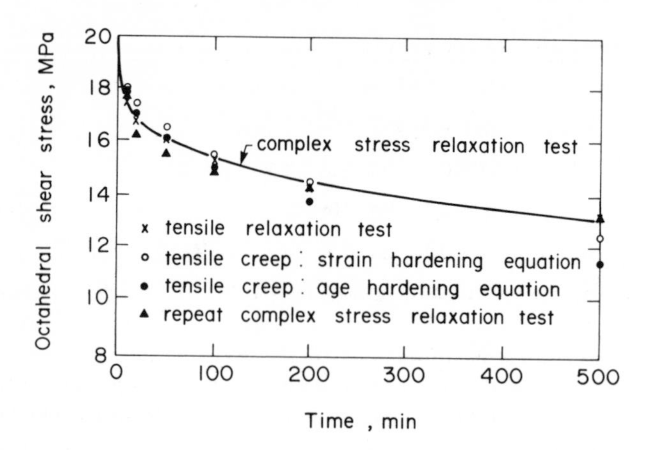

Fig. 11.16 Complex stress relaxation curve with predicted points based on tensile relaxation test. After Henderson and Snedden [*91*]

geometrically similar. Mackenzie [99] developed a method to estimate the appropriate reference stress in a component, which could then be used in making a uniaxial creep test, and this has been made more general by the analysis of Sim [*100*]. A detailed description of the reference stress method is given by Penny and Marriott [*101*], while its applicability to the common (and much analyzed) design problem of creep in thick cylinders has been reported by Fairbairn and Mackie [*102*], who showed a good correlation to exist between measured creep in the cylinder and predictions from reference stress analysis. Although the reference stress method has been used to predict the behavior of components subject to variable or cyclic loading [*103*], there is a need for more experimental work in this area, particularly for work concerned with materials aspects, but taking cognizance of the analytical methods available.

Here, we shall discuss a few examples only of multiaxial creep, without introducing problems relating to stress redistribution.

Consider a thin-walled tube of wall thickness h and radius R, with closed ends and subject to internal pressure p. The radial stress (σ_r) will vary from p at the inside wall to 0 at the outside, and may be neglected in many cases. Thus, we may write the principal stresses in the tangential, axial and radial directions, respectively, as

$$\sigma_t = pR/h$$

$$\sigma_a = pR/2h$$

$$\sigma_r = 0$$

From Eqs. (11.110) and (11.111) the creep rates in these principal directions are

$$\left.\begin{aligned} \dot{\epsilon}_t &= \dot{\epsilon}(3/4)^{(n+1)/2} \\ \dot{\epsilon}_a &= 0 \\ \dot{\epsilon}_r &= -\dot{\epsilon}_t = -\dot{\epsilon}(3/4)^{(n+1)/2} \end{aligned}\right\} \quad (11.120)$$

where $\dot{\epsilon}$ is the uniaxial tensile creep rate under a stress $\sigma = \sigma_t$. For steady state conditions, the total secondary creep strain is obtained by multiplying by time, t.

The foregoing was a particular case of biaxial stressing with σ_1 $(=\sigma_t) = 2\sigma_2$ $(=2\sigma_a)$, and we shall consider the more general case of biaxial stressing. As pointed out by Finnie and Heller [*103*], little loss in generality from the triaxial case is involved, since a hydrostatic component of stress of magnitude σ_3 can be subtracted from each of the principal stresses, to give principal stresses of magnitude $(\sigma_1 - \sigma_3)$, $(\sigma_2 - \sigma_3)$, and 0, for which the strains are calculated. The hydrostatic component will not influence the strains. Thus, the principal strain rates from Eq. (11.111) become

$$\left.\begin{aligned} \dot{\epsilon}_1 &= B\,(\sigma_1^2 - \sigma_1\sigma_2 + \sigma_2^2)^{(n-1)/2}\,(\sigma_1 - \sigma_2/2) \\ \dot{\epsilon}_2 &= B\,(\sigma_1^2 - \sigma_1\sigma_2 + \sigma_2^2)^{(n-1)/2}\,(\sigma_2 - \sigma_1/2) \\ \dot{\epsilon}_3 &= B\,(\sigma_1^2 - \sigma_1\sigma_2 + \sigma_2^2)^{(n-1)/2}\,(-\sigma_1 - \sigma_2)/2 \end{aligned}\right\} \quad (11.121)$$

Putting $\sigma_2/\sigma_1 = \alpha$ (where $-1 \leqslant \alpha \leqslant 1$), we may write

$$\left.\begin{aligned} \dot{\epsilon}_1/\dot{\epsilon} &= (1 - \alpha + \alpha^2)^{(n-1)/2}\,(1 - \alpha/2) \\ \dot{\epsilon}_2/\dot{\epsilon} &= (1 - \alpha + \alpha^2)^{(n-1)/2}\,(\alpha - 1/2) \\ \dot{\epsilon}_3/\dot{\epsilon} &= (1 - \alpha + \alpha^2)^{(n-1)/2}\,(-1 - \alpha)/2 \end{aligned}\right\} \quad (11.122)$$

where $\dot{\epsilon}$ is again the uniaxial tensile creep rate under a stress of σ_1.

So far, we have assumed the material to be isotropic. However, in the general case this is not so, and although it may be taken as a reasonable approximation in many practical situations, considerable error may be introduced in particular cases. One example where considerable preferred crystallographic orientation combined with a limited number of slip systems can lead to erroneous predictions of creep rates when isotropy is assumed is a zirconium alloy reactor pressure tube discussed by Ross-Ross *et al.* [*104, 105*].

In considering anisotropic material, we assume it to be orthotropic, which is generally the case for metals, and that the creep rate again depends on an invarient characterizing the multiaxial stressing. Considering the principal axes of anisotropy as the axes of the system, we may write the invariant or effective stress in the generalized von Mises form proposed by Hill [*106*] as

$$\sigma_{eff} = [F\,(\sigma_{xx} - \sigma_{yy})^2 + G\,(\sigma_{yy} - \sigma_{zz})^2 + H\,(\sigma_{zz} - \sigma_{xx})^2 + 6L\tau_{xy}^2 + 6M\tau_{yz}^2 + 6N\tau_{zx}^2]^{1/2} \quad (11.123)$$

where *F*, *G*, etc. are constants. The shear stress components must be included as we are dealing with axes of anisotropy which are not necessarily the axes of principal stress and strain. The strain rates are assumed to depend on a function of σ_{eff}, as before, and from Eq. (11.109).

$$\dot{\epsilon}_{ij} = (\partial \sigma_{eff} / \partial \sigma_{ij}) \dot{\epsilon}_{eff}$$

For a power law dependency of creep rate on stress, the principal strain rates are

$$\left.\begin{aligned} \dot{\epsilon}_1 &= B(\sigma_{eff})^{n-1} [F(\sigma_1 - \sigma_2) - H(\sigma_3 - \sigma_1)] \\ \dot{\epsilon}_2 &= B(\sigma_{eff})^{n-1} [G(\sigma_2 - \sigma_3) - F(\sigma_1 - \sigma_2)] \\ \dot{\epsilon}_3 &= B(\sigma_{eff})^{n-1} [H(\sigma_3 - \sigma_1) - G(\sigma_2 - \sigma_3)] \end{aligned}\right\} \quad (11.124)$$

where *B* is defined by Eq. (11.110). The constants *F*, *G*, and *H* may be evaluated from tests made in mutually perpendicular directions to the material's most natural orientations, while the remainder may be obtained from tests of obliquely cut specimens.

Where the axes of anisotropy correspond with those of principal stress, as will be the case for a tube, for example, we may simplify the equation for the invarient to

$$\sigma_{eff} = [F(\sigma_1 - \sigma_2)^2 + G(\sigma_2 - \sigma_3)^2 + H(\sigma_3 - \sigma_1)^2]^{1/2} \quad (11.125)$$

Considering the case of a tube, let the principal stresses in the axial, transverse and radial directions be given by σ_a, σ_t and σ_r, replacing σ_1, σ_2 and σ_3, respectively, in Eq. (11.125). For a uniaxial creep test in the axial direction, $\sigma_t = \sigma_r = 0$, and it is easily shown that the principal strain rates are given by

$$\left.\begin{aligned} \dot{\epsilon}_a &= (\dot{\epsilon}_{eff}/\sigma_{eff}) (F + H)\sigma_a \\ \dot{\epsilon}_t &= -(\dot{\epsilon}_{eff}/\sigma_{eff}) F\sigma_a \\ \dot{\epsilon}_r &= -(\dot{\epsilon}_{eff}/\sigma_{eff}) H\sigma_a \end{aligned}\right\} \quad (11.126)$$

For a creep test on the tube with closed ends where $\sigma_t = 2\sigma_a$, and $\sigma_r = 0$, we obtain

$$\left.\begin{aligned} \dot{\epsilon}_t &= (\dot{\epsilon}_{eff}/\sigma_{eff}) (G + 0.5F)\, \sigma_t \\ \dot{\epsilon}_a &= (\dot{\epsilon}_{eff}/\sigma_{eff}) (0.5H - 0.5F)\, \sigma_t \end{aligned}\right\} \quad (11.127)$$

For isotropic material, $F = G = H = 0.5$, and for anisotropic material $F + G + H$ is taken as 1.5 for consistency. By careful measurement of the creep rates in tube tests as above, the constants may be evaluated, and the corresponding creep rates in the different directions have been correlated with crystallographic texture for Zircaloy tubing [*104*].

Although the theory has been developed to some extent for anisotropic materials, what is lacking are good experimental data, and careful studies of multiaxial creep in well-characterized orthotropic materials are required, rather than refinement of theories.

11.6.2 Creep-Rupture Under Multiaxial Stress

As in the case of creep deformation under multiaxial stress, the majority of the experimental data available concerning creep-rupture under multiaxial stress are due to Johnson and his co-workers at NEL. They showed that two criteria for creep-rupture exist, namely the maximum principal stress criterion, where materials show general or continuous cracking in tertiary creep, and the octahedral shear stress criterion (which is equivalent to a criterion based on σ_{eff} or the von Mises second order invarient) where a single major crack (or, perhaps two) develops in the region of fracture [*87, 89, 107*]. It was found that the fracture mode of a material did not change over its working range of temperature; thus, a tensile creep-rupture test and subsequent metallographic examination of the specimen serve to determine the failure criterion for multiaxial stressing. Finnie and Abo el Ata [*108*] have suggested a mixed criterion, with crack initiation controlled by maximum tensile stress and propagation governed by σ_{eff}. However, it has been pointed out by Henderson and Snedden [*109*] that this has no particular advantage, and the results of the latter authors on widely different materials correlate well on the basis of a single criterion governed by fracture mode.

These observations may be rationalized on the basis of the operative mechanisms for fracture. Nucleation of voids at triple points and other grain boundary regions will be caused by sliding, and this, in turn, will be controlled by shear stresses, which are dependent on σ_{eff} under conditions of multiaxial stress. Void growth is controlled by the condensation of vacancies arising from deformation or transferred by a Nabarro-Herring mechanism. Thus, growth will be controlled by the maximum tensile stress.

A number of attempts have been made to apply phenomenological theories of creep to multiaxial creep-rupture, but they have met with limited success only, and do not, as yet, appear to be of particular help to the designer. Odqvist [*110*] has discussed the application of Kachanov's theory of creep-rupture [*111*] to multiaxial stressing; however, Henderson and Snedden [*109*] have shown this theory to be inapplicable to their results.

There is undoubtedly a need for more, careful experimental studies of multiaxial creep fracture, and it appears that at present prediction of rupture life may be made, to a first approximation, using the method of uniaxial tensile testing proposed by Johnson. Such studies should endeavor to provide correlations between multiaxial stress, fracture mode, and ductility, including Manjoine's analysis [*112*] discussed in Section 3.2.

11.7 CREEP-FATIGUE INTERACTION

In many practical situations components operating at elevated temperature may be subject to cyclic stress or strain and the basis of the design cannot be assumed to be creep or fatigue only, but the two processes are likely to interact, contributing to accumulated damage in the component. Such situations may arise in gas turbines, nuclear reactor components and chemical processing plants, for example. Reviews of creep-fatigue interaction have been made by Coffin [*113*] and Manson [*114*], several symposia and conference proceedings dealing with the topic have been published in recent years [*115-117*], and an extensive overall review has also been prepared [*118*]. In the following, the different approaches which may be taken to design for creep-fatigue interaction are examined briefly, and then the phenomenon of strain accumulation or *racheting* is discussed.

11.7.1 Interaction Theories

The principal approaches which have been proposed for design for creep-fatigue interactive conditions are as follows:

- Linear and modified linear damage accumulation rules;
- Modified relationships for fatigue developed for low temperature conditions;
- Strainrange partitioning;
- Ductility exhaustion.

Damage accumulation

In this approach damage incurred under operative fatigue and creep conditions considered separately is summed, with failure being predicted when the total reaches the limit of damage which the material can accept. This is the basis of ASME Code Case N-47 for nuclear pressure vessel components [*119*], and the criterion used for a design to be acceptable is that creep and fatigue damage satisfy the equation [cf. Eq. (3.37)] representing damage accumulation for "stress effects" and "strain effects"),

$$\sum_{j=1}^{p} (n/N_d)_j + \sum_{k=1}^{q} (t/T_d)_k \leqslant D \tag{11.128}$$

where

D = total allowable creep-fatigue damage,
n = the number of applied cycles at loading condition j,
N_d = the number of design allowable cycles at loading condition j, obtained from fatigue curves corresponding to the maximum metal temperature during the cycle for the equivalent strain range,
t = the time load condition k is applied,
T_d = the allowable time at a given stress level (for elastic analysis) or at a given effective stress (for inelastic analysis) at load k. T_d values are obtained by entering the stress-to-rupture curve at a stress equal to the calculated stress (from load k) divided by the factor $k' = 0.9$ for the ASME Code Case.

D may be either a linear or a modified linear damage summation depending on the material, and on the loading conditions.

Eq. (11.128) is a representation of the classic Palmgren-Miner damage rule, and the allowable damage may be taken as 1 or some lower quantity depending on experience. For example, ASME Code Case N-47 utilizes $D = 1$ as the criterion for elastic analysis and takes D from Fig. 11.17 for inelastic analysis. The creep component in Eq. (11.128) may be replaced by $\int_0^t (1/T_d)dt$ in inelastic analysis. For multiaxial stress-strain situations, an equivalent strain range approach is used, with fatigue strain range being given [from Eq. (1.61)] by

$$\Delta\epsilon_{eff} = (\sqrt{2}/3)\,[(\Delta\epsilon_1 - \Delta\epsilon_2)^2 + (\Delta\epsilon_2 - \Delta\epsilon_3)^2 + (\Delta\epsilon_3 - \Delta\epsilon_1)^2]^{1/2} \qquad (11.129)$$

The equivalent creep life is given in terms of the effective stress as defined in Eq. (1.60).

The background to ASME Code Case N-47 and the supportive experimental data demonstrating its general validity are detailed in a separate ASME publication [*120*]. However, questions remain concerning the adequacy of the simple procedures involved when short time tests must be used to predict designs for long time service conditions, when strain ranges are small, and for multiaxial conditions and situations involving discontinuities and strain concentrations. In addition, there is the very real problem that metallurgical changes occurring during prolonged elevated temperature service may cause changes in mechanical properties which are not accounted for by the Code. Hopefully, these are accounted for in utilizing the correct extrapolation or data smoothing procedure as discussed in Section 11.5.

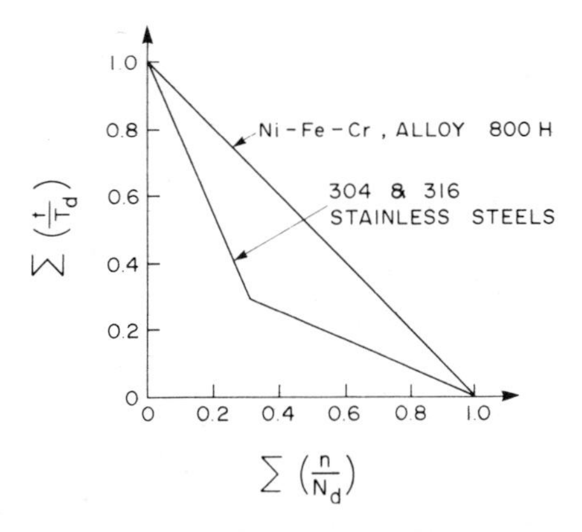

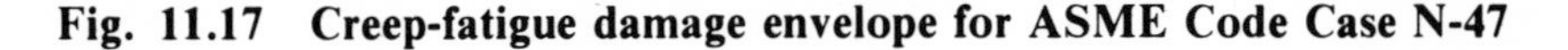

Fig. 11.17 Creep-fatigue damage envelope for ASME Code Case N-47

Modified low temperature relationship

The Coffin-Manson relationship for low cycle fatigue was given in Eq. (10.25). At elevated temperature a time dependency must be introduced, and Coffin [*113*] has argued that, because of the importance of environmental effects, frequency or cycle time should be introduced rather than strain rate. Thus, Coffin uses an equation of the form

$$\nu^k t_f = \text{constant} = f(\Delta\epsilon_p) \tag{11.130}$$

where ν is frequency, t_f is the time to failure, and k is a constant depending on temperature. Eq. (11.130) may be rewritten as

$$\nu^k t_f = \nu^k (N_f/\nu) = N_f\nu^{k-1} \tag{11.131}$$

and this quantity is termed the frequency modified fatigue life. Substituting this in the Coffin-Manson relation [Eq. (10.25)], we obtain

$$(N_f\nu^{k-1})^{1/c}\Delta\epsilon_p = C_2 \tag{11.132}$$

where C_2 is a constant. This relation provides a method of taking time effects into account when dealing with plastic strain.

Stress range, plastic strain range and frequency can be related by the equation [*121*]

$$\Delta\sigma = A\,(\Delta\epsilon_p)^{n'}\,\nu^{k'} \tag{11.133}$$

where A, n' and k' are coefficients determined by regression analysis of test data, n' being the cyclic strain hardening exponent. For a triangular waveform, $\dot{\epsilon}_p = 2\Delta\epsilon_p\nu$, and Eq. (11.133) becomes

$$\Delta\sigma = B(\Delta\epsilon_p)^{n'_1}\,\dot{\epsilon}_p{}^m \tag{11.134}$$

where $B = A(2)^{-k'}$, $n'_1 = n' - k'$, and $m=k'$. Eliminating $\Delta\epsilon_p$ from Eqs. (11.132) and (11.133), a high temperature equivalent of the Basquin equation for low temperature fatigue [Eq. (10.30)] is obtained in the form

$$\Delta\epsilon_e = \Delta\sigma/E = (A'/E)N_f^{-b}\,\nu^{k''} \tag{11.135}$$

where $A' = AC_2^{n'}$, $b = (1/c)n'$, $k'' = (1/c)n'\,(k\text{-}1) + k'$.

Fatigue at high temperature may be represented conveniently by Eqs. (11.132) and (11.135), the various coefficients being determined by regression analysis, or the two components of strain may be combined to give

$$\Delta\epsilon_t = \Delta\epsilon_e + \Delta\epsilon_p = C_2\,(N_f\nu^{k-1})^{-1/c} + (A/E)\,C_2{}^{n'}\,N_f^{-(1/c)n'}\,\nu^{-(1/c)n'\,(k-1)+k'} \tag{11.136}$$

which is similar in form to Manson's Universal Slopes equation [Eq. (10.4)], and identical to it if $k = 1$, $k' = 0$, $C_2 = D^{0.6}$, $(1/c) = 0.6$, $n' = 0.2$ and $A = 3.5\sigma_u/D^{0.12}$.

These relations have been applied to a number of alloys by Coffin and his co-workers [*113, 122, 123*], and the approach is a useful one for many high temperature situations involving fatigue. However, its applicability to more complex cycles is less certain, and the large number of constants, which must be determined at the temperature level of interest, represent considerable experimental time for data generation.

Strainrange partitioning

This method of describing creep-fatigue interaction was introduced by Manson and his associates [*124*]. It is postulated that a strain cycle can be considered in terms of two different types of strain designated "plasticity" and "creep", respectively, and which may operate in either a positive or a negative manner. The four basic types of cycle are shown in Fig. 11.18, and are designated $\Delta\epsilon_{pp}$, $\Delta\epsilon_{cp}$, $\Delta\epsilon_{pc}$, $\Delta\epsilon_{cc}$, where the first subscript designates the type of deformation in the tensile part of the cycle and the second designates that in the compressive part. Although the use of the terms "creep" and "plasticity" is inexact and misleading in designating time-dependent and time-independent inelastic deformation respectively, they are retained here, but in inverted commas, in order to conform with Manson's terminology.

In an actual strain cycle several of these strainrange components will occur, and Fig. 11.19 shows a cycle involving $\Delta\epsilon_{pp}$, $\Delta\epsilon_{cc}$, and $\Delta\epsilon_{pc}$. In a complex cycle four quantities may be recognized, viz. tensile "creep", tensile "plasticity", compressive "creep" and compressive "plasticity". In any cycle $\Delta\epsilon_{cp}$ or $\Delta\epsilon_{pc}$ can exist, not both, depending on whether "creep" or "plasticity" is greater in each direction.

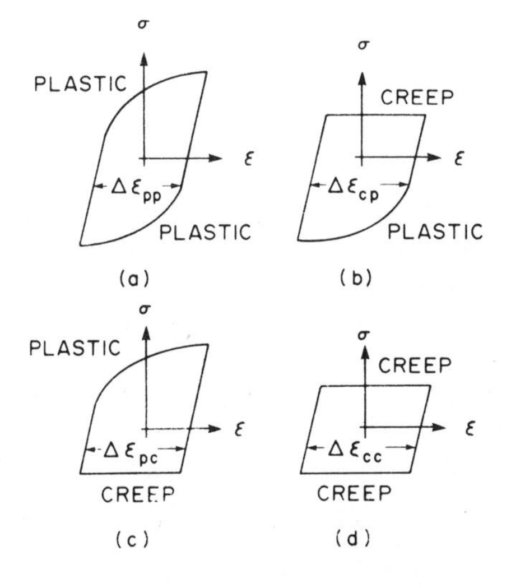

Fig. 11.18 The four type of partitioned strainrange cycles: (a) PP-type cycle; (b) CP-type cycle; (c) PC-type cycle; (d) CC-type cycle

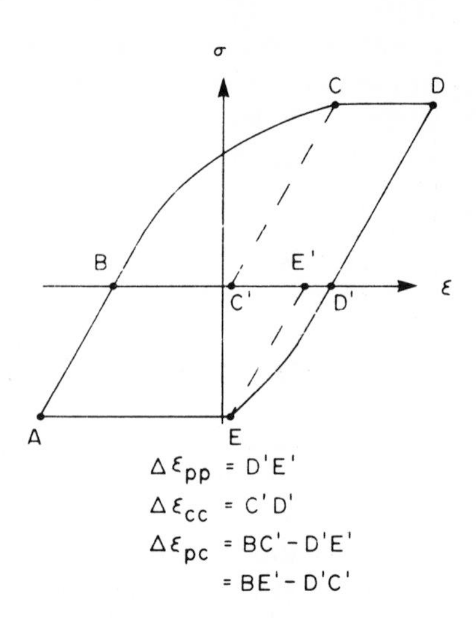

Fig. 11.19 Hysteresis loop involving several components

For each of the basic strainranges life may be set down in terms of a Coffin-Manson relationship as

$$\left.\begin{aligned} N_{pp} &= A_1\,(\Delta\epsilon_{pp})^{a_1} \\ N_{cp} &= A_2\,(\Delta\epsilon_{cp})^{a_2} \\ N_{pc} &= A_3\,(\Delta\epsilon_{pc})^{a_3} \\ N_{cc} &= A_4\,(\Delta\epsilon_{cc})^{a_4} \end{aligned}\right\} \quad (11.137)$$

where the values of A and a are specific for each material, being relatively independent of temperature for materials whose ductility is relatively temperature independent for both tensile and creep deformation. Hence, plots of strainrange versus life can be constructed for each strain component as shown in Fig. 11.20.

For a cycle of total inelastic strainrange $\Delta\epsilon_i$, the component strainranges can be written as fractions, i.e.,

$$F_{pp} = \Delta\epsilon_{pp}/\Delta\epsilon_i,\ F_{cp} = \Delta\epsilon_{cp}/\Delta\epsilon_i,\ F_{pc} = \Delta\epsilon_{pc}/\Delta\epsilon_i,\ F_{cc} = \Delta\epsilon_{cc}/\Delta\epsilon_i$$

and the *interaction damage rule*, which has been found to be valid [*114*], is

$$1/N_f = F_{pp}/N_{pp} + F_{cp}/N_{cp} + F_{pc}/N_{pc} + F_{cc}/N_{cc} \quad (11.138)$$

N_{pp}, etc., designating the lives determined from Eq. (11.137) or Fig. 11.19 when $\Delta\epsilon_i$ is used in the appropriate relation.

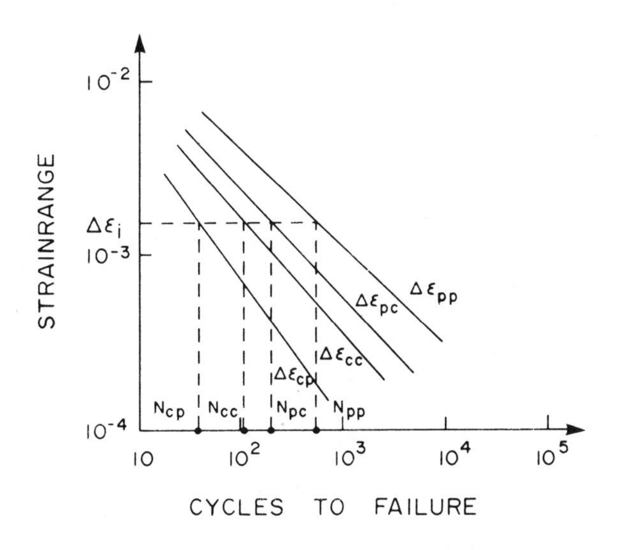

Fig. 11.20 Coffin-Manson plots for strainrange components

For continuous cycling at constant frequency and temperature, the deformation can be approximated as all $\Delta\epsilon_{pp}$ at high frequency and all $\Delta\epsilon_{cc}$ at low frequency. At intermediate frequencies the strain cycle consists of $\Delta\epsilon_{cc}$ and $\Delta\epsilon_{pp}$, $\Delta\epsilon_{cp}$ or $\Delta\epsilon_{pc}$ not being present from symmetry, and $\Delta\epsilon_{pp}$ can be separated by making a rapid cycle under stress control between limits appropriate to the hysteresis loop. As shown in Fig. 11.21, the magnitude of $\Delta\epsilon_{pp}$ is obtained from the width of the rapidly cycled loop, and $\Delta\epsilon_{cc}$ is determined by subtraction from the width of the total hysteresis loop. Good correspondence between experimental data and the results of application of the interaction damage rule has been reported [*125*].

In more complex nonsymmetrical cycles, the strainrange can also be partitioned using procedures which have been developed for this purpose [*125*] and it has been found that there is good correlation between predicted and experimental data in terms of life to failure for the limited number of cases examined.

Ductility exhaustion

The importance of ductility was emphasized in Section 3.2, and a component can be considered to fail when a plastic strain is imposed on it after it has reached its ductility limit. In the ductility exhaustion model for combined creep and fatigue it is assumed that the reduction in ductility resulting from strain cycling can be represented as an incremental creep extension, with failure taking place when the total reduction reaches its

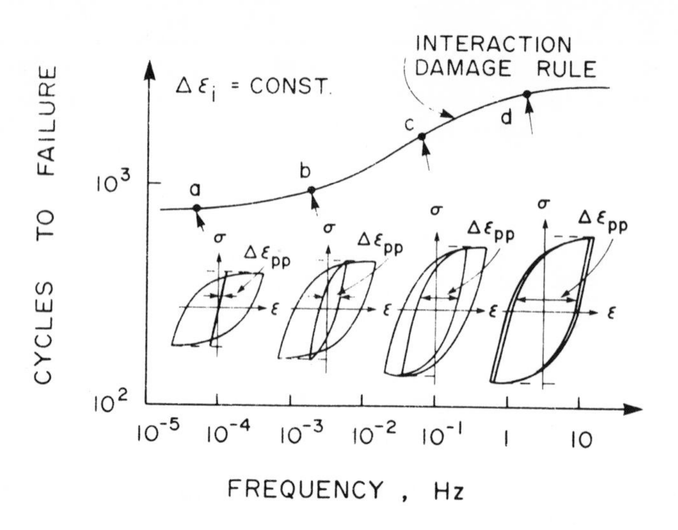

Fig. 11.21 Partitioning of hysteresis loops generated at high temperature at different frequencies by means of rapid cycling to produce PP-type loops

limiting value. The approach used by Polhemus *et al.* [*126*] in analyzing ductility exhaustion is described in the following.

The residual ductility in a specimen which has undergone N cycles at a strain range $\Delta\epsilon$ is

$$\epsilon_F = \epsilon_{Fo}\,(1 - N/N_f)^{1/a} \tag{11.139}$$

where ϵ_{Fo} is the ductility of the virgin material and a is a material constant equal to the slope of the fatigue curve plotted on the basis of $\log\Delta\epsilon$ versus $\log N_f$. The ductility reduction is

$$\epsilon_{Fo} - \epsilon_F = \epsilon_{Fo}\,[1 - (1 - N/N_f)^{1/a}] \tag{11.140}$$

and, substituting the power law relation in the form [*127*]

$$(\Delta\epsilon)^a N_f = (2\epsilon_{Fo})^a/4$$

into Eq. (11.140), we obtain

$$\epsilon_{Fo} - \epsilon_F = \epsilon_{Fo}\{1 - [1 - 4N\,(\Delta\epsilon/2\epsilon_{Fo})^a]^{1/a}\} \tag{11.141}$$

For combined creep and strain cycling, the accumulated decrements of available ductility can be considered in terms of a stepwise movement along a graph representing creep deformation at constant temperature, this having the form [cf. Eq. (11.72)],

$$\epsilon_{cr} = A_1 t^m \tag{11.142}$$

Creep and fatigue are assumed to contribute to the same failure mechanism, thus damage during the first increment of cycles is equivalent to a creep strain increment, with ductility loss given by Eq. (11.141). From Eq. (11.142) and the graph representing creep deformation, the equivalent creep time is determined, this being the time in creep to cause the same ductility decrement as in the cycling, i.e.,

$$t_{eq} = (\Delta\epsilon / A_1)^{1/m} \qquad (11.143)$$

The actual creep time increment (t_{inc}) is added, and the ductility after one creep and one cyclic increment is

$$\epsilon_{FR} = \epsilon_{Fo} - A_1 (t_{eq} + t_{inc})^m \qquad (11.144)$$

Solution is obtained by addition of increments of cycles and, as the equations are non-linear, computer solution is required in a practical case. The method, which has been applied to several gas turbine alloys with a reasonable degree of success [*126*], appears preferable to a linear cumulative damage summation when dealing with alloys exhibiting lower levels of ductility, in which the damage fraction summation may be considerably less than unity.

11.7.2 Ratcheting

Ratcheting may be defined as progressive cyclic inelastic deformation occurring in a component subjected to cyclic mechanical stress, cyclic thermally induced stress, or both, in the presence of a sustained primary stress. It is of particular concern in reactor systems operating at elevated temperature because frequent and rapid changes in temperature can produce the appropriate cyclic stress conditions, and is treated in ASME Code Case N-47 [*119*].

At lower temperature conditions, where a cyclic stress is added to a steady loading, the situation is essentially one of low cycle fatigue. However, the more normal situation occurring at elevated temperature is more complex.

Consider a component carrying a sustained tensile stress σ_m and subjected to periodic and rapid cooling on one wall from T_1 to T_2, while the other wall does not vary appreciably in temperature. As a first approximation consider the temperature distribution to be of the simplified form shown in Fig. 11.22, with a rectangular form.

Because of the rapid temperature drop, high thermal stresses are induced which may, if sufficiently large, lead to plastic yielding at either or both wall surfaces. This, in turn, produces incremental plastic strain and a residual stress distribution after subsequent reheating. The latter will cause an increase in the tensile stress in part of the component leading to accelerated creep during the high temperature hold period, because of the nonlinear dependence of creep on stress. Thus, ratcheting of the component can be caused both by the inelastic deformation caused by time-independent plasticity and by the enhanced creep deformation. These two effects are termed *plastic ratcheting*

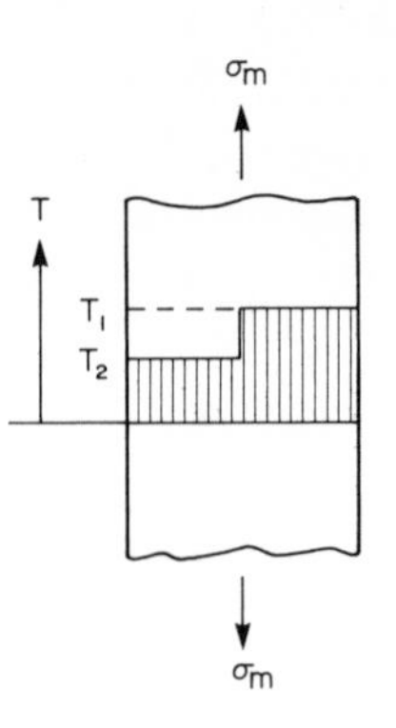

Fig. 11.22 Cyclically applied temperature variation applied to stressed plate component

or *thermal ratcheting* and *creep ratcheting*, respectively.

Figure 11.23 shows a two-bar representation of the component with areas A_1 and A_2 corresponding to the regions remaining at high temperature (T_1) and being cycled between T_1 and T_2 respectively. Taking $A_1 = A_2$, the stress after a temperature drop is $E\alpha\,(T_1 - T_1)/2$ (α being the coefficient of thermal expansion of the material), and for this greater than σ_Y, yielding will occur. For elastic-perfectly plastic material, the deformation modes for thermal ratcheting are shown in Fig. 11.24, the *shakedown* region representing conditions where yield occurs in both elements of the model during the first half cycle and the deformation settles down to purely elastic behavior during subsequent cycles. Figure 11.24 is plotted for different values of normalized thermal stress, $E\alpha\,(T_1 - T_2)/\sigma_Y$, and of normalized mean stress, σ_m/σ_Y. Similar

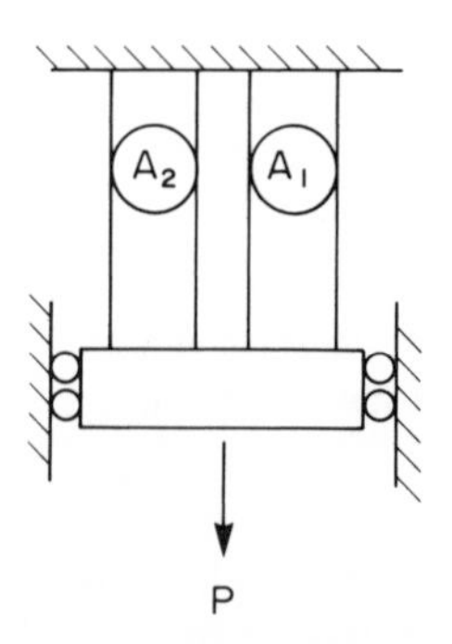

Fig. 11.23 Two-bar model of component subject to temperature cycling

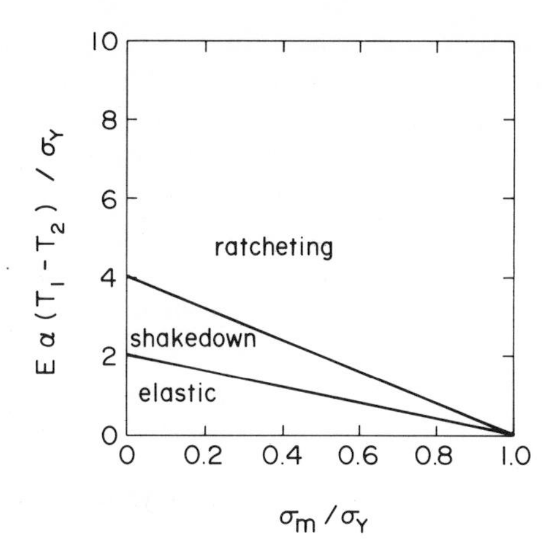

Fig. 11.24 Deformation modes for elastic-perfectly plastic solid

plots can be drawn up for different temperature distributions [*128*], different area ratios of the two members, and for material with kinematic strain hardening and Bauschinger effect [*129-131*].

When two dissimilar materials comprising a member undergo a temperature cycle, a similar analysis can be applied, taking into consideration the different values of α and E of the two materials.

11.8 ENVIRONMENTAL EFFECTS

11.8.1 Effects of Atmosphere

It has long been recognized that the gaseous environment in which creep takes place can cause alteration in the creep behavior and rupture time. However, there is a lack of clear understanding of the complex effects of environment in producing gas-metal reactions which may lead, for example, to increased strength in air as compared to tests in vacuum at one stress level and reduced strength at different stress levels [*132*]. Strengthening may be produced by a surface film in a number of ways: for example, it may arise from the intrinsic mechanical strength of the film or it may be produced by the film blocking the egress of dislocations (see Section 4.3.5); while weakening from the presence of an oxide film may arise because of excess vacancies caused by the diffusional processes leading to film growth and their facilitating dislocation climb near the surface [*133*].

In recent years much of the study of effects of gaseous environments on creep has been directed to tests in impure helium relating to gas-cooled nuclear reactor design and operation. A comprehensive review of environmental effects is provided by Cook and Skelton [*134*].

Although most of the observed environmental effects can be accommodated within the scatter bands arising from material and test variation, it is important that environmental effects be considered at an early stage in high temperature design as severe reduction in high temperature properties does arise in specific cases, for example in superalloys when exposed to oxygen or hydrogen under particular conditions [*135*]. Such effects can cause large differences in service life from that predicted by extrapolation from air or vacuum tested specimen data. Generally, the effect is most severe in thin sections where the ratio of surface area to volume is large, as it is a surface-related phenomenon. However, at the present time it appears that there is no substitute for testing the material of interest in the reactive atmosphere to be met in operation, when there is a lack of specific prior data.

An area of particular importance appears to be high temperature fatigue which should perhaps be considered as involving creep-fatigue-environment interactions, rather than simply creep-fatigue interaction. This point has been made strongly by Coffin [*113*].

11.8.2 Effect of Radiation

Radiation can accelerate creep deformation by the displacement of atoms from interaction with high energy particles. However, creep can also be produced in a reactor core at temperatures which are normally too low for creep to take place. This occurs because internal stresses can be produced under irradiation, these being greater than the applied stress in some cases. Thus, in a high energy neutron flux an unstressed single crystal of uranium or zirconium changes shape in a manner controlled by the formation of interstitials in one specific set of planes and the agglomeration of vacancies on another. Unstressed, randomly oriented polycrystalline material does not change shape because each grain is restrained by its neighbors and it yields (or creeps) to balance the irradiation induced deformation. When an applied stress is superimposed, considerable creep can occur at relatively low temperatures.

In austenitic stainless steel and nickel-based superalloys no shape change takes place in a single crystal, however such polycrystalline materials still display accelerated creep in a high neutron flux. This arises because the voids and excess vacancies produced cause swelling. To accommodate this, creep takes place, and the presence of the excess vacancies will aid in dislocation climb.

A detailed discussion of the available data for irradiation creep and of the mechanisms is provided by Gittus [*3*].

REFERENCES

1. Ahlquist, C. N., and Nix, W. D., *Acta Met.*, **19**, 373 (1971).
2. Le May, I., and White, W. E., in *Proc. 3rd Int. Conf. on Pressure Vessel Tech.*, p. 861, ASME, New York (1977).
3. Gittus, J., *Creep, Viscoelasticity and Creep Fracture in Solids*, Halsted Press, Wiley, New York (1975).
4. Garofalo, F., *Fundamentals of Creep and Creep Rupture in Metals*, Macmillan, New York (1965).
5. Lagneborg, R., *International Metall. Revs.*, **17**, No. 165, 130 (1972).
6. Evans, A. G., and Rawlings, R. D., *Phys. Stat. Solidi*, **34**, 9 (1969).
7. Nabarro, F. R. N., in *Bristol Conf. on Strength of Solids*, p. 75, Physical Soc., London (1948).
8. Herring, C., *J. App. Phys.*, **21**, 437 (1950).
9. Coble, R. L., *J. App. Phys.*, **34**, 1679 (1963).
10. Lifshitz, I. M., *Soviet Physics*, **17**, 909 (1963).
11. Mohamed, F. A., and Langdon, T. G., *Trans. ASME, J. Eng. Matls. and Technology*, **98**, 125 (1976).
12. Bailey, R. W., *J. Inst. Met.*, **35**, 27 (1926).
13. Orowan, E., *J. West Scot. Iron Steel Inst.*, **54**, 45 (1946-47).
14. McLean, D., *Rep. Progress Phys.*, **29**, 1 (1966).
15. Gibbs, G. B., *Phil. Mag*, **23**, 771 (1971).
16. Ahlquist, C. N., Gasca-Neri, R., and Nix, W. D., *Acta. Met.*, **18**, 663 (1970).
17. Lagneborg, R., Forsen, B. H., and Wiberg, J., in *Creep Strength in Steel and High-Temperature Alloys*, p. 1, The Metals Soc., London (1975).
18. Lagneborg, R., and Forsen, B. H., *Acta Met.*, **21**, 781 (1973).
19. Lagneborg, R., *Metal Sci. J.*, **6**, 127 (1972).
20. Ostrom, P., and Lagneborg, R., *Trans. ASME, J. Eng. Matls. and Technology*, **98**, 114 (1976).
21. Gittus, J. H., *Phil. Mag.*, **21**, 495 (1970).
22. Gittus, J. H., *Phil. Mag.*, **23**, 1281 (1971).
23. Gittus, J. H., *Phil. Mag.*, **24**, 1423 (1971).
24. Gittus, J. H., *Trans. ASME, J. Eng. Matls. and Technology*, **98**, 52 (1976).
25. McLean, D., in *Mechanical Behavior of Materials*, Special Volume, p. 91, The Society of Materials Science, Japan (1972).
26. Friedel, J., *Dislocations*, Pergamon Press, Oxford (1964).
27. Gittus, J. H., in *Proc. Int. Conf. on Creep and Fatigue in Elevated Temperature Applications*, Vol. 1, p. 115.1, I. Mech. E., London (1975).
28. McVetty, P. G., *Mech. Eng.*, **56**, 149 (1934).
29. Li, J. C. M., *Acta. Met.*, **11**, 1269 (1963).
30. Morrison, J., Ph.D. Thesis, McMaster Univ., Hamilton, Ont. (1975).
31. Morrison, J., Sargent, C. M., and Embury, J. D., *Scripta Met.*, **12**, 513 (1978).
32. Davies, P. W., Nelmes, G., Williams, K. R., and Wilshire, B., *Metal Sci. J.*, **7**, 85 (1973).
33. Williams, K. R., and Wilshire, B., *Metal Sci. J.*, **7**, 176 (1973).

34. PARKER, J. D., and WILSHIRE, B., *Metal Sci.*, **9**, 248 (1975).
35. EVANS, W. J., and HARRISON, G. F., NGTE Rept. No. R. 340, NGTE, Pyestock, Hants (1976).
36. RAJ, R., and ASHBY, M. F., *Met. Trans.*, **2**, 1113 (1971).
37. HARPER, J. G., and DORN, J. E., *Acta Met.*, **5**, 654 (1957).
38. ASHBY, M. F., and VERRALL, R. A., *Acta Met.*, **21**, 149 (1971).
39. WEERTMAN, J., *Trans. ASM*, **61**, 681 (1968).
40. ASHBY, M. F., *Acta Met.*, **20**, 887 (1972).
41. MOHAMED, F. A., and LANGDON, T. G., *Met. Trans.*, **5**, 2339 (1974).
42. SHERBY, O. D., and DORN, J. E., *Trans. AIME*, **194**, 959 (1952).
43. ANDRADE, E. N. DA C., *Proc. Roy. Soc., A*, **84**, 1 (1910).
44. ANDRADE, E. N. DA C., *Proc. Roy. Soc., A*, **90**, 329 (1914).
45. KENNEDY, A. J., *Processes of Creep and Fatigue in Metals*, Oliver and Boyd, Edinburgh (1962).
46. BAILEY, R. W., *Proc. I. Mech. E.*, **131**, 131 (1935).
47. GRAHAM, A., *Research*, **6**, 92 (1953).
48. WYATT, O. H., *Proc. Phys. Soc. London, B.*, **66**, 495 (1953).
49. DORN, J. E., *J. Mech. Phys. Solids*, **3**, 85 (1955).
50. NORTON, F. H., *The Creep of Steel at High Temperature*, McGraw-Hill, New York (1929).
51. MCVETTY, P. G., *Trans. ASME*, **65**, 761 (1943).
52. SODERBERG, C. R., *Trans. ASME*, **58**, 735 (1936).
53. LUBAHN, J. D., and FELGAR, R. P., *Plasticity and Creep of Metals*, John Wiley, New York (1961).
54. GOLDHOFF, R. M., in *Advances in Creep Design: The A. E. Johnson Memorial Volume*, p. 81 (A. I. Smith and A. M. Nicolson, eds.), Applied Science Publishers, London (1971).
55. PAO, Y. H., and MARIN, J., *Trans. ASME, J. Applied Mech.*, **20**, 245 (1953).
56. MILLER, A., *Trans. ASME, J. Eng Matls. and Technology*, **98**, 97 (1976).
57. LUBAHN, J. D., *Trans. ASME*, **45**, 787 (1953).
58. NIELD, B. J., and QUARRELL, A. G., *J. Inst. Metals*, **85**, 480 (1956-57).
59. JEFFRIES, Z., *J. Inst. Metals*, **11**, 300 (1917-18).
60. MILLER, D. A., and LANGDON, T. G., *Met. Trans.*, **10A**, 1635 (1979).
61. ASHBY, M. F., Report No. CUED/C/MATS/TR.34, Cambridge University Engineering Department, Cambridge (1977).
62. RAJ, R., and ASHBY, M. F., *Acta Met.*, **23**, 653 (1975).
63. RAJ, R., SHIH, H. M., and JOHNSON, H. H., *Scripta Met.*, **11**, 839 (1977).
64. PAVINICH, W., and RAJ, R., *Met. Trans.*, **8A**, 1917 (1977).
65. RAJ, R., *Acta Met.*, **26**, 341 (1978).
66. WRAY, P. J., *J. Appl. Phys.*, **40**, 4018 (1969).
67. ASHBY, M. F., and RAJ, R., in *The Mechanics and Physics of Fracture*, p. 148, The Metals Society, London (1975).
68. GOLDHOFF, R. M., *J. of Testing and Evaluation*, **2**, 387 (1974).
69. CONWAY, J. B., *Stress-Rupture Parameters: Origin, Calculations and Use*, Gordon and Breach, New York (1969).
70. MANSON, S. S., in *ASM Publication D-8-100*, p. 1, ASM, Metals Park (1968).

71. HOLLOMON, J. H., and JAFFE, L. D., *Trans. AIME*, **162**, 223 (1945).
72. LARSON, F. R., and MILLER, J., *Trans. ASME*, **74**, 765 (1952).
73. ORR, R. L., SHERBY, O. D., and DORN, J. E., *Trans. ASM*, **46**, 113 (1954).
74. MANSON, S. S., and HAFERD, A. M., NACA TN 2890 (1953).
75. GOLDHOFF, R. M., and HAHN, G. J., in *Publication D-8-100*, p. 199, ASM, Metals Park (1968).
76. WHITE, W. E., and LE MAY, I., *Trans. ASME, J. Eng. Matls. and Technology*, **100**, 319 (1978).
77. MANSON, S. S., and ENSIGN, C. R., NASA TM.X-52999 (1971).
78. GOLDHOFF, R. M., in *Proc. Int. Conf. on Creep and Fatigue in Elevated Temperature Applications*, Vol. 1, p. 171.1, I. Mech. E., London (1975).
79. WHITE, W. E., and LE MAY, I., *Trans. ASME, J. Eng. Matls. and Technology*, **100**, 333 (1978).
80. MANSON, S. S., and ENSIGN, C. R., in *Characterization of Materials for Service at Elevated Temperatures*, p. 299 (G. V. Smith, ed.), ASME, New York (1978).
81. MONKMAN, F. C., and GRANT, N. J., *Proc. ASTM*, **56**, 593 (1956).
82. LE MAY, I., *Trans. ASME, J. Eng. Matls and Technology*, **101**, 326 (1979).
83. WHITE, W. E., and LE MAY, I., in *Mechanical Behaviour of Materials, ICM 3*, Vol. 2, p. 293 (K. J. Miller and R. F. Smith, eds.), Pergamon, Oxford (1980).
84. ODQVIST, F. K. G., *Roy. Swed. Acad. Eng. Research Proc.*, **141**, 31 (1936).
85. SODERBERG, C. R., *Trans. ASME*, **58**, 733 (1936).
86. SMITH, A. I., and NICOLSON, A. M., eds., *Advances in Creep Design: The A. E. Johnson Memorial Volume*, Applied Science Publishers, London (1971).
87. JOHNSON, A. E., HENDERSON, J., and KHAN, B., *Jt. Int. Conf. on Creep, Proc. I. Mech. E.*, **178**, (Pt. 3A), 2-27 (1963).
88. JOHNSON, A. E., *Proc. I. Mech. E.*, **164**, 432 (1951).
89. JOHNSON, A. E., HENDERSON, J., and KHAN, B., *Complex-stress Creep, Relaxation, and Fracture of Metallic Alloys*, HMSO, Edinburgh (1962).
90. JOHNSON, A. E., HENDERSON, J., and MATHUR, V. D., *Aircr. Eng.*, **31**, (361), 75; **32**, (362), 113 (1959).
91. HENDERSON, J., and SNEDDEN, J. D., in *Advances in Creep Design; The A. E. Johnson Memorial Volume*, p. 163 (A. I. Smith and A. M. Nicolson, eds.), Applied Science Publishers, London (1971).
92. JOHNSON, A. E., HENDERSON, J., and KHAN, B., *Proc. I. Mech. E.*, **175**, 1043 (1961).
93. JOHNSON, A. E., HENDERSON, J., and KHAN, B., *Engineer*, **212**, 1078 (1961).
94. JOHNSON, A. E., and KHAN, B., *Int. J. Mech. Sci.*, **5**, 507 (1963).
95. JOHNSON, A. E., and KHAN, B., *Proc. I. Mech. E.*, **178** (Pt. 3L), 29 (1964).
96. JOHNSON, A. E., HENDERSON, J., and KHAN, B., *Int. J. Mech. Sci.*, **4**, 195 (1962).
97. JOHNSON, A. E., HENDERSON, J., and KHAN, B., *App. Mater. Res.*, **3**, 45 (1964).
98. SMITH, E. M., *J. Mech. Eng. Sci.*, **7**, 82 (1965).
99. MACKENZIE, A. C., *Int. J. Mech. Sci.*, **10**, 441 (1968).
100. SIM, R. G., *J. Mech. Eng. Sci.*, **13**, 47 (1971).

101. PENNY, R. K., and MARRIOTT, D. L., *Design for Creep*, McGraw-Hill, London (1971).
102. FAIRBAIRN, J., and MACKIE, W. W., *J. Mech. Eng. Sci.*, **10**, 286 (1968).
103. FINNIE, I., and HELLER, W. R., *Creep of Engineering Materials*, McGraw-Hill, New York (1959).
104. ROSS-ROSS, P. A., FIDLERIS, V., and FRASER, D. E., *Can. Met. Quarterly*, **11**, 101 (1972).
105. ROSS-ROSS, P. A., and FIDLERIS, V., in *Proc. Int. Conf. on Creep and Fatigue in Elevated Temperature Applications*, Vol. 1, p. 216.1, I. Mech. E., London (1975).
106. HILL, R., *The Mathematical Theory of Plasticity*, Clarendon Press, Oxford (1950).
107. JOHNSON, A. E., *Metall. Rev..*, **5**, No. 51, 447 (1960).
108. FINNIE, I., and ABO EL ATA, M. M., in *Advances in Creep Design: The A. E. Johnson Memorial Volume*, p. 329 (A. I. Smith and A. M. Nicolson, eds.), Applied Science Publishers, London (1971).
109. HENDERSON, J., and SNEDDEN, J., in *Proc. Int. Conf. on Creep and Fatigue in Elevated Temperature Applications*, Vol. 1, p. 172.1, I. Mech. E., London (1975).
110. ODQVIST, F. K. G., in *Advances in Creep Design: The A. E. Johnson Memorial Volume*, p. 31 (A. I. Smith and A. M. Nicolson, eds.), Applied Science Publishers, London (1971).
111. KACHANOV, L. M., in *Problems in Continuum Mechanics, Contributions in Honor of Seventieth Birthday of N. I. Muskhelishvili*, p. 202 (J. R. M. Radok, ed.), Philadelphia (1961).
112. MANJOINE, M. J., *Trans. ASME, J. Eng. Matls. and Technology*, **97**, 156 (1975).
113. COFFIN, L. F., *Proc. I. Mech. E.*, **188**, 109 (1974).
114. MANSON, S. S., in *Fatigue at Elevated Temperature*, p. 774, (A. E. Carden, A. J. McEvily and C. H. Wells, eds.), STP 520, ASTM, Philadelphia (1973).
115. CARDEN, A. E., MCEVILY, A. J., and WELLS, C. H., eds., *Fatigue at Elevated Temperature*, STP 520, ASTM, Philadelphia (1973).
116. *Creep and Fatigue in Elevated Temperature Applications*, Proceedings of the International Conference, I. Mech. E., London (1975).
117. CURRAN, R. M., ed., *1976 ASME-MPC Symposium on Creep-Fatigue Interaction*, MPC-3, ASME, New York (1976).
118. COFFIN, L. F., *et al.*, ORNL-5073, ERDA, Oak Ridge, Tennessee (1977).
119. Code Case N-47, *ASME Boiler and Pressure Vessel Code, Code Cases*, ASME, New York (1977).
120. *Criteria for Design of Elevated Temperature Class 1 Components in Section III, Division 1, of the ASME Boiler and Pressure Vessel Code*, ASME, New York (1976).
121. BERLING, J. T., and SLOT, T., in *Fatigue at High Temperature*, p. 3, STP 459, ASTM, Philadelphia (1969).
122. COFFIN, L. F., in *Fatigue at Elevated Temperatures*, p. 5 (A. E. Carden, A. J. McEvily and C. H. Wells, eds.), STP 520, ASTM, Philadelphia (1973).

123. Henry, M. F., Solomon, H. D., and Coffin, L. F., in *Creep and Fatigue in Elevated Temperature Applications*, p. 182.1, I. Mech. E., London (1975).
124. Manson, S. S., Halford, G. R., and Hirschberg, M. H., in *Design for Elevated Temperature Environment*, p. 12, ASME, New York (1971).
125. Manson, S. S., Halford, G. R., and Nachtigall, A. J., NASA Tech. Memo. TM X071737 (1975).
126. Polhemus, J. F., Spaeth C. E., and Vogel, W. H., in *Fatigue at Elevated Temperatures*, p. 625, (A. E. Carden, A. J. McEvily and C. H. Wells, eds.), STP 520, ASTM, Philadelphia (1973).
127. Ohji, K., Miller, W. R., and Marin, J., ASME Paper 65-WA/met-5 (1965).
128. Burgreen, D., in *Fatigue at Elevated Temperatures*, p. 535, (A. E. Carden, A. J. McEvily and C. H. Wells, eds.), STP 520, ASTM, Philadelphia (1973).
129. Mulcahy, T. M., *Trans. ASME, J. Engineering Materials and Technology*, **96**, 214 (1974).
130. Mulcahy, T. M., *Trans. ASME, J. Engineering Materials and Technology*, **98**, 264 (1976).
131. Brunsvold, A. R., Ahmed, H. U., and Stone, C. C., *Trans. ASME, J. Engineering Materials and Technology*, **98**, 256 (1976).
132. Shahinian, P., and Achter, M. R., in *High Temperature Materials*, p. 448, (R. F. Hehemann and G. M. Ault, eds.), Wiley, New York (1959).
133. Le May, I., Truss, K. J., and Sethi, P. M., *Trans. ASME, J. of Basic Eng.*, **91**, 575 (1969).
134. Cook, R. H., and Skelton, R. P., *Int. Metall. Rev.*, **19**, 199 (1974).
135. Woodford, D. A., in *Proc. Int. Conf.: Engineering Aspects of Creep*, I. Mech. E., London (1980).

APPENDIX
Sample Problems

Chapter 1

1.1 At a point in a flat plate the stresses are as follows: $\sigma_{xx} = -100$ MPa, $\sigma_{yy} = 300$ MPa, $\tau_{xy} = -150$ MPa. Determine the magnitude and direction of the principal stresses and of the maximum shear stresses.

1.2 A point is acted upon by the stresses, $\sigma_{xx} = 100$ MPa, $\sigma_{yy} = \sigma_{zz} = 60$ MPa, $\tau_{yz} = -\tau_{xy} = -\tau_{zx} = 20$ MPa. Determine the magnitude and orientation of the principal stresses and of the maximum shear stresses.

1.3 Three strain gages are arranged on a flat surface with gages a and b at right angles and the intermediate gage, c, at $\pi/4$ to the others. Show that the principal strains are specified by the relation:

$$\epsilon_{1,2} = (\epsilon_a + \epsilon_b)/2 \pm [2(\epsilon_a - \epsilon_c)^2 + 2(\epsilon_b - \epsilon_c)^2]^{1/2}/2$$

Chapter 2

2.1 Plot the following points (or their projections) on the π-plane:

$$\sigma_1 = 20,\ \sigma_2 = 14,\ \sigma_3 = -10;\ \sigma_1 = 9,\ \sigma_2 = 0,\ \sigma_3 = 0.$$

2.2 An element is subjected to the following stresses: $\sigma_1 = 30$, $\sigma_2 = 6$, $\sigma_3 = -15$. What are the magnitudes of the hydrostatic and deviatoric stresses?

2.3 A well lubricated cube of metal of side 80 mm is loaded in compression and yielding commences at a load of 3.2 MN. What load would be required if the other sides were constrained by forces of 0.8 MN and 1.6 MN?

2.4 A metal is tested in compression with $\sigma_1 = 3\sigma_2 = -2\sigma_3$. Yield is observed at $\sigma_2 = 145$ MPa. Determine the yield stress in simple tension, using the von Mises yield criterion. If the material were tested under conditions where $\sigma_1 = -\sigma_3$, $\sigma_2 = 0$, at what value of σ_3 would yield occur?

2.5 The von Mises criterion for yield can be expressed in the following ways. Demonstrate that all are equivalent:

$$-(\sigma_1'\sigma_2' + \sigma_2'\sigma_3' + \sigma_3'\sigma_1') = [(\sigma_1')^2 + (\sigma_2')^2 + (\sigma_3')^2]/2$$
$$=[(\sigma_1 - \sigma_2)^2 + (\sigma_2 - \sigma_3)^2 + (\sigma_3 - \sigma_1)^2]/6 = k^2 = Y^2/3$$

2.6 Pressurized liquid is contained in a brittle ceramic tube which has tensile and compressive strengths of 42 MPa and 280 MPa, respectively. The tube is 50 mm in diameter and has a wall thickness of 2.5 mm. Determine the pressure it can stand (a) if the ends are open and free to slide through O-ring joints in tubeplates, (b) if the ends are closed.

2.7 a) A white cast iron component is loaded in biaxial compression with $\sigma_1 = 2\sigma_2$. The tensile and compressive fracture strengths are respectively 205 MPa and 1520 MPa. If the working stresses are kept such that a safety factor of 1.5 is introduced, estimate the magnitude of an unexpected tensile stress in the third principal direction to cause failure.
b) A gray cast iron pipe is of 0.8 m diameter and 10 mm wall thickness. It is clamped between two end platens to which it is sealed by gaskets, the clamping being such that the initial axial compressive stress in the cast iron is 140 MPa. The cast iron has tensile and compressive strengths of 290 MPa and 965 MPa respectively.

The pipe is then pressurized internally until failure by fracture or leakage occurs. At what pressure would this be expected and would failure be expected to start at the outside or the inside? Explain your answer.

2.8 Steel specimens have a fatigue limit in rotating bending of 150 MPa and, when tested in 0-to-tension loading the fatigue limit is found to be 100 MPa. A shaft made from this steel carries a steady torque of 300 N.m, a fluctuating torque of ± 50 N.m and an in-phase fluctuating axial load of ± 500 N. Determine the minimum diameter to avoid fatigue failure.

2.9 a) Discuss the probable effects of cyclic compressive load on the life of a component.
b) A compression coil spring is made from music wire (cold drawn 0.8% C steel) of diameter (d) 5.21 mm. Tests on the wire in 0-to-tension loading have shown the value of $\Delta\sigma$ for $N_f = 10^6$ cycles to be 700 MPa. The UTS is 2000 MPa.

If the spring is to carry a compressive load of 2000 N together with a fluctuating load of ± 1000 N and the design is for 10^6 cycles, determine the appropriate diameter of the coil, D.
c) Comment on a proposal that, for reasons of safety, designers must avoid any possibility of fatigue failure occurring in the vehicles to be employed in a new rapid transit system.

Chapter 3

3.1 Determine the engineering strain, e, the true strain, ϵ, and the reduction in area, q, for each of the following: (i) extension from l to $1.1l$; (ii) compression from h to $0.9h$; (iii) extension from l to $2l$; (iv) compression from h to $h/2$; (v) compression to zero thickness.

3.2 The initial diameter of a tensile specimen is 12.8 mm. At a specified load the diameter is 9.9 mm. Compute the true and engineering strains at this point, stating any assumptions made.

3.3 A true stress-true strain curve is plotted for an alloy and the slope measured at various points, so that a plot of $d\sigma/d\epsilon$ versus σ can be drawn. From this it is found that $d\sigma/d\epsilon = \sigma$ when $\sigma = 334$ MPa. The corresponding true strain is 0.45. Determine the UTS of the alloy and the uniform strain at the start of necking.

3.4 A wire is reduced in cross-section by passing it through seven successive dies, the cross-sectional area being reduced by 35% during each operation. Determine the increase in length in terms of (a) the engineering strain; (b) the true strain.

3.5 True stress-true strain data for a metal are plotted as a straight line on a $\log\sigma$ - $\log\epsilon$ plot. The end points of the line are defined by $\sigma_1 = 172$ MPa, $\epsilon_1 = 0.01$; and $\sigma_2 = 483$ MPa, $\epsilon_2 = 0.20$. Determine the equation of the true stress-true strain line between these points and plot the curve on σ-ϵ coordinates.

3.6 Draw a true stress-strain diagram assuming that the curve for strain hardening obeys the relationship $\Delta\sigma = 690\epsilon^{0.5}$ (MPa). Start with a yield stress of 69 MPa and end at $\epsilon = 100\%$. Convert your plot to a conventional engineering stress-strain diagram. Comment on what occurs for $\epsilon > 0.8$.

3.7 Write a concise account of the role of ductility in facilitating deformation during metal forming. You should consider both ductile and brittle materials.

3.8 You have a supply of normalized steel bar 25 mm in diameter, with $n = 0.2$ and a true stress at maximum load of 600 MPa. Required are rods of 6.25 mm diameter with a yield strength of 520 MPa. Devise a suitable procedure to produce the latter if you have hot rolling, cold rolling and annealing facilities available.

3.9 An annealed metal cylinder is tested in compression. The friction can be considered negligible and the cylinder's initial dimensions are 20 mm diameter, and 25 mm long. The following load-compression points are obtained:

Load (kN)	Compression, Δh(mm)
15.2	0.1
22.8	0.7
30.4	1.9
38.0	3.6
45.5	5.1
53.1	6.6
60.7	8.0

Evaluate the UTS and the engineering strain at necking for tensile loading of the material. What is the value of the strain hardening exponent?

Chapter 4

4.1 What stress must be applied in the [110] direction to produce slip on the (111)$[01\bar{1}]$ system of an Al crystal, if the CRSS is 1 MPa?

4.2 A single crystal of Zn is oriented with the normal to the basal plane making an angle of 1.042 rad with the tensile axis, and the three slip directions making angles of 0.663 rad, 0.785 rad and 1.466 rad with the tensile axis. Plastic deformation is observed at a tensile stress of 2.275 MPa. Determine the CRSS for Zn.

4.3 A tensile specimen of cross-sectional area 19.6 mm^2 is cut from a single crystal of Mg. The plane on which slip takes place and the slip direction have true angles to the tensile axis of 0.47 rad and 0.67 rad, respectively. Yielding in tension takes place at 45 N. Determine the CRSS.

Chapter 5

5.1 (a) A material is represented by the standard linear solid. It is pulled in a tensile testing machine at constant cross-head speed, until the slope of the force versus time curve tends to its limiting value. At this point the cross-head's motion is reversed. Sketch the stress-strain curve.
(b) A Maxwell element is tested as in a stress-strain test, i.e., the strain rate is maintained approximately constant at $\dot{\epsilon}_0$. Evaluate the stress-strain relationship, and sketch this as a function of strain rate.

5.2 A polymer behaves approximately as a Maxwell solid. When subjected to tensile loading, it is found that the strain rate is 1.1/s when $\sigma = 15$ MPa. If the elastic modulus is determined to be 2.5 GPa, determine its viscosity and the relaxation time.

5.3 The viscosity of glass varies from 10^{21} N.s/m^2 at 20°C to 1.1 x 10^{12} N.s/m^2 at 575°C, and it may be assumed to vary linearly with temperature. Given that E and ν have values of 13.8 GPa and 0.25, respectively, and that these do not change appreciably with temperature, plot the variation in relaxation time with temperature.

Chapter 6

6.1 (a) A metal contains strong second phase particles of spacing $100b$, where b is the Burgers vector of the maxtrix. Determine the ratio of the actual strength to the theoretical strength for slip.
(b) An Al alloy is strengthened by means of dispersed particles of Al_2O_3, 10 μm in diameter, 2 w/o being present. Estimate the materials's strength. G for Al = 28 GPa; ρ(Al) = 2.7 Mg/m^3; $\rho(Al_2O_3)$ = 3.96 Mg/m^3.

6.2 A metal, whose shear modulus is 10 GPa, contains a fine dispersion of hard second phase particles of average diameter 10 nm, with an average interparticle spacing of 50 nm. If the shortest distance between atoms in the lattice is 0.2 nm, what additional strengthening increment will be provided by the particles?

Chapter 7

7.1 (a) A composite consists of aligned carbon fibers in a epoxy resin matrix, with a composition 60% fibers by weight. The density of the fibers is 1.3 Mg/m^3, and their elastic modulus is 270 GPa. The density and elastic modulus of the epoxy are 1.12 Mg/m^3, and 4.5 GPa.

Determine the elastic modulus of the composite in the direction of the fibers.

(b) Discontinuous fibers of Al_2O_3 of diameter 12.5 μm are dispersed in a metallic matrix. The fibers have UTS and E values of 2 GPa and 500 GPa, respectively, while the true yield stress versus strain curve for the matrix is given by $\sigma = 30\ \epsilon^{0.2}$ MPa.

For a volume fraction of fibers, V_f, = 0.3, determine the required fiber length if the fracture strength of the composite is to be 60% of one having continuous fibers.

7.2 A composite material contains 40% by volume of aligned high-strength steel wires of fracture strength 2.8 GPa. If the flow curve for the matrix aluminum is given by $\sigma = 150\epsilon^{0.2}$ MPa, and for loading along the axis of the wires, determine the composite's longitudinal strength. Fracture of the composite occurs when the wires break and isostrain conditions are assumed throughout.

7.3 (a) A composite sheet material consists of an aluminum matrix reinforced with continuous graphite fibers. The fibers have UTS = 2 GPa and E = 686 GPa, and the aluminum matrix has a true stress/true strain curve which can be represented by $\sigma = 150\epsilon^{0.2}$. The fiber volume fraction is 20%.

The sheet is loaded in biaxial tension with $\sigma_1 = 2\sigma_2$, and the fibers are aligned in the two directions of principal stress such that failure in both directions would occur at exactly the same point in loading.

Calculate the values of the applied principal stresses at failure.

(b) If discontinuous fibers were used instead, and the strength was to be the same as before, calculate the required increase in fiber content for the following conditions: fiber length, 50 mm; fiber diameter, 1 mm. It is found by experiment that the force required to pull out a fiber which has 10 mm of its length in the matrix is 1.25 kN.

Chapter 8

8.1 (a) A glass rod of 5 mm diameter contains three small sharp cracks, two being 1 μm in length, the third 0.8 μm. Determine the load it can carry given that $E = 70$ GPa, $\gamma = 0.3$ J/m^2, $\nu = 0.25$.
(b) Repeat (a) for a brittle metallic rod which shows local plastic deformation at the crack tips. In this case $E = 200$ GPa, γ_p (effective surface energy) = 1 kJ/m^2, $\nu = 0.28$.
(c) Comment on the validity of the statement: "The larger the specimen the lower its strength", as applied to brittle solids.

8.2 The work of fracture, $\mathcal{G}_c$, for a steel plate in a bridge has a value of 35 kN/m. Determine the load to cause fracture in the plate which is 0.3 m wide, 10 mm thick and contains a central crack 10 mm long. If the crack were 50 mm long, estimate the fracture load.
With increase in plate thickness, $\mathcal{G}_c$ falls to a constant value of 17 kN/m. Determine the fracture stress for a 10 mm crack in this case. Take $E = 207$ GPa.

8.3 A high strength steel has a UTS of 1950 MPa and a sheet of it contains a crack 4 mm long at right angles to the loading direction. What is the % reduction in strength due to this crack, given that $E = 200$ GPa, $\gamma_s = 2$ J/m^2; and γ_p, the work of plastic deformation per unit area of crack surface, = 20 kJ/m^2? If the crack were in line with the stress axis, what reduction would then be caused?

8.4 BCC metals generally fracture in a brittle manner at low temperature, particularly under impact loading, but FCC metals are ductile even to the lowest temperature. Explain why this occurs.
It was observed that when a steel specimen made of two halves joined by brazing with a layer of silver (FCC) was pulled in tension, necking started in the silver layer and then fracture took place *by cleavage* of the silver. The steel had not yet reached its failure condition. Explain why cleavage failure could occur in the silver.

8.5 Quenched and tempered 4140 steel has $\sigma_Y = 1500$ MPa and $K_{Ic} = 66$ MPa m$^{1/2}$. Plot the envelope describing the safe regime where a component will neither yield nor will a crack propagate. This should be drawn on the basis of σ versus defect size, a.

8.6 A consulting engineer has examined a large shaft driving a piece of heavy rotating machinery which failed in service causing loss of life. He is retained as an expert witness in a lawsuit brought against the manufacturer and his report includes the following comments.
"The fracture was at right angles to the axis of the shaft, indicating failure under torsional loading. It was generally smooth and shiny without any noticeable surface markings. This smooth, shiny appearance indicates fatigue failure took place owing to the periodically varying torsional loading, and the absence of any noticeable plastic deformation on the fracture surface indicates that the material broke in a brittle or cleavage fracture mode, again indicative of a fatigue crack running through the component. Because the component failed by fatigue and because the service loads were not excessive (as is admitted by all parties) I examined the microstructure of the steel shaft. Speaking as an expert in the area of physical metallurgy, I consider that the heat treatment was 'unsatisfactory' and that the manufacturers are at fault."
You are retained to advise legal counsel for the manufacturers. Suggest some ways in which he might demolish the credibility of the expert witness in the eyes of the jury on the basis of the evidence he has offered. (NOTE: You are not asked to provide contrary evidence of failure mechanisms, as you have not had access to the shaft, this being in the hands of the claimant.)

Chapter 9

9.1 A cylindrical steel pressure vessel with closed ends is to contain a pressure (p) of 15 MPa and the steel is to be chosen to give minimum mass. It is to be fabricated from steel plates welded together at their edges. The volume of the vessel is to be 1000 m^3, the maximum stress is to be $\leqslant \sigma_Y/2$ and a safety factor of $\geqslant 1.1$ is called for, based on the ratio of fracture stress to working stress.

Three steels are available in different thicknesses as follows:

Steel	Thickness (m)	σ_Y (MPa)	K_{Ic} (MPa$\sqrt{m}$)
A	0.08	965	280
B	0.06	1310	66
C	0.04	1700	40

Defects may be present in the form of thumbnail cracks at the inner surface of the welds, having an aspect ratio, $a/2c = 0.2$, while the stress intensity factor is given by

$$K_{Ic} = 1.12\,(\sigma + p)\,(\pi a/Q)^{1/2}$$

where Q is a flaw shape parameter which can be evaluated from Fig. 9.25, and the other symbols have their usual meanings. The longest defect which may be present without detection is found to have a length ($2c$) of 20 mm.

Determine the appropriate steel for the vessel.

9.2 One engineer indicates that stress-relief annealing is necessary in a welded structure as the residual stresses induced by welding are tensile in nature. Discuss the statement and comment on its relevance to design on the basis of fracture mechanics.

9.3 A large plate of alloy steel is to be loaded in tension. The minimum flaw size for detection is 2 mm in length and the design stress is taken as 0.5 x UTS. In order to cut down on the mass of the structure it is proposed to alter the heat treatment cycle to raise UTS from 1500 MPa to 2000 MPa with consequent reduction in K_{Ic} from 70 to 35 MPa$\sqrt{m}$. Is the proposal a sound one? (Assume plane strain conditions to hold).

Chapter 10

10.1 A cold-worked mild steel rod is stressed cyclically at $\pm\sigma_a$ while a mean tensile stress, σ_m, is applied. The loading is such that $\sigma_a + \sigma_m < \sigma_Y$, its monotonic tensile yield stress. It is observed that the rod starts to elongate after several hundred load cycles have been applied. Explain.

10.2 A group of components have a fatigue life of 20 x 10^3 cycles at $\pm$414 MPa and 1 x 10^6 cycles at $\pm$276 MPa, and the S-logN_f relation can be considered a straight line between these points. If a component were subjected to 20 x 10^3 cycles at 360 MPa and 30 x 10^3 cycles at 574 MPa, estimate the number of cycles to failure at $\pm$ 310 MPa.

10.3 A smooth bar, cyclically-strained steel specimen fails after 1000 cycles. Determine the total strain range to which it has been subjected if the base material has the following properties: E = 205 GPa; σ_f = 1850 MPa, ϵ_f = 0.7; n' = 0.15. State any assumptions made.

10.4 Explain whether a high or low cyclic strain hardening exponent is appropriate for fatigue resistance under (a) high cycle, low stress conditions; (b) low cycle, high strain conditions.

Chapter 11

11.1 Experiment has shown that, for a particular bolting material, $\dot{\epsilon}_s = 4.0 \times 10^{-8}$/h at 30 MPa and 550°C, and that the strain rate exponent, $n = 3.8$. Taking E to be 200 GPa and considering it to be unaffected by temperature, determine the stress in a bolt of this material after one year, if the initial tightening stress used in clamping two plates together at 550°C were 70 MPa.

11.2 (a) In designing a component for high temperature operation under constant loading it is important that a limiting strain be specified which may occur during the projected 20 year life of the part. Creep curves are available for a range of values of stress and temperature, all leading to such a strain at much shorter times. It is suggested that an extrapolation procedure be used, based on the short-time data, even although such procedures are normally used for rupture time. Is the proposal justified?
(b) Comment of the stress dependency of activation energy during creep and how this may determine to some extent the relevant parameter for extrapolation.

11.3 A thin walled tube at high temperature is under tensile axial load and internal pressure. The axial load is additional to that from the effect of pressure, and of magnitude three times the latter. Determine the creep rates in the principal directions of strain as a function of uniaxial tensile creep rate for the material at the same temperature.

11.4 The wall of a high temperature system under tensile loading is subjected to periodic cooling from T_1 to T_2 on one side, the other remaining at approximately constant temperature T_1. For a square wave temperature variation, and for a sharp variation in temperature through the wall, determine the regimes for ratcheting, shakedown and elastic behavior as a function of normalized thermal and mean stresses. The inner portion of the wall, subject to temperature variation, can be considered to be one-third of the total wall thickness.

11.5 A gas turbine runs at constant speed, the centrifugal stress along the blades being ~ 40 MPa. There is 1 mm clearance between the blade tips and the shroud for a blade length of 100 mm.
Laboratory tests have shown the steady state creep rate of the blading alloy to be as follows at 40 MPa: $T = 800°C$, $\dot{\epsilon} = 7.3 \times 10^{-8}$/s; $T = 950°C$, $\dot{\epsilon} = 1.29 \times 10^{-5}$/s.
Estimate the safe life of the blades at an operating temperature of 750°C.

11.6 Data are given in the table for the stress-rupture of Type 316 stainless steel. On the basis of one or other of the parametric methods of extrapolation, draw up graphs showing stress versus log (rupture time) at 500, 600 and 700°C for times ranging from 1 h to 10^5 h.

T (°C)	σ (MPa)	Time to rupture (h)
832	46.3	130
803	46.3	525
850	46.3	95
950	46.3	2.5
850	46.3	97
801	61.8	255
770	61.8	808
825	61.8	95
850	61.8	37
899	61.8	5.9
782	123.6	9.1
735	123.6	70.4
685	123.6	710.5
800	123.6	4.3
715	123.6	139
685	247.1	4.0
615	247.1	458
615	247.1	228
675	247.1	5.6
650	247.1	30.3

AUTHOR INDEX

SUBJECT INDEX